Nanowire

This book comprehensively:

- Covers material for nanowires, device models, and applications for circuit design based on nanowire field-effect transistors
- Explains a wide spectrum for designing robust multi-gate nanowire devices and low-power circuits
- Highlights the conventional and novel nanomaterials, devices, and circuits, leakage current mitigation techniques, and important trade-offs along with exhaustive analysis
- Includes nanowire device structures and modeling along with MATLAB® codes
- Offers an in-depth insight into the temporal degradation awareness and interconnects delay, which provides inside exposure to the impact of reductions in speed and consumption of more power

The text is primarily written for senior undergraduates, graduate students, and academic researchers in electrical engineering, electronics and communications engineering, materials science, nanoscience, and nanotechnology.

Editors

Shailendra Singh is currently working as an Assistant Professor, Department of Electronics Engineering at Madhav Institute of Technology & Science, Gwalior (M.P.), India. His research interest includes modeling and simulation of semiconductor devices, especially Vertical TFET for low-power applications. He has published more than 39 SCI and Scopus-indexed research papers with 10 book chapters and many national/International conferences. His Google Scholar profile lists 1047 citations, h-index 18, and i10-index 27. Dr. Singh actively collaborates with researchers from IIT Bhubaneswar, IIIT Jabalpur, NIT Sikkim, and NIT Jalandhar, and has delivered expert lectures in IEEE and FDP programs on VLSI and Nanoelectronics.

Balwinder Raj (MIEEE'2006) currently working as an associate professor, in the Department of Electronics and Communications Engineering, at the National Institute of Technology Jalandhar, India. For further research work, the European Commission awarded him the "Erasmus Mundus" Mobility of Life research fellowship for postdoc research work at the University of Rome, Tor Vergata, Italy, in 2010-2011. His areas of research interest are classical/nonclassical nanoscale semiconductor device modeling, nanoelectronics and their applications in hardware security, sensors, and circuit design, FinFET-based memory design, low-power VLSI design, digital/analog VLSI design, field programmable gate array (FPGA) implementation. He has published more than a hundred research papers in peer-reviewed international/national journals and conferences.

Arun Kumar Manoharan is working as an associate professor, in the Department of Electrical, Electronics and Communication Engineering, at GITAM (Deemed to be University), Bengaluru, Karnataka, India. His current research area is the fabrication of sensor devices for biosensor applications. His research areas include energy storage applications. He has published research papers in journals and conferences.

Dr. Girdhar Gopal is working as an Assistant Professor in the Department of Electronics and Communication Engineering at the National Institute of Technology Patna (Bihar), India. His research areas include VLSI design, semiconductor device engineering, organic optoelectronic devices, thin-film technologies, and VLSI-DSP.

He has a total of 24 publications, which include 12 international journal papers, 9 conference papers, and 3 book chapters. Dr. Gopal also serves as a reviewer for several well-known journals and is a member of the Editorial Board of *Scientific Reports*.

Nanowire
Materials, Devices, Circuits, and Applications

Edited by
Shailendra Singh
Balwinder Raj
Arun Kumar Manoharan
Girdhar Gopal

CRC Press
Taylor & Francis Group
Boca Raton London New York

CRC Press is an imprint of the
Taylor & Francis Group, an **informa** business

Designed cover image: © 2026 Shutterstock

First edition published 2026
by CRC Press
2385 NW Executive Center Drive, Suite 320, Boca Raton FL 33431

and by CRC Press
4 Park Square, Milton Park, Abingdon, Oxon, OX14 4RN

CRC Press is an imprint of Taylor & Francis Group, LLC

ISBN: 978-1-041-06267-7 (hbk)
ISBN: 978-1-041-06272-1 (pbk)
ISBN: 978-1-003-63464-5 (ebk)

DOI: 10.1201/9781003634645

Typeset in Sabon
by SPi Technologies India Pvt Ltd (Straive)

Contents

List of Contributors vii
Preface ix
Acknowledgment xv

1 III–V Semiconductor Nanocrystal formation using aqueous state superlattice in silicon nanowires 1
SHAKUN SRIVASTAVA AND ANSHU SRIVASTAVA

2 Flexible energy storage devices 21
TANYA SINGH, AKRITI DEWANGAN, PUJA KUMARI, BALWINDER RAJ, AND MANDEEP SINGH

3 2D Material-based sensor design 62
BALWINDER RAJ, VINAY CHANDEL, RAVI JYOTI, MANISH AGRAHARI, BALWANT RAJ, AND MANDEEP SINGH

4 Machine learning in quantum material 93
MANDEEP SINGH, PAWAN KUMAR MAURYA, MANAS BAISWAR, GANDEPALLI SAI TEJA, AND BALWINDER RAJ

5 Nanowire-based biosensors: Unveiling growth to applications 118
MOHIT KUMAR SRIVASTAVA, PREETI AGARWAL MITTAL, SHAILENDRA SINGH, VIVEK KUMAR, ARTI SAXENA, AND MAN MOHAN SHUKLA

6 Nanowire-FET: Role in healthcare applications 149
ANITA SHUKLA, ANKIT JAIN, SHAILENDRA SINGH, PUSPRAJ SINGH CHAUHAN, AND IMRAN ULLAH KHAN

7 Enhancing nanowire TFET performance through hetero-dielectric and hetero-material at the device and circuit level 171

SUGANDHA TEJESWEE, MOHIT SRIVASTAVA, PREETI AGARWAL MITTAL, APARNA DIXIT, AND ARTI SAXENA

8 Reconfigurable nanowire transistors and their applications in novel electronic circuits and systems 203

PREETI AGARWAL MITTAL, MOHIT KUMAR SRIVASTAVA, MANMOHAN SHUKLA, SHAILENDRA SINGH, AND VIVEK KUMAR

9 Nanowire field effect transistors-applications in integrated circuits and sensing 235

PREETI AGARWAL MITTAL, MOHIT KUMAR SRIVASTAVA, SHAILENDRA SINGH, APARNA DIXIT, ARTI SAXENA, AND VIVEK KUMAR

10 IoT-based energy meter for performance evaluation of nanowire-enhanced lithium-ion batteries 265

ANKIT JAIN, ANITA SHUKLA, SHAILENDRA SINGH, PUSPRAJ SINGH CHAUHAN, AND IMRAN ULLAH KHAN

11 A comprehensive review on nanowires: Synthesis, utility, and technical hurdles 290

ANKIT JAIN, ANITA SHUKLA, VIVEK KUMAR, PUSPRAJ SINGH CHAUHAN, SHAILENDRA SINGH, AND IMRAN ULLAH KHAN

12 Nano-enhanced biosensors 309

ANITA SHUKLA, ANKIT JAIN, PUSPRAJ SINGH CHAUHAN, SHAILENDRA SINGH, AND IMRAN ULLAH KHAN

Index 329

Contributors

Manish Agrahari
NIT Jalandhar
Jalandhar, Punjab

Manas Baiswar
NIT Jalandhar
Jalandhar, Punjab

Vinay Chandel
NIT Jalandhar
Jalandhar, Punjab

Puspraj Singh Chauhan
IIT Indore
Indore, India
Pranveer Singh Institute of Technology
Kanpur, India

Akriti Dewangan
NIT Jalandhar
Jalandhar, Punjab

Aparna Dixit
Pranveer Singh Institute of Technology
Kanpur, India

Ankit Jain
IIT Indore
Indore, India
Pranveer Singh Institute of Technology
Kanpur, India

Ravi Jyoti
NIT Jalandhar
Jalandhar, Punjab

Imran Ullah Khan
IIT Indore
Indore, India
Pranveer Singh Institute of Technology
Kanpur, India

Vivek Kumar
Pranveer Singh Institute of Technology
Kanpur, India

Puja Kumari
NIT Jalandhar
Jalandhar, Punjab

Pawan Kumar Maurya
NIT Jalandhar
Jalandhar, Punjab

Preeti Agarwal Mittal
Pranveer Singh Institute of Technology
Kanpur, India

Balwant Raj
NIT Jalandhar
Jalandhar, Punjab

Balwinder Raj
NIT Jalandhar
Jalandhar, Punjab

Arti Saxena
Pranveer Singh Institute of Technology
Kanpur, India

Anita Shukla
IIT Indore
Indore, India
Pranveer Singh Institute of Technology
Kanpur, India

Manmohan Shukla
Pranveer Singh Institute of Technology
Kanpur, India

Man Mohan Shukla
Pranveer Singh Institute of Technology
Kanpur, India

Mandeep Singh
NIT Jalandhar
Jalandhar, Punjab

Shailendra Singh
Pranveer Singh Institute of Technology
Kanpur, India
IIT Indore
Indore, India

Tanya Singh
NIT Jalandhar
Jalandhar, Punjab

Anshu Srivastava
Pranveer Singh Institute of Technology
Kanpur, India

Mohit Srivastava
Pranveer Singh Institute of Technology
Kanpur, India

Mohit Kumar Srivastava
Pranveer Singh Institute of Technology
Kanpur, India

Shakun Srivastava
Pranveer Singh Institute of Technology
Kanpur, India

Gandepalli Sai Teja
NIT Jalandhar
Jalandhar, Punjab

Sugandha Tejeswee
Pranveer Singh Institute of Technology
Kanpur, India

Preface

This book exhibits a unique way to explain nanowire materials, devices, and analysis of its design parameters to meet the sub-nano-regime challenges for low-power chip design. Since process variability, device sizing, and power supply scaling are ongoing challenges in large VLSI circuits design, this book highlights the conventional and novel nano materials, devices and circuits, leakage current mitigation techniques, and other important trade-offs along with exhaustive analysis. More focus has been placed throughout the book on the various trade-offs for high-speed and low-power chip design using nanowire device technology.

This book provides a complete and precise introduction to nanowire materials, devices design and analysis, emphasizes different trade-offs for achieving the best possible nanoelectronic devices design for high-performance integrated circuits design. Further, this book offers in-depth insight into the temporal degradation aware and interconnects delay, which provides the inside exposure on the impact of various reductions in speed and consumes more power. The discussion on various high mobility materials and multi-gate devices mechanisms will be useful for the researchers to appositely extrapolate the aging phenomenon. Since the lifetime of the circuit has severely degraded due to the temporal degradation of materials in the devices. Better tolerance to aging phenomenon will provide more robustness and extend the IC lifetime.

Chapter 1: The primary obstacle to improving compatible oxide of metal semiconductor (CMOS) achievement and increasing the direct combination of excessive potency of III–V compounds with current Si-dependent CMOS clarifying capabilities. As was initially shown utilizing molecular beam epitaxy methods, heterowires containing silicon with compatible III–V fragments are among the most favorable possibilities for sophisticated nano photonics. Here, employing consecutive ion thruster insemination and photonic sintering, photonic curing, and intense pulsed light,we provide a unique method for fabricating in arsenic silicon cross bread nanowires through aqueous superlattice, regrowth in the millisecond range. We demonstrate that a single nanowire can monolithically integrate such severely mismatched systems. High-quality hetero-nanowire manufacturing and the creation of microscopic or infinitesimal, knife-like connections linking indium and silicon fragments are confirmed by optical and microstructural studies. Upcoming rapid velocity multipurpose microelectronics gadgets made only with one chip have new avenues to follow thanks to such hybrid systems.

Chapter 2: Flexible energy storage technologies, such as flexible batteries and supercapacitors, have advanced as a result of the growing demand for wearable, flexible, and sustainable electronics. Flexible displays, wearable technology, and next-generation portable electronics all depend on these devices because they can bend, stretch, and fold without losing their functionality. It has been seen that how flexible devices have changed over time, how their structures have changed to store energy, and the essential components that give them flexibility and durability—such as carbon-based nanofibers and conductive polymers. Modern applications require mechanical resilience and energy storage capacity, which are optimized by a variety of structures such as fiber, island bridges, and interdigital designs. Whereas issues such as low energy density, expensive production, and environmental impact are discussed, highlighting the necessity for more innovation. The development of flexible energy storage devices ensures for improving energy efficiency and facilitating the shift to a resilient, flexible, and sustainable energy ecosystem.

Chapter 3: When the bulk material is only a few layers thick, within a range of 1 nanometer to almost 1 micrometer (a millionth of a meter), it turns into what are called two-dimensional materials, or the common jargon term "2D materials." Examples include graphene, a conductive, strong and flexible material formed by carbon atoms arranged in hexagonal lattices, insulating, thermally stable h-BN with mechanical strength that is similar to graphene; electronic effective TMDs such as molybdenum disulfide (MoS_2); and phosphorene geometry of black phosphorus known as a semiconductor. Showing high surface area, tunable properties (such as electrical conductivity), mechanical strength, and transparency, these materials are ideal candidates for applications in electronics, energy storage, and nanotechnology. Those properties are why they hold promise for future technologies, such as flexible electronics and wearables, their high conductivity, low density, and ease of production yield screens that can be made thin and flexible.

Chapter 4: In recent years, the tech world has seen a rapid shift toward quantum devices, in order to push the limits of classical technology. For decades, metal-oxide-semiconductor field-effect transistors have been the backbone of electronic systems, fueling remarkable advances in computing and data processing, largely thanks to Moore's Law. This principle held that the number of transistors on a chip would double every couple of years, leading to exponential growth in processing power. But now, as these transistors approach the scale of just a few nanometers, fundamental challenges have started to crop up. When transistor sizes drop below 10 nanometers, they face short-channel effects like unwanted leakage currents and weakened gate control, which both affect performance and reliability.

Chapter 5: The further developments of nanotechnology catalyze critical innovations in other fields, especially in biosensing. Further investigation into the anomalous properties of nanomaterials, more specifically nanowires, hastened the advance of highly sensitive and efficient biosensors. Such nanoscale materials, on account of a structure with an enormously high ratio of surface-to-volume, find high utility to interact with targeted analytes as applied in monitoring real-time events in different settings. This nanowire-based biosensor novel represents a critical step forward toward multifunctional devices for rapid detection. Such applications address critical healthcare, environmental, and food security needs effectively. Wearable technology will be a promising area of future opportunities in personalized medicine and proactive health management. This introduction will set the context for analyzing growth trajectories and practical implications of nanowire-based biosensors, which can help to establish their transformative potential in current research and industry.

Chapter 6: Nanomaterials like carbon nanotubes (CNTs), nanowires, including nanoparticles, have revolutionized biosensing because of their unique quantum properties, which enhance the sensitivity of sensors. Their small size, comparable to that of biomolecules, allows for precise detection, making them invaluable in various applications. Nanowires are important for sensor technologies because of their electrical controllability and chemically reactive surfaces, which allow for sensitivity and multiple sensors measuring properties associated with a detection target. Field-effect transistors (FETs) have existed since the 1930s and have a prominent role in today's technology; they are primarily used in a metal-oxide-semiconductor (MOSFET) transistor. These FETs are especially pervasive in digital technology because they are easy to design and can be produced at low cost. Transistor technology has seen scales between 1 and 2 μm and even smaller. The art and science of producing smaller transistors does not necessarily mean the performance has continued to improve, which has led to research and development in novel, advanced technology. Nanowire field-effect transistors (NW-FETs) are an exciting choice to produce unique nanomaterial sensors in healthcare applications that have the highest sensitivity and flexibility. Nanowire FETs could allow researchers to find biomarkers, infectious viruses, and even DNA. An extremely small yet highly sustainable sensor has potential for diagnostic use, drug discovery, and real-time monitoring. NW-FETs may dominate future thinking and capabilities in medical diagnostics and personalized medicine. This chapter looks at how NW-FETs can be exploited for NANOWIRE. FET technology can be valuable in healthcare for virus detection, finding biomarkers, and DNA and drug development. We will show how NW-FET technology can marry theoretical principles to practical application via improved diagnostics and therapeutic advantages from improved sensing that is more accurate, reliable, and fast as we need for healthcare technologies.

Chapter 7: Tunnel field-effect transistors (TFETs) have experienced a fair amount of research awareness as an alternative to the regular metal-oxide-semiconductor field effect transistors (MOSFETs) for less power electronic scenarios, because they can achieve steep subthreshold swings (SS) and function at lower supply voltages. Unlike conventional FETs, where charge carriers travel over a potential barrier through thermionic emission, TFETs exhibit band-to-band tunneling (BTBT) over a tunneling junction from the source to the channel regions. This unique quantum mechanical transport yields a SS that can exceed the thermodynamically determined confinement as 60 mV/decade at room temperature for MOSFETs. This feature means TFETs can reduce static power dissipation and are ideal for energy-constrained devices. The recent advancements in material engineering such as heterostructures, high-k gate dielectrics, and low-bandgap semiconductors (e.g., InAs and GaSb) have all led to significant improvements in ION and switching efficiency. Comparative images of TFET and MOSFET operation showcase the differences in charge injection mechanisms as well as the benefits of tunneling-based transport as future nanoscale and low-power device technology.

Chapter 8: The development of electronic circuits requires innovative materials and devices to overcome various performance and integration issues. Probably, the most notable innovation is in reconfigurable nanowire transistors, which really open up avenues for great circuit design flexibility and functionality. They exploit the intrinsic properties of the nanowire, which promise applications in such areas as neuromorphic computing and complex logic gates. The two-gate design materials can utilize materials like transition metal dichalcogenides that might increase the efficiency and result in smaller designs. On the other hand, transistors whose ability is in their reconfiguration ability will offer highly notable power-use reductions as well as greater speeds in processing since energy consumption problems in electronic systems have continued increasing. By considering these points, the discussion will now outline why reconfigurable nanowire transistors are crucial for the future of electronic circuits and systems.

Chapter 9: The search for better materials and designs has become increasingly pressing as electronics get smaller and more efficient. A highly promising alternative option for use of integrated circuits and sensing is a type of the nanowire FETs. The chips based on these transistors are faster, consume less electric power, and are able to carry faster charge. Recent experiments have shown that the on/off current ratio of silicon nanowire-FET can exceed 10^6 and can be used for various types of sensing, including biomolecular detection. A diagram explaining the multiple uses ascribed to these instruments demonstrates how important they are in current electronics and biosensing and, thus, for new directions in health research.

Chapter 10: Nanowire-enhanced lithium-ion batteries have significantly advanced energy storage technology by addressing key challenges such as limited capacity, poor cycling stability, and thermal management. The integration of nanowires boosts surface area, electrical conductivity, and mechanical flexibility, directly enhancing energy density, charge–discharge cycles, and overall battery lifespan. These batteries use a solid electrolyte composed of nanowires, usually created by electrospinning. By compressing this electrolyte between the anode and cathode, performance is enhanced in terms of durability, energy efficiency, and charging speed. Because of these developments, nanowire-enhanced lithium-ion batteries are perfect for Internet of Things (IoT) applications where long-term performance, energy density, and efficiency are critical. IoT-based energy meters have been created to further improve the performance of these cutting-edge batteries. Battery performance can be continuously monitored thanks to these meters' real-time data on important parameters such as voltage, current, energy density, and charge–discharge cycles. Predictive maintenance and remote monitoring are made possible by the meters' ability to send the data they collect to cloud platforms via IoT connectivity. This guarantees that the batteries run as efficiently as possible and can be modified or serviced as needed. These energy meters' sophisticated sensors and analytics capabilities provide insightful information about battery health and energy efficiency, extending the life and maximizing the performance of battery systems in IoT ecosystems. For real-time data collection in real-world applications, the system incorporates an ESP32 Wi-Fi module, SCT-013 current sensors, and ZMPT101B voltage sensors. After collection, the data is sent to the Blynk server for worldwide access and analysis. The system offers comprehensive energy audits in addition to allowing the monitoring of energy consumption in household appliances. Wearables, smart homes, and industrial automation are just a few of the industries that stand to gain from this arrangement, which combines nanotechnology in energy storage with IoT-based monitoring to help develop smarter, more sustainable energy solutions.

Chapter 11: One-dimensional nanostructures are commonly referred to as nanowires (NWs), nanofibers, nanotubes, etc. One of the most interesting features of these materials is their electrical conductivity and when they scale to nano-alloys (<100 nm), the conductivity is not subjected to increased resistance like traditional physics the conductivity of the wire enhances as the diameter of the wire decreases. These materials and their qualities, such as optical, magnetic, biological, electrical conductivity, and the ability to sense through environment, have a wide variety of industrial applications. NWs, along with various industrial sectors, have the power to influence electronics, computers, memory, data storage, communications, manufacturing, health, governance, and national security. The converted large surface-to-volume ratio of NWs means lower defect density, greater optical output, and tunable n-type conductivity, thereby making them suitable and more sensitive for applications such as sensing. NWs can be used for chemical sensors, gas sensors, electrochemical sensors, and nano biosensor applications due to their ultrasensitive, real-time response. In photovoltaic devices, NWs have been used to improve optical absorption and collection efficiency. NWs are also the basis for large circuits that can be fabricated at the scale of commercial planar technologies, and they are very small, lightweight, and inexpensive for mass production. NWs are one-dimensional (1D) nanostructures that remain conductive even though their cross-section is reduced, unlike ordinary electrical wires. NWs can be conducting materials such as Ni, Pt, and Au; semiconducting materials such as Si, InP, and GaN; and insulating materials such as SiO_2, TiO_2, and more. NWs have properties including mechanical, electrical, chemical, and optical and thermal properties.

Chapter 12: Nanowire biosensors are a subclass of highly sensitive sensors in which the primary sensing elements are nanometer-scale wires or fibers with assemblies sometimes built from biological molecules including DNA, polypeptides, fibrin proteins, and filamentous bacteriophages. They offer distinct benefits based on the nanoscale size of the materials, particularly a high surface area-to-volume ratio, which provides significantly higher sensitivity than conventional, larger-scale materials. Nanowire biosensors have promising potential in a range of biomedical applications, including diagnostics of diseases, drug discovery, and detection of biomarkers, virions, and DNA, and can be highly accurate in detecting and classifying chemical and biological species. Their sensitivity is useful for early-stage diagnosis, personalized medicine, and monitoring patients in real time and makes potential timely and life-saving medical treatments much more likely. Recent developments in nanowire biosensors have also improved their functionality since their initial implementation: self-powered systems, stability in high ionic strength environments, reusability, and long-lasting operation time are some recent developments that address a lot of previously mentioned limitations of the biosensor. One of the most essential facts that influences the performance of nanowire biosensors is the Debye effect. The Debye effect can be decisive in the detection and sensitivity of signals from biosensors, especially in physiological media. In this review, we describe recent developments in nanowire biosensors, which include methods of fabrication and basic principles of operation. The focus here will be on practical applications to detect the most important biomolecules for human health. Due to their overall sensitivity, specificity, and versatility, nanowire biosensors have the potential to completely transform healthcare diagnostics. Combined with their efficiency, adaptability, and inherent potential for small size and portability that can allow complete integration, nanowires have received considerable attention in the literature and will continue to be a hotbed for research, commercialization, and future directions that will be an important component of clinical and biomedical applications.

Acknowledgment

We would like to extend our sincerest thanks to Prof. Binod Kumar Kanaujia, Director of National Institute of Technology Jalandhar for providing us with this opportunity and helping to create the environment for the start of this book project. We are also thankful to Dr. Ramesh Kumar Sunkaria, Head of the Department of Electronics and Communication Engineering, NIT Jalandhar, for guidance and support. We have been fortunate in receiving support, assistance, and encouragement from our colleagues and scholars in the Department of Electronics and Communication Engineering at NIT Jalandhar.

We would like to thank and acknowledge the efforts and support of all contributors for completing the book. This book is based on the research work carried out by a number of authors. Many people have contributed greatly to this book on "Nanowire: Materials, Devices, Circuits and Applications." We are grateful for the number of friends and colleagues in encouraging us to start the work on this book. We as the editors would like to acknowledge all of them for their valuable help and generous ideas in improving the quality of this book. With our feelings of gratitude, we would like to introduce them in turn. The first mention is the authors and reviewers of each chapter of this book. Without their outstanding expertise, constructive reviews, and devoted effort, this comprehensive book would become something without contents. The second mention is the CRC Press, Taylor and Francis staff for their constant encouragement, continuous assistance, and untiring support. Without their technical support, this book would not have been completed. The third mention is the editor's family for being the source of continuous love, unconditional support, and prayers not only for this work, but throughout our life. Last but far from least, we express our heartfelt thanks to the almighty for bestowing upon us the courage to face the complexities of life and complete this work.

July 2025
Shailendra Singh
Balwinder Raj
Arun Kumar Manoharan
Girdhar Gopal

Chapter 1

III–V Semiconductor Nanocrystal formation using aqueous state superlattice in silicon nanowires

Shakun Srivastava and Anshu Srivastava

1.1 INTRODUCTION

Nanowires (NWs) of III–V semiconductors and their capsized properties of (self-addition) production are discussed in this work for use in innovative and highly efficient between-band photovoltaic cells (BBPVCs). Nanowires-BBPVCs formed by this technology III-Nanowires-V mixture have the potential to remove maximum drawbacks that have been come across thus distant semiconductor nanocrystals or flat (planar) heterojunction structure BBPVCs. In fact, photovoltaic cells with previously unheard levels of power transformation strength, basic understanding of their geometry, lessened constructional limitations, minimum compound use, and manufacturing price should originate, thanks to the combination of BBPVCs' specific practicality, which is due to the presence of physical characteristics with nanowire-constituent gadgets. In spite of the very keen engineering of the gadgets having inner nanowire architectures, it is necessary for the manufacturing of III to V NWs-BBPVCs in order to meet the unique physical features of III–V semiconductor amalgams as well as the strict operating requirements of BBPVCs. As we are using this initially, we suggest the III–V main nanowires have more than one-cell heterostructure because of viable options for BBPVCs use. However, their manufacturing necessitates carefully regulated self-assembly methods. Reviewing the current state of the field's research, with an emphasis on forming a huge structure development of III–V NWs by the attributes of as-grown nanostructures and several of the atomic rays and metal carbon aqueous phase superlattice techniques, might be overcome by nanowire-BBPVCs, which depend on III-N-V compounds. The benefits and drawbacks of using bottom-up techniques are the formation of very low concentration III–V-based nanowire-BBPVCs, which are given and reviewed, along with significant findings from recent research and unresolved issues. Semiconductor NWs, a type of quasi one-dimensional nanostructure, have enormous promise for three-dimensional nanoelectronics in the future [1–10] to integrate the well-known complementary. The engineering based on CMOS requires realization by high-quality III–V/Si heterojunctions in order to benefit from the execution benefits of III–V semiconductors.

DOI: 10.1201/9781003634645-1

Gated resistors [8], LEDs [11], electronic and junction transistors [12–14], photodetectors [15], and photovoltaic devices [16] have all been manufactured making use of semiconductor NWs. Complete controlled fabrication of longitudinal III–V NWs [2, 17] or main shell NW heterostructures on Si reactants[18] is now possible thanks to recent advancements in chemical vapor deposition (CVD). Specifically, III–V compound semiconductors have every attribute needed for high-speed logic applications of the future. The majority of III–V binary semiconductors are ideal for use in optoelectronic devices because they have straight bandgaps and substantially better electron mobilities than Si. With one of the greatest electron mobilities among III–V semiconductors, As is a perfect choice for the channel material of tunnel field effect transistors (TFETs), which can operate at low voltages (<0.5 V) without experiencing a loss of performance. The challenges are the straightforward development of III–V semiconductors on Si, which are the primary barriers to the mass manufacture of such hybrid devices. These include substantial lattice mismatch, counter/counterbalancing superficial incongruity, and the variation in warm enlargement coefficient in the middle of Si and III–V substances. The ion beam impregnation method, followed by intense pulsed light (IPL) sintering "or" photonic curing, was previously employed to fabricate Si/III–V, two different semiconductors meeting in Si (silicon) at insulator (SOI) wafers [19] and bulk silicon [20, 21]. Here, we have investigated whether the ion beam impregnation method is appropriate for creating silicon-dependent prominent hetero-NW. In order to form we can go through treated samples that had each [12] aligned because –developed on silicon nanowires or nanowires used on shockproof(insulating) Si wafer with uppermost part of SiO_2 that was 100 nm thick. After being implanted with In and As ions, these samples were strengthened and hardened for 20 milliseconds. Si/in As/Si hetero-branched NWs with (11) oriented zinc blende in arsenic parts or fragments, or the SiNW in the As quantum dot (QD) system, were created, depending on the annealing temperature. We have demonstrated that the transition from rigid appearance to crystalline and thin film to liquid appearance to crystalline (LA) development during post-implantation annealing is the cause of the shift in the recrystallization process in the middle of the silicon NWs in arsenic QD structure and the Si/InAs/Si structure.

1.2 SYNTHESIS

The SiNWs with a <11> orientation were produced epitaxially on (1,000) Si compounds using a little-pressure substance utilizing a thin (<3 nm) thermionic vacuum tube sputtered gold coating as a trigger (catalyst) and silane (SiH_4) as a precursor gas in a vapor deposition (LPCVD) technique through the vapor–liquid–solid (VLS) development process [22]. The as-grown NWs had a diameter of between 50 and 200 nm. Aqua regia was used to remove the Au following NW growth. In essence, two kinds of samples were examined: flat and as-grown SiNWs placed on Si wafers with a thermally generated SiO_2 layer of 100 nm. Prior to implantation, a 20-nm SiO_2 shell was formed by rapidly thermally oxidizing the planar NWs. As and In ions were inserted into both kinds of SiNWs with a fluence of 1 × 1,016, 2 × 1,016, or 3.1 × 1,017 ions per cm^2 for every constituent. For the purpose of supplying a superimposed profundity dispensation of indium and arsenic atoms through the silicon NWs, the ion impregnation energy was classified as 101 keV for arsenic cation and 142 keV for indium cation. A Ni solid stiffen mask was obtained using electron rays lithography and electron ray evaporation of 85 Å Ni, and launch sequence and take off process using propanone, withdraw from impregnation channels of around 459 nm × 119 nm nickel coated nanowires through the use of micro-Raman spectroscopy, the optical characteristics were examined. Reverse Raman spectroscopy, also known as back scattering, was performed using an Alpha 300 phase contrast or centripetal microscope system that included a wide-aperture 100 × objective and a 532 nm Yttrium aluminum garnet doped with neodymium (Nd: YAG) laser [23–29]. At room temperature, between 150 and 1,000 cm^{-1}, the Raman microscopy was recorded at the sample surface using a low-level laser therapy (LLLT) of around 50 μW to avoid heating the specimen. Energy scatterers X-ray technique (ESX), measuring energy loss in the analysis process by the electron spectroscopy (ES), very elevated resolution transition electron microscopy (HRTEM), and elevated High-angle annular dark-field scanning transmission electron microscopy" (HAADF-STEM), which is often simply called "Z-contrast imaging" due to its sensitivity to atomic number (Z) variations in a sample. HAADF-STEM was carried out in a broad brush structure using an field electron and ion transmission electron microscope (FEI) transmission electron microscope process, F21 running at 200 keV. Lamellas for the cross-structures having cross-section transmission electron microscopy analysis were made within the lamellas, which were generated by focused ion beam milling within the axis or central line of silicon NWs for the cross-section transmission electron microscope (TEM) study. Following implantation, "a pulsed light annealing system," "photonic sintering system," or "intense pulsed light curing system" was used to anneal the samples at various temperatures. The plasma surface activation (PSA) was supplied for 20 ms after three minutes of preheating at 600°C [23]. The fractional crystallization of silicon NWs and the huddling or bunching of As were initiated by the pre-warm-up or before heating. We will just mention the flash lamp annealing (FLA) conditions in the discussion that follows because the preheating conditions were maintained at the same level for every sample. During FLA, the energy density collected over the sample layer ranged from 46 to 67 J/cm^2, which corresponds to a final temperature between 1,000 and 1,300°C. The annealing process was carried out in an environment with continuous argon flow. Additionally, the planar SiNWs underwent a specific area of ion beam impregnation, which was again used by PSA. In order to do this, NWs were distributed using a 100 nm SiO_2 to layer on an insulating Si substrate. Two methods were determined to be appropriate [30–37]. The first method included using plasma-enhanced CVD (PECVD) at 300°C to deposit 120 nm SiO_2. The SiNW core was revealed by buffered hydrofluoric acid (HF) etching after implantation of window-centered NWs, which were coated in a Perspex or Plexiglass (PMMA) surface using e-beam lithography. A 177 Å oxide coating for impregnation masking beyond the impregnation channel was then produced by PECVD after 54 nm SiO_2 was once again deposited (Figure 1.4a). For the second method, PECVD was used to deposit 50 nm SiO_2 at 300°C.

1.3 RESULT (FINDINGS) AND CONVERSATION

The scanning electron microscope (SEM) pictures of silicon nanowire (SiNWs) executed with various influences and annealed at the values for 20 ms are 58.1 J/cm^2 ((a) and (b)), 66.5 J/cm^2 (c), or 54.2 J/cm^2 ((d) and (e)). In all NWs, the InAs precipitates (bright region) are easily visible because of the Z-imparity between silicon and impregnated atoms. Arsenic nanodots are sporadically scattered across the length of the flat silicon NWs that have been reinforced beneath the melting temperature of the unstructured or powdered silicon bond (see Figure 1.1a and b). As implantation fluence rises, so does the size of the As nanodots. Conglomerate silicon in arsenic NWs with isolated arsenic fragments was generated in a specimen ingrafted with 2.9 × 1,016 ions/cm^2 and reinforced with a higher melting temperature of a-Si (see Figure 1.1c). The coefficient of absorption of the reinforced specimen has a significant impact on the ultimate temperature attained after FLA treatment [47–55]. Since the dense vertical SiNW-grown samples absorb significantly greater light in the visible spectrum than planarly scattered silicon NWs, they are used for the purpose of achieving identical NWs. The absorption coefficient of the annealed specimen has a significant impact on the ultimate temperature attained after FLA treatment. The quantity of energy imparted to the sample during FLA has to be elevated in order to achieve the same temperature. As the samples were grown, they exhibited a dense vertical structure. Silicon NWs capture significantly more light of greater intensity in the visible spectrum region than flatly distributed silicon NWs. It turns out that the peak temperature attained in the flat silicon NWs reinforced at 57.9 KJ/m^2 was lower by 101°C, while the ultimate temperature achieved for the flat silicon NWs subjected to reinforcing at 67.9 KJ/m^2 for 19.9 milliseconds is equivalent to that of the as-grown silicon NWs reinforced at 53.9 J/cm^2, which corresponds to a peak temperature of roughly 1,190°C. The silicon NWs are completely amorphous right after high-fluence ion implantation [56–64]. Given that bulk InAs melts at around 940°C and that the melting temperature of a silicon bond is roughly 199 ± 50 K. The value is less than that of crystalline silicon [23], we may presume that the NWs shown in Figure 1.1c–e. The material was liquefied during the FLA process.

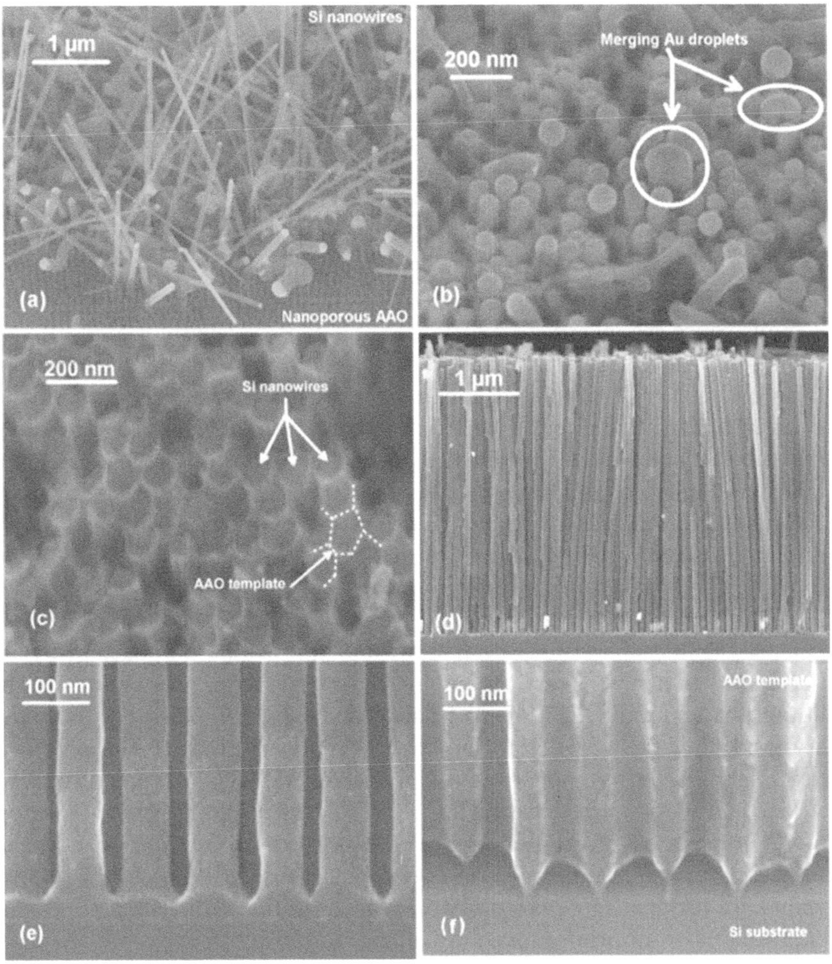

Figure 1.1 A picture of the planarization of a silicon nanowire array captured by scanning electron microscopy. (a) Following growth, silicon nanowires that proliferated from the anodic aluminum oxide (AAO) template fill it. (b) The post-growth AAO surface is revealed when the sample is sonicated, breaking the exterior nanowires. (c) Inductively coupled plasma (ICP) silicon etching and I-KI gold etching cause the nanowire array to become planar. (d) After alumina is removed by HF etching, a cross-section reveals the "top-down-like" nanowire array with excellent length homogeneity and high density. (e) A close-up of the interface between the customized nanowires and the Si (100) substrate. (f) In accordance with the geometry of (g), empty the AAO template prior to gold catalyst electrodeposition [38–46].

As a result, conglomerate silicon in arsenic NWs is obtained owing to the aqueous state superlattice in the millisecond range. Aggressive diffusion of In and As occurs when the system is operating in the liquid phase. At a speed of 10–20 m/s, the unstructured aqueous/structured solid boundary travels down the NW throughout this process, combining the implanted elements in a floating manner toward the tip of the silicon NWs, where they separate to create arsenic fragments or parts (see Figure 1.1c–e) [23]. Moreover, compared to the solid state, the liquid Si diffusion coefficient for In and As (~10–4 $cm^2 \bullet s^{-1}$) is around five orders of magnitude greater [25].

This translates to an implanted element's diffusion length (LD) of around 2 m/ms. Since the investigated NWs have an average length of around 2 m, the implanted elements can flow across the entire length of the nanowire throughout the FLA process, as long as they are in the form of a liquid silicon phase, which is visible only for one millisecond. Silicon NWs are not molten in samples that have been annealed at 58.1 J/cm^2 (about 1,100°C) and recrystallize as a result of solid-phase epitaxy (Figure 1.1a). A hybrid nanowire is created in the millisecond range. An explosive crystallization process (ECP) is a good way to explain FLA treatment [26, 27]. According to the annealing settings, Geile et al. [26] suggested four forms of ECP: (i) explosive processes include (i) solid-phase nucleation, (ii) explosive solid-phase superlattice, (iii) explosive liquid-phase nucleation, and (iv) explosive liquid-phase superlattice. It is most likely the result of explosive aqueous state super lattice with in liquid phase epitaxy (LPE)or artificial seeded substrate (ASS) regulates the crystallization process, that hybrid nanowires with distinct. The provided text illustrates the scanning electron microscopy [65–72].

Plasmon mapping reveals the SiO_2 shell in image (f) and the Si plasmon mode in image (g), corresponding to the as-grown SiNWs shown in image (d). The images reflect the ion fluencies applied during the process. InAs segments are formed. Hybrid nanowire creation in the millisecond range. An explosive crystallization process (ECP) is a good way to explain FLA treatment [26, 27]. According to the annealing settings, Geile et al. [26] suggested four forms of ECP: (i) explosive solid-phase nucleation, (ii) explosive solid-phase epitaxy, (iii) explosive liquid-phase nucleation, and (iv) explosive liquid-phase epitaxy. The production of hybrid nanowires with distinct InAs segments is most likely the result of explosive liquid-phase epitaxy, where the crystallization process is regulated by LPE. In contrast, explosive solid-phase epitaxy controls the crystallization of the SiNWs with InAs QDs. We assume that before the SiNWs melt, a silicon dioxide shell is formed, consuming residual oxygen present in the annealing chamber. Based on our simulation, the SiNWs are liquid for approximately 3 ms. The 25-nm thick oxide layer is deemed adequate for stabilizing the silicon nanowire during the explosive recrystallization process. The temperature simulation was conducted in accordance with the wave parameters. At a fluence of 54.2 J/cm^2 for a duration of 20 ms, a temperature of 1,000°C is attained after 10 ms, which is sufficient for the effective formation of the oxide layer (Figure 1.2).

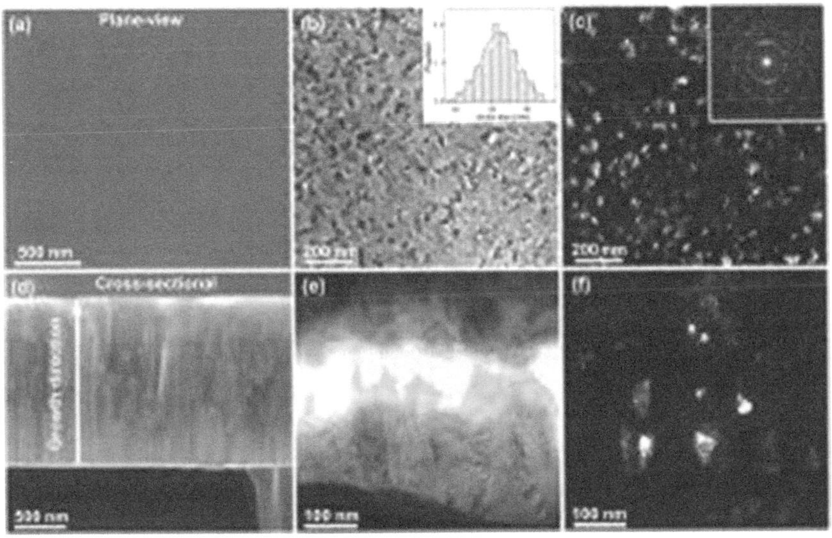

Figure 1.2 Typical nanocrystalline (NC) Mo film made by magnetron sputtering is shown in SEM (a, d), bright-field TEM (b, e), dark-field TEM (c, f), photographs, and (d–f) from cross-sectional view (a–c) from plane view. The grain size distribution based on 500 grains is displayed in the inset in (b). 44 nm is the average grain size. The matching selected area electron diffraction (SAED) pattern is seen in the inset in (c).

The estimated implantation range for In and As is around 70 nm below the surface, where they are generated. Micro-Raman spectroscopy was used to examine the production of crystalline InAs segments and the fractional crystallization procedure for the establishment of silicon NWs [73–87]. The micro-Raman spectra from a single hybrid Si/InAs NW are displayed in Figure 1.3. The prolonged optical phonon mode in crystalline silicon is represented by a single peak at 520 cm[1] in the micro-Raman spectroscopy taken from the crystalline SiNW (curve 1 in Figure 1.3) [29]. The tabulated values for bulk InAs crystals are well-fitted by the crest locations of arsenic connected to transverse optical (TO) and longitudinal optical (LO) phonon modes [30]. This demonstrates that the material is strain free and validates the good, standard arsenic fragments or parts produced in silicon NWs. In addition, the TO phonon modes predominate in the micro-Raman spectra, suggesting that the InAs crystals are oriented (111) in silicon NWs, which is consistent with HRTEM findings. The length of the examined InAs segment is normally around 150 nm, while the lateral spatial resolution of the four Raman systems is more than 400 nm. As a result, the Raman spectrum includes peaks that belong to both crystalline arsenic and silicon [88–96]. Tabulated values for bulk InAs crystals are well-fitted by the peak locations of the InAs-related TO and LO phonon modes [30]. This demonstrates that the material is strain-free and validates the good standard using the aqueous

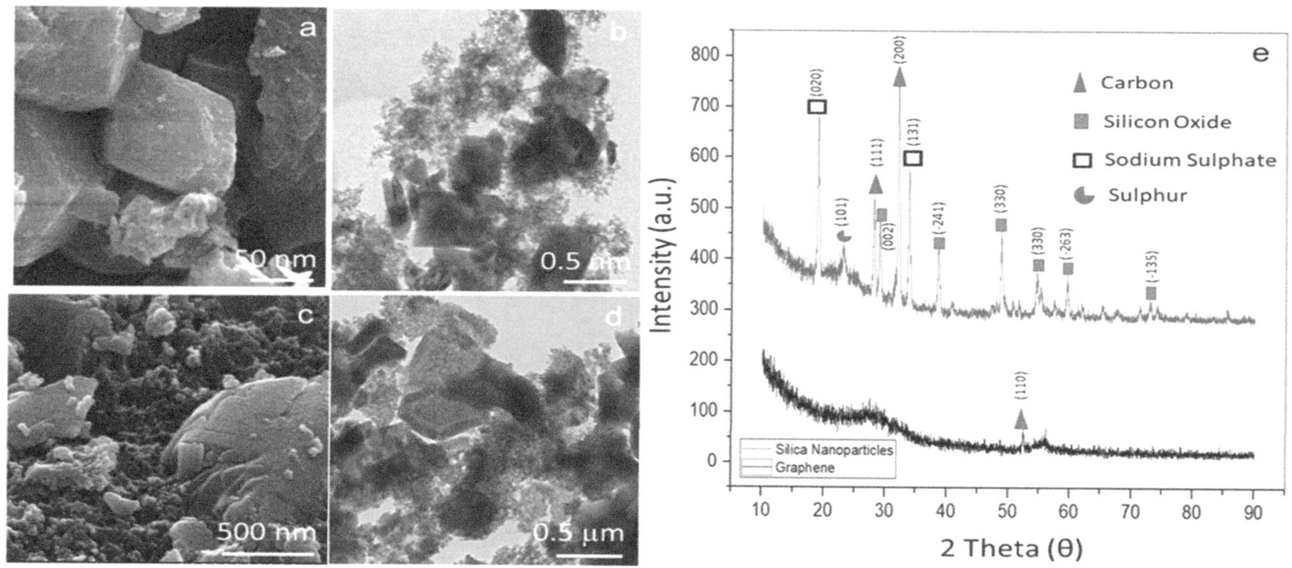

Figure 1.3 Using a newly designed current-volt aptasensor, silica and graphene mediate the detection of arsenic in mature rice grains.

state superlattice segment produced in silicon NWs using the aqueous state superlattice. Additionally, the TO phonon modes predominate in the micro-Raman spectra, suggesting that the InAs crystals are oriented (111) inside the SiNW, which is consistent with HRTEM findings [88–96]. The InAs segments are now randomly distributed along the SiNW. Particular area ion grafting was employed to create AsQDs at well-defined locations. To do this, e-beam lithography was used to define the implantation windows using either a 5,000 Å-thick Ni mask placed over the SiNWs (Figure 1.4b) or a 1,730 Å-thick PECVD SiO_2 layer (Figure 1.4a).

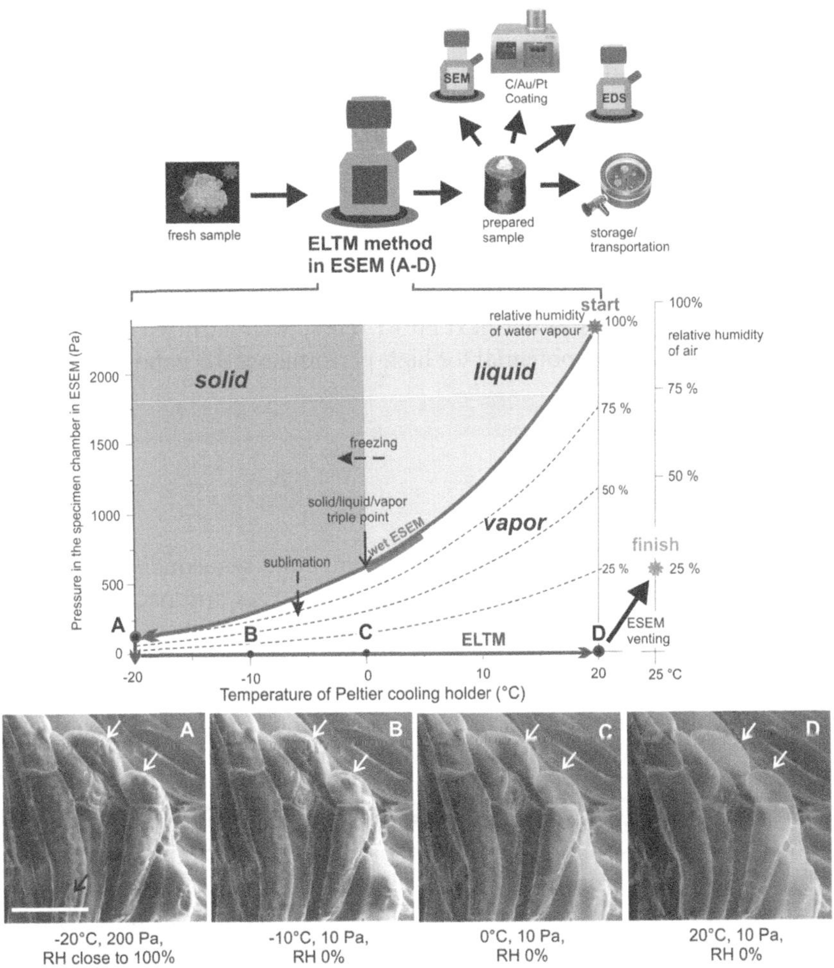

Figure 1.4 Plant samples are prepared in situ in an environmental scanning electron microscope (ESEM) for energy-dispersive X-ray microanalysis and repeated SEM and ESEM observations.

This is because the explosive liquid-phase recrystallization process begins on both borders of grafted channels at the same time [97–105]. The extent via the grafting channel, the ion grafting articulation, and the diameter of SiNWs all affect the size of InAs in the In arsenic part, shown having a length of around 600 Å and a width of 1,200 Å, which is the same as the implanted SiNW. This indicates that the average number of indium atoms in the InAs crystals is about 12 × 106. The engrafted portion of the silicon NWs has around 15.99 × 106 In elements if we assume that all of the indium that has been implanted (via a grafting channel of 5,000 Å× 1,200 Å) has collected within the SiNW [106–115]. We can determine that at least 75% of grafted atoms, including indium and arsenic, are agglomerated into single arsenic parts by comparing the number of grafted atoms with the number of atoms composing the arsenic nanocrystal. It can be inferred that the SiO_2 layer exhibits greater properties resistant to the ion graft at the ion technique based on the effectiveness of the PECVD SiO_2 and Ni masking techniques. The uppermost part underneath the nickel and the oxide shell around the silicon NWs were co-implanted because the Ni layer was partly sputtered during ion grafting. This results in the process of roughening the surface during the FLA or ASS and chemical milling procedures. The precise placement of scattered SiNW on CMOS circuits is undoubtedly difficult. Therefore, we will investigate top-down methods for creating horizontal NWs on e-SOI wafers in our next effort. These well-localized III–V heterojunctions have a lot of potential for high-performance 3D nanoelectronics in the future [116–125].

1.4 CONCLUSION

We have shown how to integrate III–V compound semiconductor material with CMOS technology in a revolutionary way. Our approach to the production of hetero NWs is based on well-accepted technology in the microelectronic sector [126–138]. The ion impregnation and post-implantation tempering and reinforcing environment completely govern the crystallographic conformations and form the crystals on indium and silicon implanted in SiNWs. The ion-impregnated silicon NWs change the silicon NWs to the indium arsenic QD technique to Si/In As/Si hetero NWs for a fixed ion control as the reinforcing temperature rises. The method of recrystallization by the process of post-impregnation reinforcing determines the various types of NWs that are produced. Atomically crisp connections among the zinc blende and silicon in arsenic indium fragments are seen in the Si/InAs/Si hybrid NWs produced by aqueous state superlattice [139–150]. Lastly, selective ion beam implantation was used to create Si/InAs/Si hybrid NWs with well-localized mono indium as parts or fragments. We have demonstrated that this technique works for both horizontal and epitaxial SiNWs. We have demonstrated that this technique works for both horizontal and epitaxial SiNWs, and then a pure top-down strategy via the proper e-beam. We will investigate dry etching and patterning the arrangement of silicon or silicon-on-insulator wafers' lateral distribution.

REFERENCES

1. Hocevar, M., Immink, G., Verheijen, M., Akopian, N., Zwiller, V., Kouwenhoven, L., Bakkers, E. Growth and optical properties of axial hybrid III–V/silicon nanowires. *Nature Communications*, 3, p. 1266, 2012.
2. Otnes, G., Borgstrom, M. T., *Nano Today*, 12, p. 31, 2016.
3. Polman, A., Knight, M., Garnett, E. C., Ehrler, B., Sinke, W. C., Photovoltaic materials: Present efficiencies and future challenges. *Science*, 352, p. aad4424, 2016.
4. Tomioka, K., Yoshimura, M., Fukui, T. A III–V nanowire channel on silicon for high-performance vertical transistors. *Nature*, 488, pp. 189–192, 2012.
5. Tang, Y., Chen, Z., Song, H., Lee, C., Cong, H., Cheng, H., Zhang, W., Bello, I., Lee, S., Vertically aligned p-type single-crystalline GaN nanorod arrays on n-type Si for heterojunction photovoltaic cells. *Nano Letters*, 8, p. 4191, 2008.
6. Justice, J., Bower, C., Meitl, M., Mooney, M. B., Gubbins, M. A., Corbett, B. Wafer-scale integration of group III–V lasers on silicon using transfer printing of epitaxial layers. *Nature Photonics*, 6, pp. 610–614, 2012.
7. Minot, E. D., Kelkensberg, F., Van Kouwen, M., Van Dam, J. A., Kouwenhoven, L. P., Zwiller, V., Borgstrom, M. T., Wunnicke, O., Verheijen, M. A., Bakkers, E. P., Single quantum dot nanowire LEDs. *Nano Letters*, 7, p. 367, 2007.
8. Bessire, C. D., Björk, M. T., Schmid, H., Schenk, A., Reuter, K. B., Riel, H., Trap-assisted tunneling in Si–InAs nanowire heterojunction tunnel diodes. *Nano Letters*, 11, pp. 4195–4199, 2011.
9. Ng, K. W., Ko, W. S., Tran, T. T., Chen, R., Nazarenko, M. V., Lu, F., Dubrovskii, V. G., Kamp, M., Forchel, A., Chang-Hasnain, C. J. Unconventional growth mechanism for monolithic integration of III–V on silicon. *ACS Nano*, 7, pp. 100–107, 2013.
10. Lugstein, A., Steinmair, M., Hyun, Y. J., Hauer, G., Pongratz, P., Bertagnolli, E. Pressure-induced orientation control of the growth of epitaxial silicon nanowires. *Nano Letters*, 8, pp. 2310–2314, 2008.
11. Ohdaira, K., Fujiwara, T., Endo, Y., Nishizaki, S., Matsumura, H. Explosive crystallization of amorphous silicon films by flash lamp annealing. *Journal of Applied Physics*, 106, 044907, 2009.
12. Skorupa, W., Gebel, T., Yankov, R. A., Paul, S., Lerch, W., Downey, D. F., Arevalo, E. A. J. Advanced thermal processing of ultrashallow implanted junctions using flash lamp annealing. *Journal of the Electrochemical Society*, 152, pp. G436–G440, 2005.
13. Thompson, M. O., Galvin, G. J., Mayer, J. W., Peercy, P. S., Poate, J. M., Jacobson, D. C., Cullis, A. G., Chew, N. G. Melting temperature and explosive crystallization of amorphous silicon during pulsed laser irradiation. *Physical Review Letters*, 52, p. 2360, 1984.
14. Garandet, J. P. New determinations of diffusion coefficients for various dopants in liquid silicon. *International Journal of Thermophysics*, 28, 1285–1303, 2007.

15. Geiler, H.-D., Glaser, E., Götz, G., Wagner, M. J. Explosive crystallization in silicon. *Journal of Applied Physics*, 59, 3091, 1986.
16. Smith, M., McMahon, R. A., Voelskow, M., Skorupa, W. Modeling and regrowth mechanisms of flash lamp processing of SiC-on-silicon heterostructures. *Journal of Applied Physics*, 96, 4843, 2004.
17. Martienssen, W., Warlimont, H. *Springer Handbook of Condensed Matter and Materials Data*, Springer, 2005.
18. Prucnal, S., Turek, M., Drozdziel, A., Pyszniak, K., Zhou, S. Q., Kanjilal, A., Skorupa, W., Zuk, J. Formation of InAs quantum dots in silicon by sequential ion implantation and flash lamp annealing. *Applied Physics B: Lasers and Optics*, 101, 315–319, 2010.
19. Polman, A., Atwater, H. A., *Nature Materials*, 11, 174, 2012.
20. Conibeer, G., *Materials Today*, 10, 42, 2007.
21. Tsakalakos, L., *Materials Science and Engineering Reports*, 62, 175, 2008.
22. Beard, M. C., Luther, J. M., Nozik, A. J., *Nature Nanotechnology*, 9, 951, 2014.
23. Maeda, S., Tomioka, K., Hara, S., Motohisa, J., *Japanese Journal of Applied Physics*, 51, 02BN03, 2012.
24. del Alamo, J. A. Nanometre-scale electronics with III–V compound semiconductors. *Nature*, 479, 317–323, 2011.
25. Lubyshev, D., Fastenau, J. M., Liu, W. K. Integration of III–V and Si CMOS devices by molecular beam epitaxy. *ECS Transactions*, 19, 295–308, 2009.
26. Colinge, J.-P., Lee, C.-W., Afzalian, A., Akhavan, N. D., Yan, R., Ferain, I., Razavi, P., O'Neill, B., Blake, A., White, M. et al. Nanowire transistors without junctions. *Nature Nanotechnology*, 5, 225–229, 2010.
27. Razavi, P., Fagas, G. Electrical performance of III–V gateall- around nanowire transistors. *Applied Physics Letters*, 103, 063506, 2013.
28. Colli, A., Fasoli, A., Ronning, C., Pisana, S., Piscanec, S., Ferrari, A. C. Ion beam doping of silicon nanowires. *Nano Letters*, 8, 2188–2193, 2008.
29. Cui, Y., Zhong, Z. H., Wang, D. L., Wang, W. U., Lieber, C. M. High performance silicon nanowire field effect transistors. *Nano Letters*, 3, 149–152, 2003.
30. Xiang, J., Lu, W., Hu, Y. J., Wu, Y., Yan, H., Lieber, C. M. Ge/Si nanowire heterostructures as high performance fieldeffect transistors. *Nature*, 441, 489–493, 2006.
31. Lu, W., Xie, P., Lieber, C. M. Nanowire transistor performance limits and applications. *IEEE Transactions on Electron Devices*, 55, 2859–2876, 2008.
32. Fan, P. Y., Chettiar, U. K., Cao, L. Y., Afshinmanesh, F., Engheta, N., Brongersma. M. L. An invisible metal– semiconductor photodetector. *Nature Photonics*, 6, 380–385, 2012.
33. Limbach, F., Hauswald, C.,Lähnemann, J., Wölz, M., Brandt, O., Trampert, A., Hanke, M., Jahn, U., Calarco, R., Geelhaar L. et al. Current path in light emitting diodes based on nanowire ensembles. *Nanotechnology*, 23, 465301, 2012.
34. Gao, Q., Saxena, D., Wang, F., Fu, L., Mokkapati, S., Guo, Y., Li, L., Wong-Leung, J., Caroff, P., Tan, H. H., *Nano Letters*, 14, 5206, 2014.
35. Peng, K., Parkinson, P., Fu, L., Gao, Q., Jiang, N., Guo, Y.-N., Wang, F., Joyce, H. J., Boland, J. L., Tan, H. H., *Nano Letters*, 15, 206, 2014.
36. Li, Z., Yuan, X., Fu, L., Peng, K., Wang, F., Fu, X., Caroff, P., White, T. P., Tan, H. H., Jagadish, C., *Nanotechnology* 26, 445202, 2015.

37. Krogstrup, P., Jorgensen, H. I., Heiss, M., Demichel, O., Holm, J. V., Aagesen, M., Nygard, J., Fontcuberta i Morral, A., *Nature Photonics* 2013, 7, 306.
38. Yadav, D., Raj, B., Raj, B., "Design and simulation of low power microcontroller for IoT applications", *Journal of Sensor Letters*, 18, pp. 401–409, May 2020.
39. Singh, S., Raj, B., "A 2-D analytical surface potential and drain current modeling of double-gate vertical t-shaped tunnel FET", *Journal of Computational Electronics*, 19, pp. 1154–1163, April 2020a.
40. Singh, J., Raj, B., "An accurate and generic window function for non-linear memristor model", *Journal of Computational Electronics*, 18, No. 2, pp. 640–647, June 2019a.
41. Kaur, Manjit, Gupta, Neena, Kumar, Sanjeev, Raj, B., Singh, Arun Kumar, Comparative RF and crosstalk analysis of carbon based nano interconnects", *IET Circuits, Devices and Systems*, 15, No. 6, pp. 493–503, February 2021.
42. Kandasamy, Nehru, Ahmad, Firdous, Ajitha, D., Raj, B., Telagam, Nagarjuna "Quantum dot cellular automata based scan flip flop and boundary scan register", *IETE Journal of Research*, 66, pp. 535–548, 2020.
43. Singh, Jeetendra, Raj, B., "Design and investigation of 7T2M NVSARM with enhanced stability and temperature impact on store/restore energy", *IEEE Transactions on Very Large Scale Integration Systems*, 27, No. 6, pp. 1322–1328, June 2019b.
44. Bhardwaj, Anil Kumar, Gupta, Sumeet, Raj, B., Singh, Amandeep, "Impact of double gate geometry on the performance of carbon nanotube field effect transistor structures for low power digital design", *Computational and Theoretical Nanoscience*, 16, pp. 1813–1820, 2019.
45. Jain, Neeraj, Raj, B., "Thermal stability analysis and performance exploration of asymmetrical dual-k underlap spacer (ADKUS) SOI FinFET for security and privacy applications", *Indian Journal of Pure & Applied Physics (IJPAP)*, 57, pp. 352–360, May 2019a.
46. Jain, Neeraj, Raj, Balwinder, "Dual-k spacer region variation at the drain side of asymmetric SOI FinFET structure: Performance analysis towards the analog/RF design applications", *Journal of Nanoelectronics and Optoelectronics*, 14, pp. 349–359, March 2019b.
47. Zhong, Z., Li, Z., Gao, Q., Li, Z., Peng, K., Li, L., Mokkapati, S., Vora, K.,Wu, J., Zhang, G., *Nano Energy*, 28, 106, 2016.
48. Mokkapati, S., Jagadish, C., *Optics Express*, 24, 17345, 2016.
49. Anttu, N., *ACS Photonics*, 2, 446, 2015.
50. Fukui, T., Yoshimura, M., Nakai, E., Tomioka, K., *Ambio*, 41, 119, 2012.
51. Gutsche, C., Lysov, A., Braam, D., Regolin, I., Keller, G., Li, Z. A.,Geller, M., Spasova, M., Prost, W., Tegude, F. J., *Advanced Functional Materials*, 22, 929, 2012.
52. Aberg, I., Vescovi, G., Asoli, D., Naseem, U., Gilboy, J. P., Sundvall, C., Dahlgren, A., Svensson, K. E., Anttu, N., Bjork, M. T., *IEEE Journal of Photovoltaics*, 6, 185, 2016.
53. Xu, S., Qin, Y., Xu, C., Wei, Y. G., Yang, R. S., Wang, Z. L. Self-powered nanowire devices. *Nature Nanotechnology*, 5, 366–373, 2010.
54. Dick, K. A., Kodambaka, S., Reuter, M. C., Deppert, K., Samuelson, L., Seifert, W., Wallenberg, L. R., Ross, F. M. The morphology of axial and branched nanowire heterostructures. *Nano Letters* 2007, 7, 1817–1822.

55. Lauhon, L. J., Gudiksen, M. S., Wang, D. L., Lieber, C. M. Epitaxial core–shell and core–multishell nanowire heterostructures. *Nature*, 420, 57–61, 2002.
56. Prucnal, S., Facsko, S., Baumgart, C., Schmidt, H., Liedke, M. O., Rebohle, L., Shalimov, A., Reuther, H., Kanjilal, A., Mücklich, A. et al. n-InAs nanopyramids fully integrated into silicon. *Nano Letters*, 11, 2814–2818, 2011.
57. Prucnal, S., Zhou, S. Q., Ou, X., Reuther, H., Liedke, M. O., Mücklich, A., Helm, M., Zuk, J., Turek, M., Pyszniak, K. et al. InP nanocrystals on silicon for optoelectronic applications. *Nanotechnology*, 23, 485204, 2012.
58. Prucnal, S., Zhou, S. Q., Ou, X., Facsko, S., Liedke, M. O., Bregolin, F., Liedke, B., Grebing, J., Fritzsche, M., Hübner, R. et al. III–V/Si on silicon-on-insulator platform for hybrid nanoelectronics. *Journal of Applied Physics*, 115, 074306, 2014.
59. Bhardwaj, Anjana, Pradeep Kumar, B. Raj, Sunny Anand, "Design and performance optimization of doping-less vertical nanowire TFET using gate stack technique", *Journal of Electronic Materials (JEMS)*, 41, No. 7, pp. 4005–4013, 2022.
60. Singh, Jeetendra, Raj, B., "Tunnel current model of asymmetric MIM structure levying various image forces to analyze the characteristics of filamentary memristor", *Applied Physics A*, 125, No. 3, pp. 203.1–203.11 February 2019a.
61. Goyal, Candy, Ubhi, Jagpal Singh, Raj, B., "Low leakage zero ground noise nanoscale full adder using source biasing technique," *Journal of Nanoelectronics and Optoelectronics*, 14, pp. 360–370, March 2019.
62. Singh, Gurmohan, Sarin, R. K., Raj, B., "A novel robust exclusive-OR function implementation in QCA nanotechnology with energy dissipation analysis", *Journal of Computational Electronics*, 15, No. 2, pp. 455–465, June 2016.
63. Wadhwa, Girish, Kamboj, Priyanka, Raj, Balwinder "Design optimisation of junctionless TFET biosensor for high sensitivity", *Advances in Natural Sciences: Nanoscience and Nanotechnology*, 10, pp. 045001, 2019.
64. Bansal, Priya, Raj, B., Memristor modeling and analysis for linear dopant drift kinetics", *Journal of Nanoengineering and Nanomanufacturing*, 6, pp. 1–7, 2016.
65. Ashima, D. V., Raj, B., "Performance analysis of charge plasma induced graded channel Si nanotube", *Journal of Engineering Research (JER)*, pp. 146–154, 2021.
66. Tomar, A. S., Magraiya, V. K., Raj, B., "Scaling of access and data transistor for high performance DRAM cell design", *Quantum Matter*, 2, pp. 412–416, 2013.
67. Jain, Neeraj, Raj, B., "Parasitic capacitance and resistance model development and optimization of raised source/drain SOI FinFET structure for analog circuit applications", *Journal of Nanoelectronics and Optoelectronins*, 13, pp. 531–539, April 2018a.
68. Singh, Shradhya, Vishvakarma, S. K., Raj, B., "Analytical modeling of split-gate junction-less transistor for a biosensor application", *Sensing and Bio-sensing*, 18, pp. 31–36, April 2018.
69. Gopal, Maisagalla, Raj, Balwinder, "Low power 8T SRAM cell design for high stability video applications", *ITSI Transaction on Electrical and Electronics Engineering*, 1, No. 5, pp. 91–97, 2013.
70. Raj, Balwinder, Mitra, Jatin, Bihani, Deepak Kumar, Rangharajan, V., Saxena, A.K., Dasgupta, S., "Analysis of noise margin, power and process

variation for 32 nm FinFET based 6T SRAM cell", *Journal of Computer (JCP)*, 5, No. 6, pp. 1–8, 2010.

71. Sharma, Divya, Mehra, Rajesh, Raj, B., "Comparative analysis of photovoltaic technologies for high efficiency solar cell design", *Superlattices and Microstructures*, 153, pp. 106861, May 2021.
72. Kaur, Pawandeep, Buttar, A. S., Raj, B., "A comprehensive analysis of nanoscale transistor based biosensor: A review", *Indian Journal of Pure and Applied Physics*, 59, pp. 304–318, April 2021.
73. Singh, Jeetendra, Sharma, Sanjeev, Raj, B., Khosla, Mamta "Analysis of barrier layer thickness on performance of In1-xGaxAs based gate stack cylindrical gate nanowire MOSFET", *JNO*, 13, pp. 1473–1477, 2018.
74. Jain, Neeraj, Raj, B., "Analysis and performance exploration of high-k SOI FinFETs over the conventional low-k SOI FinFET toward analog/RF design", *Journal of Semiconductors (JoS)*, 39, No. 12, p. 124002-1-7, December 2018b.
75. Goyal, Candy, Ubhi, Jagpal Singh, Raj, B., "A reliable leakage reduction technique for approximate full adder with reduced ground bounce noise", *Journal of Mathematical Problems in Engineering*, 2018, p. 3501041, 2018.
76. Anuradha, Jeetendra Singh, Raj, B., Khosla, Mamta, "Design and performance analysis of nano-scale memristor-based nonvolatile SRAM", *Journal of Sensor Letter*, 16, pp. 798–805, 2018.
77. Wadhwa, Girish, Raj, B., "Parametric variation analysis of charge-plasma-based dielectric modulated JLTFET for biosensor application", *IEEE Sensors Journal*, 18, No. 15, pp. 6070–6077, 2018.
78. Singh, Jeetendra, Raj, B., "Comparative analysis of memristor models for memories design", *JOS*, 39, No. 7, pp. 074006-1-12, July 2018a.
79. Yadav, Divya, Chouhan, Shailesh Singh, Vishvakarma, Santosh Kumar, Raj, B., "Application specific microcontroller design for IoT based WSN", *Sensor Letters*, 16, pp. 374–385, May 2018.
80. Singh, Gurmohan, Sarin, R. K., Raj, B., "Fault-tolerant design and analysis of quantum-dot cellular automata based circuits", *IEEE/IET Circuits, Devices & Systems*, 12, pp. 638–664, 2018.
81. Singh, Jeetendra, Raj, B., "Modeling of mean barrier height levying various image forces of metal insulator metal structure to enhance the performance of conductive filament based memristor model", *IEEE Nanotechnology*, 17, No. 2, pp. 267–268, March 2018b.
82. Jain, Aakash, Sharma, Sanjeev, Raj, B., "Analysis of triple metal surrounding gate (TM-SG) III-V nanowire MOSFET for photosensing application", *Opto-electronics Journal*, 26, No. 2, pp. 141–148, May 2018.
83. Jain, Aakash, Sharma, Sanjeev, Raj, B., "Design and analysis of high sensitivity photosensor using cylindrical surrounding gate MOSFET for low power sensor applications", *Engineering Science and Technology, an International Journal*, 19, No. 4, pp. 1864–1870, 2016.
84. Saiphani Kumar, G., Singh, Amandeep, Raj, B., "Design and analysis of gate all around CNTFET based SRAM cell design", *Journal of Computational Electronics*, 17, No. 1, pp. 138–145, March 2018.
85. Singh, Gurinder Pal, Sohi, B. S., Raj, Balwinder, "Material properties analysis of graphene base transistor (GBT) for VLSI analog circuits", *Indian Journal of Pure & Applied Physics (IJPAP)*, 55, pp. 896–902, December 2017.

86. Kumar, Sunil, Raj, B., "Estimation of stability and performance metric for inward access transistor based 6T SRAM cell design using n-type/p-type DMDG-GDOV TFET", *IEEE VLSI Circuits and Systems Letter*, 3, No. 2, pp. 25–39, June 2017.
87. Sharma, Shashikant, Kumar, Anjan, Pattanaik, Manisha, Raj, B., "Forward body biased multimode multi-threshold CMOS technique for ground bounce noise reduction in static CMOS adders", *International Journal of Information and Electronics Engineering*, 3, No. 3, pp. 567–572, 2013.
88. Singh, Hamendra, Kumar, Pankaj, Raj, Balwinder, "Performance analysis of majority gate SET based 1-bit full adder", *International Journal of Computer and Communication Engineering (IJCCE)*, 2, No. 4, 2013.
89. Bhardwaj, Anil Kumar, Gupta, Sumeet, Raj, B., "Investigation of parameters for schottky barrier (SB) height for schottky barrier based carbon nanotube field effect transistor device", *Journal of Nanoelectronics and Optoelectronics*, 15, pp. 783–791, 2020a.
90. Bansal, Priya, Raj, B., "Memristor: A versatile nonlinear model for dopant drift and boundary issues", *JCTN*, 14, No. 5, pp. 2319–2325, May 2017.
91. Jain, Neeraj, Raj, B., "An analog and digital design perspective comprehensive approach on Fin-FET (Fin-field effect transistor) technology – A review", *Reviews in Advanced Sciences and Engineering (RASE)*, 5, pp. 1–14, 2016a.
92. Raj, Balwinder, Saxena, A. K., Dasgupta, S., "Analytical modeling for the estimation of leakage current and subthreshold swing factor of nanoscale double gate FinFET device", *MicroElectronics International*, 26, pp. 53–63, 2009a.
93. Soniya, Shailendra Singh, Wadhwa, Girish, Raj, Balwinder, "An analytical modeling for dual source vertical tunnel field effect transistor", *International Journal of Recent Technology and Engineering (IJRTE)*, 8, No. 2, pp. 603–608, July 2019.
94. Singh, Shailendra, Raj, B., "Design and analysis of hetrojunction vertical T-shaped tunnel field effect transistor", *Journal of Electronics Material*, 48, No. 10, pp. 6253–6260, 2019b.
95. Goyal, Candy, Ubhi, Jagpal Singh, Raj, B., "A low leakage CNTFET based inexact full adder for low power image processing applications", *International Journal of Circuit Theory and Applications*, 47, No. 9, pp. 1446–1458, 2019.
96. Raj, B., Saxena, A. K., Dasgupta, S., "A compact drain current and threshold voltage quantum mechanical analytical modeling for FinFETs", *Journal of Nanoelectronics and Optoelectronics (JNO)*, 3, No. 2, pp. 163–170, 2008.
97. Wadhwa, Girish, and Raj, B., "An analytical modeling of charge plasma based tunnel field effect transistor with impacts of gate underlap region", *Superlattices and Microstructures*, 142, pp. 106512, June 2020.
98. Singh, Shailendra, Raj, B., "Modeling and simulation analysis of SiGe heterojunction double gate vertical t-shaped tunnel FET", *Superlattices and Microstructures*, 142, pp. 106496, June 2020b.
99. Jain, Neeraj, Raj, Balwinder, "Device and circuit co-design perspective comprehensive approach on FinFET technology – A review", *Journal of Electron Devices*, 23, No. 1, pp. 1890–1901, 2016b.
100. Kumar, Sunil, Raj, B., "Analysis of I_{ON} and ambipolar current for dual-material gate-drain overlapped DG-TFET," *Journal of Nanoelectronics and Optoelectronics*, 11, pp. 323–333, June 2016.

101. Anjum, Naveed, Bali, Tarun, Raj, B., Design and simulation of handwritten multiscript character recognition", *International Journal of Advanced Research in Computer and Communication Engineering*, 2, No. 7, pp. 2544–2549, July 2013a.
102. Singh, Karmjit, Raj, B., "Performance and analysis of temperature dependent multi-walled carbon nanotubes as global interconnects at different technology nodes", *Journal of Computational Electronics*, 14 No. 2, pp. 469–476, June 2015a.
103. Kumar, Sunil, Raj, B., "Compact channel potential analytical modeling of DG-TFET based on evanescent–mode approach", *Journal of Computational Electronics*, 14, No. 2, pp. 820–827, July 2015.
104. Singh, Karmjit, Raj, B., "Temperature dependent modeling and performance evaluation of multi-walled CNT and single-walled CNT as global interconnects", *Journal of Electronic Materials*, 44, No. 12, pp. 4825–4835, 2015b.
105. Sharma, V. K., Pattanaik, M., Raj, B., "INDEP approach for leakage reduction in nanoscale CMOS circuits", *International Journal of Electronics*, 102, No. 2, pp. 200–215, 2014a.
106. Singh, Karmjit, Raj, B., "Influence of temperature on MWCNT bundle, SWCNT bundle and copper interconnects for nanoscaled technology nodes", *Journal of Materials Science: Materials in Electronics*, 26, No. 8, pp. 6134–6142, 2015.
107. Anjum, Naveed, Bali, Tarun, Raj, B., "Design and simulation of handwritten Gurumukhi and Devanagri numerical recognition", *International Journal of Computer Applications*, 73, No. 12, pp. 16–21, 2013b.
108. Khandelwal, S., Gupta, V., Raj, B., Gupta, R. D., "Process variability aware low leakage reliable nano scale DG-FinFET SRAM cell design technique", *Journal of Nanoelectronics and Optoelectronics*, 10, No 6, pp. 810–817, 2015.
109. Sharma, V. K., Pattanaik, M., Raj, B., "ONOFIC approach: Low power high speed nanoscale VLSI circuits design", *International Journal of Electronics*, 101, No. 1, pp. 61–73, 2014b.
110. Khandelwal, S., Raj, Balwinder, Gupta, R. D., "FinFET based 6T SRAM cell design: analysis of performance metric, process variation and temperature effect", *Journal of Computational and Theoretical Nanoscience*, 12, pp. 2500–2506, 2015.
111. Singh, Sumit, Yadav, Shekhar, Jagdeep, Rahul, Srivastava, Anurag, Raj, B., "Impact of HfO_2 in graded channel dual insulator double gate MOSFET", *Journal of Computational and Theoretical Nanoscience*, 12, No. 6, pp. 950–953, April 2015.
112. Sharma, Vijay Kumar, Pattanaik, Manisha, Raj, B., "PVT variations aware low leakage INDEP approach for nanoscale CMOS circuits", *Microelectronics Reliability*, 54, pp. 90–99, 2014.
113. Raj, B., Saxena, A., Dasgupta, S., "Quantum mechanical analytical modeling of nanoscale DG FinFET: Evaluation of potential, threshold voltage and source/drain resistance", *Elsevier's Journal of Material Science in Semiconductor Processing*, 16, No. 4, pp. 1131–1137, 2013.
114. Gopal, Maisagalla, Prasad, Siva Sankar D., Raj, Balwinder, "8T SRAM cell design for dynamic and leakage power reduction", *International Journal of Computer Applications*, 71, No. 9, pp. 43–48, June 2013.

115. Pattanaik, Manisha, Raj, B., Sharma, Shashikant, Kumar, Anjan, "Diode based trimode multi-threshold CMOS technique for ground bounce noise reduction in static CMOS adders", *Advanced Materials Research*, 548, pp. 885–889, 2012.
116. Raj, Balwinder, Saxena, A. K., Dasgupta, S., "Nanoscale FinFET based SRAM cell design: Analysis of performance metric, process variation, underlapped FinFET and temperature effect", *IEEE Circuits and Systems Magazine*, 11, No. 2, pp. 38–50, 2011.
117. Sharma, V. K., Pattanaik, M., Raj, Balwinder, "Leakage Current ONOFIC Approach for Deep Submicron VLSI Circuit Design", *International Journal of Electrical, Computer, Electronics and Communication Engineering*, 7, No. 4, pp. 239–244, 2013.
118. Kaur, Parminder, Gill, Sandeep Singh, Raj, B., "Comparative analysis of OFETs materials and devices for sensor applications", *Journal of Silicon*, 14, pp. 4463–4471, 2022.
119. Sharma, Sanjeev Kumar, Kumar, Parveen, Raj, Balwant, Raj, B., "$In_{1-x}Ga_xAs$ double metal gate-stacking cylindrical nanowire MOSFET for highly sensitive photo detector", *Journal of Silicon*, 14, pp. 3535–3541, 2022.
120. Raj, B., Saxena, A. K., Dasgupta, S., "Analytical modeling of quasi planar nanoscale double gate FinFET with source/drain resistance and field dependent carrier mobility: A quantum mechanical study", *Journal of Computer (JCP)*, 4, No. 9, pp.1–8, 2009b.
121. Bhushan, S., Khandelwal, S., Raj, B., Analyzing different mode FinFET based memory cell at different power supply for leakage reduction, seventh international conference on bio-inspired computing: theories and application, (BIC-TA 2012), *Adv. Intell. Syst. Comput.*, 202, pp. 89–100, 2013.
122. Singh, Jeetendra, Raj, B., "Temperature dependent analytical modeling and simulations of nanoscale memristor", *Journal: Engineering Science and Technology, an International Journal*, 21, pp. 862–868, 2018c.
123. Singh, Shradhya, Bala, Shashi, Raj, Balwant, Raj, B., "Improved sensitivity of dielectric modulated junctionless transistor for nanoscale biosensor design", *Sensor Letters*, 18, pp. 328–333, April 2020.
124. Kumar, Vivek, Vishvakarma, Santosh Kumar, Raj, B., "Design and performance analysis of ASIC for IoT applications", *Sensor Letters*, 18, pp. 31–38, January 2020.
125. Jaiswal, Akanksha, Sarin, R. K., Raj, B., Sukhija, Shikha, "A novel circular slotted microstrip-fed patch antenna with three triangle shape defected ground structure for multiband applications", *Advanced Electromagnetic (AEM)*, 7, No. 3, pp. 56–63, 2018.
126. Wadhwa, Girish, Raj, B., "Label free detection of biomolecules using charge-plasma-based gate underlap dielectric modulated junctionless TFET", *Journal of Electronic Materials (JEMS)*, 47, No. 8, pp. 4683–4693, 2018.
127. Singh, Gurmohan, Sarin, R. K., Raj, B., "design and performance analysis of a new efficient coplanar quantum-dot cellular automata adder", *Indian Journal of Pure & Applied Physics (IJPAP)*, 55, pp. 97–103, February 2017a.
128. Kaur, Parminder, Pandey, V., Raj, B., "Comparative study of efficient design, control and monitoring of solar power using IoT", *Sensor Letters*, 18, pp. 419–426, May 2020.

129. Bhardwaj, Anil Kumar, Gupta, Sumeet and Raj, B., "Development & analysis of compact model for double gate Schottky barrier CNTFET", *Journal of Nanoelectronics and Optoelectronics*, 15, pp. 1199–1208, August 2020b.
130. Wadhwa, Girish, Kamboj, Priyanka, Singh, Jeetendra, Raj, B., "Design and investigation of junctionless DGTFET for biological molecule recognition", *Transactions on Electrical and Electronic Materials*, 22, pp. 282–289, 2021.
131. Verma, Sachin Kumar, Singh, Shailendra, Wadhwa, Girish, Raj, B., "Detection of biomolecules using charge-plasma based gate underlap dielectric modulated dopingless TFET", *Transactions on Electrical and Electronic Materials (TEEM)*, 21, pp. 528–535, June 2020.
132. Jain, Neeraj, Raj, B., "Impact of underlap spacer region variation on electrostatic and analog/RF performance of symmetrical high k SOI FinFET at 20 nm chan nel length", *Journal of Semiconductors (JoS)*, 38, No. 12, pp. 122002, 2017.
133. Singh, Shailendra, Raj, B., "Analytical modeling and simulation analysis of T-shaped III-V heterojunction vertical T-FET", *Superlattices and Microstructures*, 147, pp. 106717, 2020c.
134. Singh, Gurmohan, Sarin, R. K., Raj, B., "Design and analysis of area efficient QCA based reversible logic gates", *Journal of Microprocessors and Microsystems*, 52, pp. 59–68, May 2017b.
135. Singh, Shailendra, and Raj, B., "Parametric variation analysis on heterojunction vertical t-shape TFET for supressing ambipolar conduction", *Indian Journal of Pure and Applied Physics*, 58, pp. 478–485, June 2020.
136. Soniya, Shailendra Singh, Wadhwa, Girish, Raj, Balwinder, "Design and analysis of dual source vertical tunnel field effect transistor for high performance", *Transactions on Electrical and Electronic Materials*, 21, pp. 74–82, 2019.
137. Kaur, Manjit, Gupta, Neena, Kumar, Sanjeev, Raj, B., Singh, Arun Kumar, "RF performance analysis of intercalated graphene nanoribbon based global level interconnects", *Journal of Computational Electronics*, 19, pp. 1002–1013, June 2020.
138. Wadhwa, Girish, Raj, B., "Design and performance analysis of junctionless TFET biosensor for high sensitivity", *IEEE Nanotechnology*, 18, pp. 567–574, 2019.
139. Singh, Jeetendra, Raj, B., "Enhanced nonlinear memristor model encapsulating stochastic dopant drift", *JNO*, 14, pp. 958–963.
140. Kumar, Mohit, Chaudhary, Tarun, Raj, Balwinder, "Design and comparative analysis of ferroelectric nanowire with dielectric HfO_2 and $Al_2O_3$for low power applications", *Journal of Electronic Materials*, 53, pp. 6785–6791, 2024a.
141. Thakur, Yogesh, Raj, Balwant, Raj, Balwinder, "Design of pentacene thin-film transistor based hydrogen gas sensor for high sensitivity", *ECS Journal of Solid State Science and Technology*, 13, pp. 047005–047012, April 24, 2024.
142. Kumar, Manoharan Arun, Ismail, Kamal Batcha Mohamed, Mahalingam, Shanmugam, Raj, Balwinder, Kim, Junghwan, Atchudan, Raji, "Recent advances in two-dimensional nanomaterials and nanostructures for healthcare monitoring", *ACS Sensors*, 9, No. 4, pp. 1–8, April 2, 2024b.
143. Batcha, Mohamed Ismail Kamal, Manoharan, Arun Kumar, Mahalingam, Shanmugam, Balwinder, Raj, "Carbon fiber-reinforced polymers for energy storage applications", *Journal of Energy Storage*, 84, No. Part B, pp. 110931–110939, April 20, 2024.

144. Hussain, Shahraan, Raj, Balwinder, Sharma, Divya, "Design and parametric optimization of lead free perovskite based tandem solar cell for high efficient", *Engineering Research Express*, 6, pp. 015315–0153122, March 13¸ 2024.
145. Kumar, Parveen, Raj, Balwinder, Wadhwa, Girish, Singh, Balwinder, Kumar, Raj "Design and analysis of junctionless based gate all around N+ doped layer nanowire TFET biosensor", *ECS Journal of Solid State Science and Technology*, 13, pp. 017002–017010, January 15, 2024c.
146. Sharma, Divya, Mehra, Rajesh, Raj, Balwinder, "Enhancement in efficiency of methyl ammonium tin iodide based perovskite solar cell using SCAPS-1 D", *Nano*, 18, No. 14, pp. 2350109, 2023.
147. Kumar, Parveen, Sharma, Sanjeev Kumar, Raj, Balwant, Raj, Balwinder, "Analysis of device parameter variations in In1-xGaxAs based gate stacked double metal surrounding gate nanowire MOSFET", *TEEM Springer*, 24, pp. 570–578, 2023.
148. Bhardwaj, Anjana, Kumar, Pradeep, Raj, Balwinder, Kumar, Naveen, Anand, Sunny, "Design and optimization of vertical nanowire tunnel FET with electrostatic doping", *Engineering Research Express*, 5, pp. 045025–045033, 2023.
149. Thakur, Yogesh, Raj, Balwinder, Gill, Sandeep Singh, "Design and performance analysis of pentacene organic field effect transistor with high-K dielectric materials", *Journal of Optoelectronics and Advanced Materials – Rapid Communications*, 17, No. 7–8, July–August 2023.
150. Kumar, Praveen, Raj, Balwinder, "Design and simulation of junctionless nanowire tunnel field effect transistor biosensor for high sensitivity", *Microelectronic Journal Elsevier*, 137, pp. 105826–105834, July 2023.

Chapter 2

Flexible energy storage devices

Tanya Singh, Akriti Dewangan, Puja Kumari, Balwinder Raj, and Mandeep Singh

2.1 INTRODUCTION

Global fossil fuel consumption is increasing day by day, which has resulted in greenhouse gas emissions, damage to the environment, and a lack of energy. It is well acknowledged that employing sustainable clean energy can help mitigate these problems and that it is essential to guaranteeing sustainable growth on a global scale. The growing need for flexible displays and wearable, portable electronics gave rise to the concept of flexible energy storage devices. The evolution of flexible energy storage technologies, such as supercapacitors and flexible batteries, includes a number of significant turning points. Flexible energy storage systems are power sources that can bend, stretch, fold, or twist while preserving their performance. These devices are meant to be lightweight, thin, and adaptable to diverse forms and surfaces, making them excellent for applications in wearable electronics, flexible displays, and portable gadgets. Energy storage technologies have become increasingly significant in relation to the electric grid in recent decades. The fact that there are now many more applications that need a specific power or voltage level during a brief period of time is one of the primary causes of this [1]. Flexible and wearable energy storage devices (ESDs) with high energy and power densities, as well as exceptional flexibility and durability, are significantly needed for the development of these electronics in order to power a variety of wearable systems. Because of its potential uses in wearable and portable electronics, sensing systems, and medical equipment, flexible energy-storage devices, or flexible energy storage devices (FESDs), have garnered a lot of attention. Therefore, a lot of work has gone into creating different kinds of wearable, flexible rechargeable batteries and supercapacitors [2–7].

2.1.1 Flexible devices

Flexible devices are a rapidly advancing field in technology, characterized by their ability to bend, fold, or stretch without losing functionality. They leverage cutting-edge materials and designs, such as flexible displays, sensors,

DOI: 10.1201/9781003634645-2

and batteries, to create electronics that are more adaptable, portable, and durable [2]. Examples include foldable smartphones, wearable health trackers, bendable solar cells, and electronic, skins, shown in Figure 2.1. These devices hold significant potential across industries, from healthcare to consumer electronics, offering enhanced functionality and innovative applications that traditional rigid electronics cannot achieve [8–13].

2.1.2 History and origins of flexible devices

The 20th century saw the beginning of the history of flexible electronics. Dr. Ken Gilleo discovered one of the first descriptions of what may be called a flex circuit in Figure 2.2 [4], which was revealed in an English patent by Albert Hansen from 1903 that described a device made of paraffin-coated paper with flat metal conductors.

Anything that is extremely long or thin can stretch. Although cables and wiring are the best examples of flexibility, silicon wafers used in satellite solar cells were not thinned until the space race in order to increase their power per weight ratio and allow for some warping [14–19]. In the 1960s, this idea made it possible to create the first flexible solar cells, and researchers A.W. Crabb and H. Treble made important advances to the field of flexible electronics in 1967 [5]. They concentrated on developing lightweight, flexible solar panels using thin-film technology, namely utilizing cadmium telluride (CdTe) and cadmium sulfide (CdS). Their inventions opened the door for later developments in flexible photovoltaics by demonstrating the possibility for incorporating solar energy solutions into a range of applications [20–26]. Further research proved to be a boon for flexible devices, and Henry Letheby used sulfuric acid and anodic oxidation of aniline to produce a partially conductive substance in 1862. Most likely, the substance was polyaniline [6]. Researchers found in the 1950s that halogens and polycyclic aromatic chemicals may produce semi-conducting charge-transfer complex salts. Specifically, in the perylene–iodine combination, a high conductivity of 0.12 S/cm was observed in 1954 [6]. This discovery suggested that organic materials could conduct electricity. Dr. John McGinness created the first gadget with an organic semiconductor in 1973 [7]. In the decades that followed, significant advancements in flexibility and processability were made possible by organic semiconductors and amorphous silicon. As a result, these materials were used as the foundation for electronic devices in applications that needed characteristics such as bending, rolling, folding, and stretching that conventional electronics could not provide. Bernard J. Lechner of RCA Laboratories was the one who first came up with the concept of a thin-film transistor (TFT)-based liquid crystal display (LCD) in 1968. At Westinghouse Electric Corporation, T. Peter Brody, Fang-Chen Luo, and colleagues created the first working active-matrix liquid-crystal display (AMLCD) using thin-film transistors in 1972. Polycrystalline silicon (poly-Si) TFTs with flexibility were constructed on plastic substrates in the

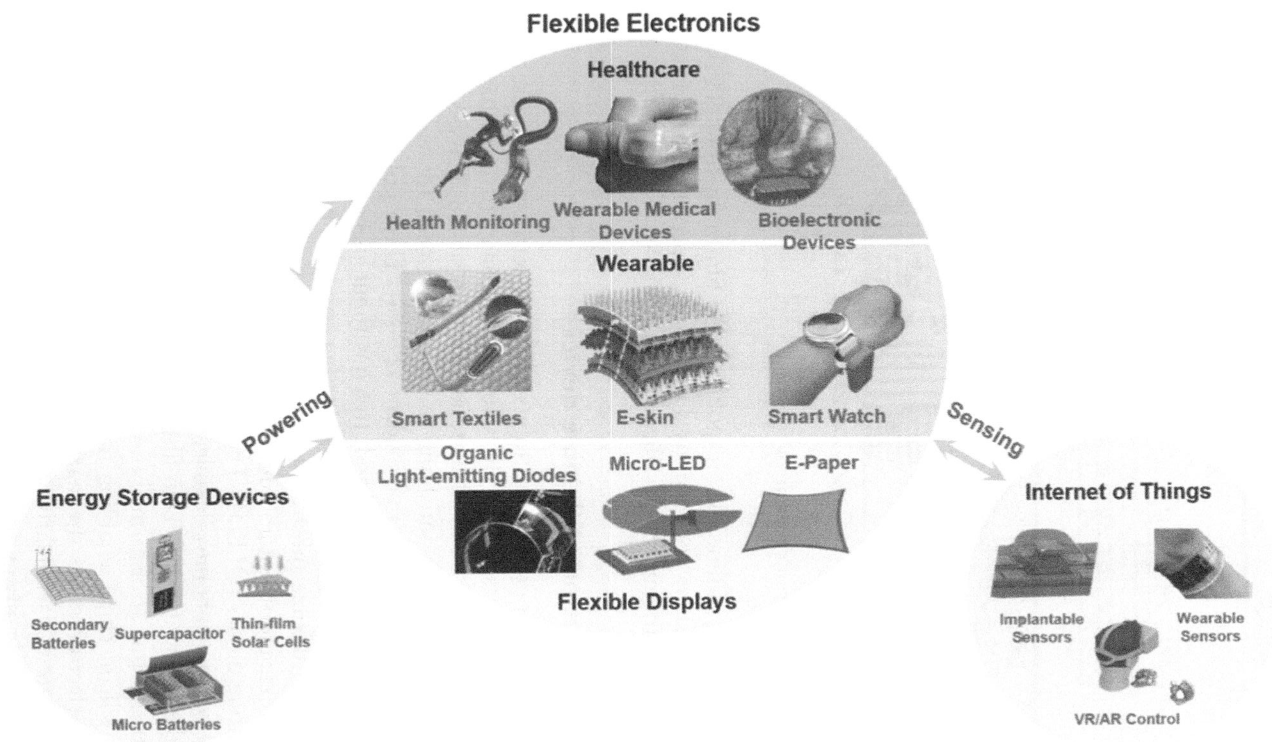

Figure 2.1 The application of flexible electronics in various fields, including wearable devices, flexible displays, healthcare and safety, the Internet of things [3].

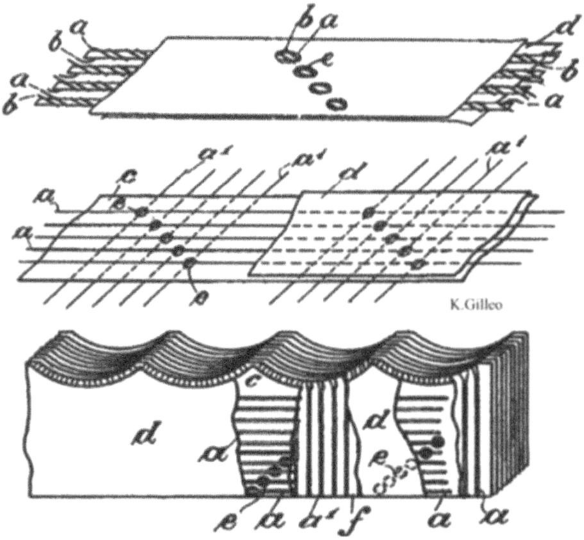

Figure 2.2 First flex circuit patent [4].

late 1990s. Polycrystalline silicon (poly-Si) thin-film transistors (TFTs) are widely used in various applications due to their superior electrical performance compared to amorphous silicon (a-Si) TFTs [8].

2.1.3 The evolution of flexible devices

The goal to get beyond the constraints of rigid electronics gave rise to the rise of flexible devices. Researchers looked to conductive polymers as a novel substitute for traditional materials such as silicon, glass, and metal as their shortcomings in satisfying the increasing need for wearable and lightweight technology became evident [27–36]. These polymers made it possible for electronics to bend and stretch without losing their operation by providing the required electrical conductivity and flexibility. Concurrently, the creation of flexible substrates—slim, robust materials like flexible glass and plastic—made it possible to produce bendable displays like organic light-emitting diodes (OLEDs). These substrates enabled the integration of electrical components into flexible, foldable, or rollable shapes while preserving their functionality. Thin-film transistors (TFTs) and organic semiconductors further transformed the sector by substituting flexible circuitry for rigid silicon chips, enabling more complex designs and functionality in flexible electronics. As the market for ergonomic, portable, and flexible technology grew, particularly in wearables, smart gadgets, and foldable screens, the sector advanced quickly. These developments in materials science transformed flexible technologies from experimental lab concepts into widely used commercial items, such as wearables, folding smartphones, and upcoming breakthroughs like stretchable electronics, all of which changed the face of contemporary technology [9]. Driven by the increasing demand for ergonomic and adaptable technology, flexible devices progressed from laboratory prototypes to commercial products as materials advanced. This technological evolution is just getting started with the continuous integration of flexible batteries, sensors, and displays into consumer electronics and healthcare systems [37–41]. To sum up, flexible devices have come a long way from their initial discoveries in conductive materials to becoming a significant area of study and industry, with the potential to make electronics more portable, flexible, and a part of everyday life in the future. These days, flexible technologies play a major role in the display business. Numerous firms and research institutions have released cutting-edge flexible technologies, such as wearables, curved-screen smartwatches, and foldable smartphones [42–48].

2.2 ENERGY STORAGE

In order to balance supply and demand in different energy systems, energy storage is an essential technology that enables excess energy to be stored and used when needed. Together with other cutting-edge materials, conductive polymers like polyaniline (PANI) are essential to the advancement of energy storage technologies, especially in batteries and supercapacitors. A vital part of a flexible operating electronic is the power source, or in this case, the energy storage component. Flexible energy storage has gained popularity recently as a means of meeting the needs for portable electronics, flexible displays, healthcare and fitness tracking devices, and other devices that are small, integrated, and self-powered [49–56]. The need to move from all-solid state to flexible energy storage has led to increased interest in carbonaceous materials and nanocarbons to enhance already efficient devices, such as Li-ion batteries and supercapacitors. Some of these issues include low volumetric energy density, high internal resistance, or poor mechanical durability [57–62].

2.2.1 Energy storage technologies and their history

2.2.1.1 Batteries

The battery storage systems (BSSs) are still the most affordable technology on the market today, even though they were the first electrical energy storage system to be introduced. Benjamin Franklin first studied the fundamentals of their operation in 1748 when he explained how to connect several Leyden jars, a device for storing static electricity. In 1873, some of the most significant scientific research on batteries was published. The first significant battery applications were not created until the 20th century [10]. Thus, in

1920, a patent was granted for a battery system for telephone circuits; other uses from that era include flashlights or even batteries that are charged using a generator powered by a gas engine [63–68] (Figure 2.3).

a. **Lead–Acid Battery**: The lead-acid battery, created in 1859 by French physicist Gaston Plante, is the most ancient type of rechargeable battery [10]. A positive electrode and a negative electrode immersed in an electrolyte make up a lead–acid battery shown in Figure 2.4. The process of electron migration from the positive electrode to the negative electrode charges the battery. Lead dioxide (PbO_2) makes up the positive plate in the charged state, while elemental lead makes up the negative plate. Here, sulfuric acid (H_2SO_4) at a concentration of roughly 33.5% is the electrolyte [69–74]. When electrons move from the negative electrode to the positive electrode, the discharging process occurs. In this instance, both the positive and negative electrodes produce grains of lead sulfate ($PbSO_4$), and the electrolyte starts to lose concentration and turn primarily into water. These grains dissolve during charging. However, if the discharge is too deep, the grains may enlarge and become insoluble, rendering the plates' active material useless. Applications for lead–acid batteries range from 1 kW (uninterruptible power supply, or UPS) to 10 MW (transport and distribution systems). They possess a low energy density, which means they store less energy relative to their size and weight compared to modern battery technologies, making them bulkier and heavier [75–82].
b. **Nickel Cadmium(Nicd)/Nickel–Metal Hydride(NiMH) Batteries**: The second kind of rechargeable batteries was NiCd batteries shown in Figure 2.5, which were created in 1899 by Swedish engineer Waldemar

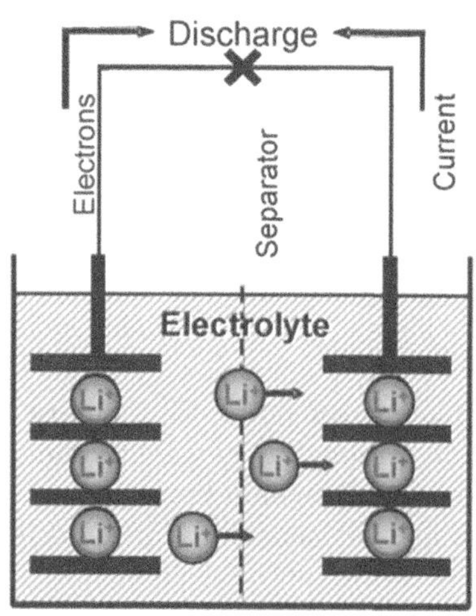

Figure 2.3 Schematic diagram of battery [13].

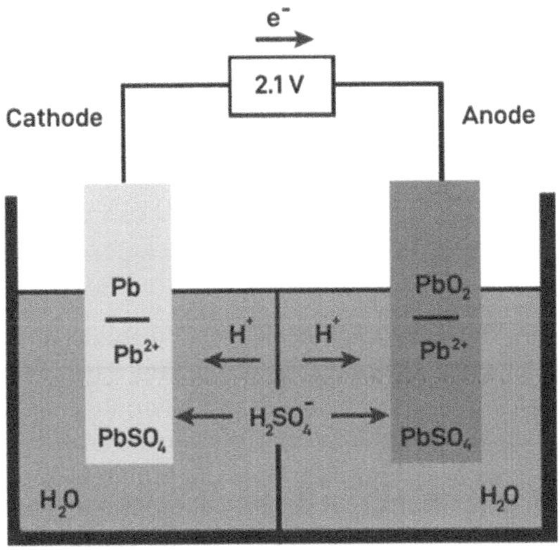

Figure 2.4 Lead–acid battery.

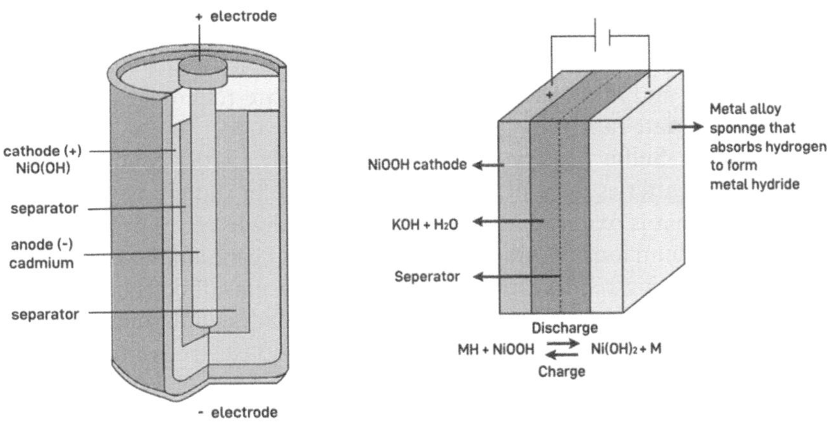

Figure 2.5 NiCd/NiMH batteries.

Jungner. NiMH batteries appeared on the market in 1989 [10]. NiCd batteries use an electrolyte (typically potassium hydroxide), a separator, a positive electrode made of nickel (III) oxide and hydrogen, and a negative electrode made of cadmium when fully charged. In NiMH, the metal M, an intermetallic compound made of a combination of lanthanum, cerium, and other rare earth elements with nickel, aluminium, cobalt, or manganese, makes up the negative electrode. Potassium hydroxide serves as the electrolyte, just like in NiCd batteries [83–89]. NiMH batteries have supplanted NiCd batteries due to their higher energy density, which is suitable for applications ranging from 1 kW to 0.5 MW, as well as the environmental risks associated with them that necessitate elaborate recycling processes. These are utilized for applications ranging from 1 kW to roughly 1 MW and have a higher energy density [10].

$$\text{Cell Reaction}: Cd(s) + 2NiO(OH)(s) + 2H_2O \rightarrow Cd(OH)_2(s) + 2Ni(OH)_2(s) \quad (2.1)$$

c. **Lithium-Ion Batteries (LIBs)**: It was introduced in 1970 by the American chemist Michael Whittingham [10]. The fundamental working principle states that during discharging and charging (Figure 2.6), the positive Li ions move in the opposite direction from the negative plate—which is often composed of graphite—to the positive plate, which is composed of an alloy containing lithium [90–95]. Ion circulation is allowed by the electrolyte, but electron conduction is not. Because lithium is highly reactive with water, the electrolyte is either nonaqueous or aprotic (it does not accept or donate hydrogen ions. These batteries have a high energy density, no memory effect, and a low discharge rate. The primary drawback is the potential for internal resistance to heat up and leads to battery failure. For this reason, overcurrent and overvoltage protection devices are required to guarantee appropriate functioning [10].
d. **Sodium–Sulfur Battery**: Beginning in the early 1970s, sodium–sulphur (NaS) batteries were developed. It consists of a sodium negative electrode that is oxidized at the surface during discharge to produce positive sodium ions that move through the solid beta-alumina electrolyte before being reduced to produce sodium pentasulfide (Na2S5). This Na2S5 progressively changes into polysulfides with more sulphur [96–101]. The electrolyte conducts nearly no electrons but conducts sodium ions, which limits the rate of self-discharge. Electricity passes through the external load as a result of electrons produced during discharge by the molten elemental sodium-positive electrode [10]. Unfortunately, a number of dangerous substances are present in these batteries, including metallic sodium, which burns when it comes into touch with water. For this reason, NaS cells also include sand, which serves to both fix them and put out any potential fires. The discovery of a high-voltage electrolyte could lead to the possibility of other chemicals. One such instance is the pyrotemperature Li/S battery, which Sion Power (Tuscon, AZ) has been developing for over ten years under a license from Polyplus Corporation. Nowadays, some military markets use commercial cells for flight power [11]. As observed in Figure 2.7 [11], they already significantly outperform the

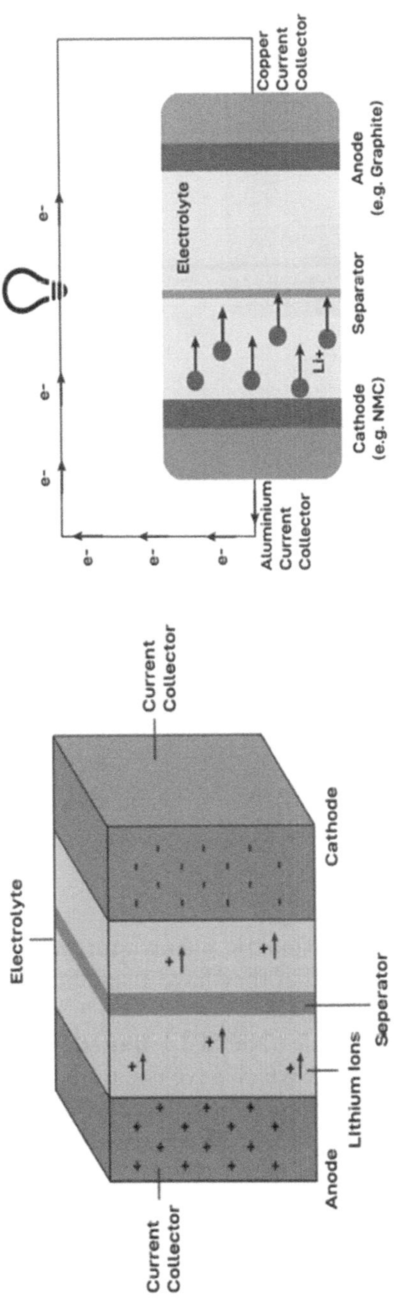

Figure 2.6 Lithium-ion batteries (discharging and charging).

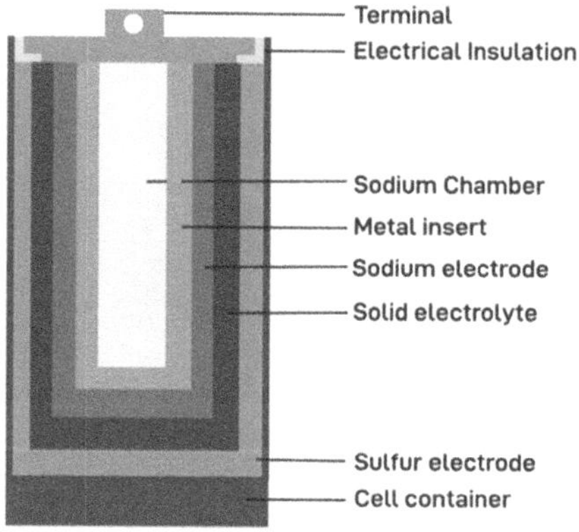

Figure 2.7 Sodium–sulfur battery.

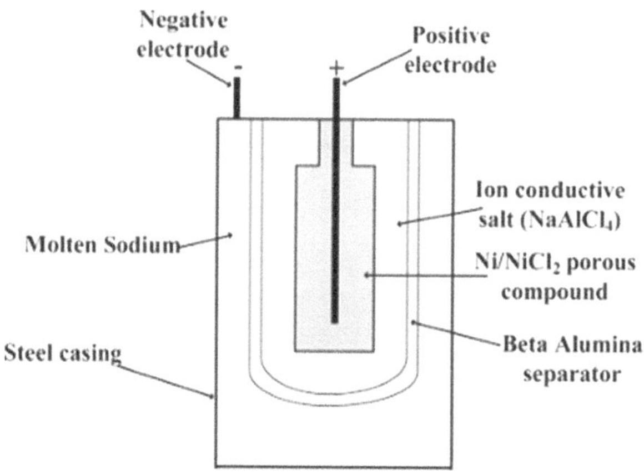

Figure 2.8 Sodium–nickel–chloride battery [10].

Li ion in terms of gravimetric energy storage capacity, but the figure also demonstrates that they have a significantly lower volumetric storage capacity [102–107].

e. **Sodium–Nickel–Chloride ($NaNiCl_2$) Batteries:** The Zeolite Battery Research Africa Project, which gave rise to the moniker ZEBRA, first launched this battery in 1985, it operates at high temperatures of approximately 270 C–350 C. Nickel (Ni) and salt (NaCl) are transformed into sodium and nickel chloride ($NiCl_2$) during the charging process. The reactions are reversed during discharge. Sodium serves as the negative electrode, whereas $NiCl_2$ in the charged state and Ni in thedischarged state serve as the positive electrodes, as shown in Figure 2.8. An electrolyte that separates the two electrodes allows ion conduction but not electron conduction. As a result, the circuit is

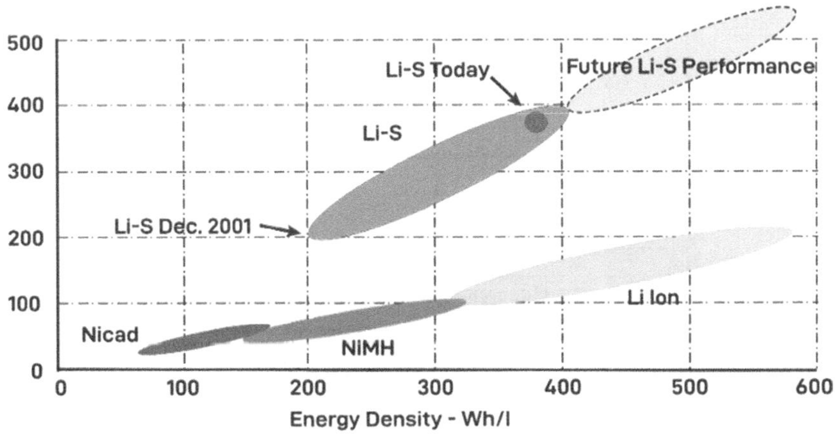

Figure 2.9 Energy storage capability of lithium–sulfur cells compared to other cells [13]

closed and current flows through the battery when an external load is connected to the terminals [10] (Figure 2.9).

2.2.1.2 Supercapacitor Storage Systems (SSSs)

Supercapacitors, referred to as ultracapacitors or electrochemical capacitors, are energy-storage devices that, unlike batteries, store and release electrical energy by electrostatic charge separation. They fill the energy storage void between capacitors and batteries with a capacitance that is far higher than that of ordinary capacitors. Compared to batteries, supercapacitors have a lower energy density. They can charge and discharge in seconds, unlike batteries which take more time [108–113]. Three main types of supercapacitors can be distinguished due to their unique structures and internal storing mechanisms: hybrid supercapacitors, electric double-layer capacitors (EDLC), and pseudo capacitors shown in Figure 2.10.

a. **Electric Double-Layer Capacitor:** By positioning two parallel conductors as electrodes with dielectric interlayers inside and providing a voltage to one, grounding the other, an electric double-layer capacitor is created as shown in Figure 2.11a. Due to this configuration, charges build up on both sides until an electrostatic equilibrium is reached. Sometimes, charges will build up on both sides of the electrodes and be stored by quick, highly reversible redox processes on the electrode charges on both sides, producing high power densities. However, the energy density is not as great as it is for batteries [12].
b. **Pseudo Capacitors:** Pseudo capacitors have a high contact surface, short electron transport channel lengths, short ions diffusion lengths,

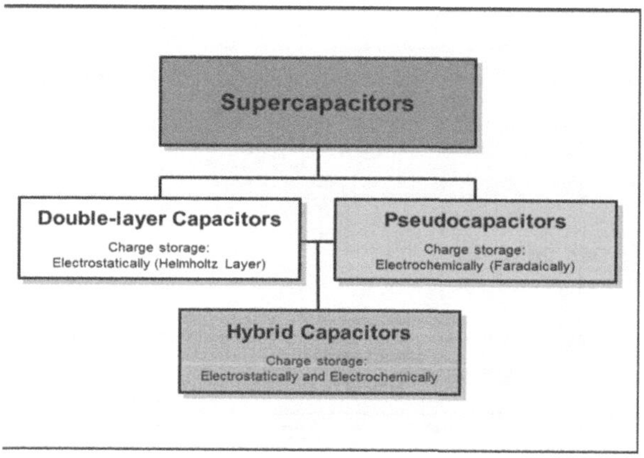

Figure 2.10 Different types of supercapacitors.

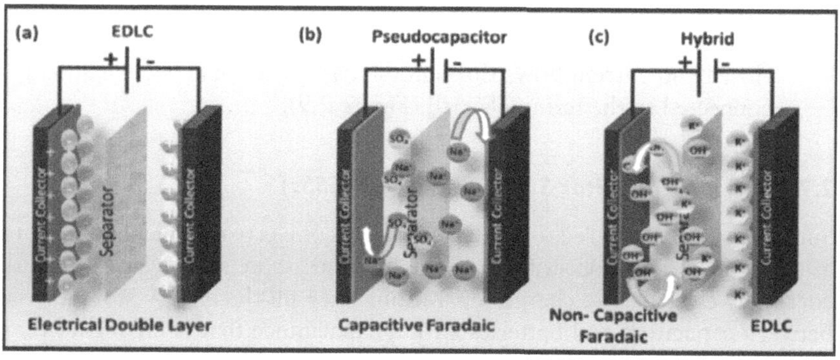

Figure 2.11 Electric double-layer capacitor, pseudo capacitors, and hybrid supercapacitors, respectively.

and even an enhanced cycle life. They store energy by pseudocapacitive or faradaic redox processes. Their energy storage mechanism operates in tandem with EDLCs. However, there is a risk of decreased electroactive surface area, adsorption sites, and performance if the electrode film's shape, structure, composition, architecture, and characteristics are not well controlled, Figure 2.11b [12].

c. **Hybrid Supercapacitors**: Hybrid supercapacitors are devices that combine an anode- or cathode-like capacitive electrode with a battery-type electrode. It combines, on electrodes, the non-Faradaic surface reaction and the Faradaic intercalation. By constructing electrodes based on 3D mesoporous structure materials, hybrid supercapacitors can preserve substantial cycling stability and affordability with both high power and energy densities, despite the poor conductivity of some types of carbon- and metal oxide-based electrodes shown in Figure 2.11 [12].

2.3 CRITERIA FOR A DEVICE TO STORE ENERGY

High efficiency, dependability, efficient power quality control, scalability, compatibility with distributed energy resources (DER), modular redundancy, and robust thermal management are the main factors considered when assessing an energy storage device for power systems. In order to reduce energy loss during conversion operations, devices should ideally achieve above 90% efficiency at different load levels. Additionally, reliability is a top concern, especially in high-tech settings where continuous availability—often up to 99.999%—is required [114–119]. This is accomplished through distributed and redundant setups that reduce disruptions. In order to maintain a steady output voltage and safeguard delicate equipment, the device must also efficiently regulate power quality by managing voltage sags, harmonics, and other variations. Scalability maximizes efficiency and optimizes costs by enabling the system to modify its power rating in response to changing load demands. The system is further strengthened by compatibility with DER, such solar or wind, which makes it possible to integrate renewable sources, boost energy resilience, and lessen reliance on the grid. For important loads during outages or voltage disturbances, modular backup and redundancy options—such as distributed configurations and embedded UPS capabilities—offer dependable, temporary power support. Finally, efficient thermal management reduces heat production, particularly in small- or high-demand settings, which enhances the system's performance and longevity. When taken as a whole, these standards guarantee that an energy storage system may satisfy the requirements of effectiveness, dependability, and flexibility for a variety of intricate power settings [13].

2.3.1 The critical role of energy storage in modern energy systems

We need energy storage for several critical reasons, primarily to ensure reliability, efficiency, and sustainability in energy systems. Initially, it aids in balancing supply and demand in the energy grid because energy output, particularly from renewable sources like solar and wind, can vary during the day, while consumption varies. Systems that store excess energy during periods of low demand and release it during periods of high demand guarantee a steady supply of electricity. Storage systems keep the flow of power constant by storing excess energy during times of high output and releasing it when the supply is low. By buffering against fluctuations, preventing blackouts, and guaranteeing a continuous power supply—especially during peak demand or unplanned outages—this capacity also improves grid stability and reliability. Energy storage can save money since it allows businesses and households to store energy during periods of low price and use it during periods of high price, which lowers overall energy expenses [120–126]. Additionally, it aids in the transition to a low-carbon energy system by promoting the use of renewable energy sources and enhancing energy efficiency. By offering vital backup power during natural catastrophes or other disturbances, energy storage improves community resilience. Energy storage will be essential for controlling growing electricity demand and guaranteeing a steady supply as industries like heating and transportation become more electrified. By lowering reliance on fossil fuels and storing extra electricity from sources like solar or wind, storage technologies further maximize the use of renewable energy. They also promote efficiency and cost savings by avoiding expensive infrastructure modifications to handle peak loads and enabling the use of stored energy during times of peak pricing. Energy storage also facilitates a more seamless switch from fossil fuels to renewable energy sources, which lowers carbon emissions and advances climate goals. Storage is valuable for resilience and sustainability in the energy environment because it supplies vital backup power to vital infrastructure in times of crisis, including data centers, hospitals, and communication services. [4, 9, 10, 12–14].

2.4 NEED OF FLEXIBLE ENERGY STORAGE DEVICES

The field of flexible electronics is a crucial driver of technological advancement with its close ties to human life and special applications in flexible electronics, wearable technology, and healthcare, and flexible electronics is a key force behind technological progress, Figure 2.12. The growing popularity of flexible energy-storage devices can be attributed to their distinct and promising features, including their flexibility, shape diversity, lightweight nature, and so on [15]. As a result, flexible energy storage devices (FESDs) are desperately needed to meet the energy storage requirements of numerous flexible product types.

Modern energy systems require flexible energy storage, particularly as we depend more and more on renewable energy sources.

Flexible energy storage is crucial for modern energy systems, especially as we increasingly rely on renewable energy sources like solar and wind.

FESDs can be classified into three categories based on spatial dimension these are as follows:

i. **Distributed Energy Storage**: These systems are located close to the point of energy use or generation, such as residential batteries or small-scale storage systems. They provide localized energy management and can help reduce transmission losses [127–130].
ii. **Centralized Energy Storage**: These larger-scale storage facilities are situated away from the point of use and are typically integrated into the grid. Examples include large battery plants or pumped hydro storage facilities that can supply energy to a wider area.

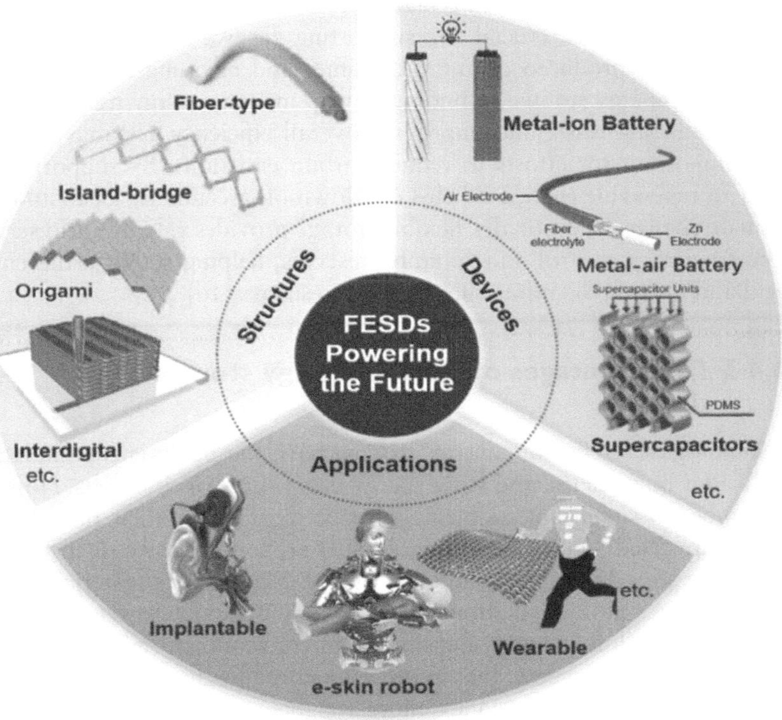

Figure 2.12 Illustration of FESDs powering the future [3].

iii. **Utility-Scale Energy Storage**: These systems are designed to provide significant capacity and are often deployed by utilities to support grid stability, peak shaving, and large-scale renewable energy integration. They can include large battery installations, compressed air energy storage, and other technologies.

Each category plays a vital role in enhancing energy flexibility and reliability, catering to different needs and applications within the energy system.

2.4.1 Advantages of flexible energy storage devices

The growing demand for flexible energy storage devices has led to major innovations that lower costs, making these technologies more affordable. Researchers at North Carolina State University have developed new manufacturing methods for flexible electronics that improve productivity while reducing prices. Companies are also working hard to make flexible circuits cheaper compared to traditional alternatives. Beyond being affordable, flexible storage solutions improve the reliability of energy supply. They can provide backup power during blackouts or when energy demand is high. These solutions are crucial for supporting renewable energy by storing excess power produced during peak times and releasing it when needed. Advanced energy storage technologies help manage energy use more effectively, reducing waste and improving overall efficiency. Flexible storage is also important for efforts to reduce carbon emissions. By supporting the shift to renewable energy, it helps cut down on greenhouse gases and fight climate change. Additionally, flexible storage provides valuable grid services like frequency control and spinning reserves, helping stabilize the energy grid and making the infrastructure more resilient. [16]

2.4.2 Disadvantages of flexible energy storage devices

Technologies for flexible energy storage have the potential to fundamentally alter how we use and store energy in a variety of situations. However, a number of significant problems must be fixed before their full potential may be realized. One of the primary issues is energy density, or the amount of energy stored per unit volume or weight. Many of the materials used in flexible energy storage solutions, such as organic or polymer-based batteries, typically have lower energy densities than conventional rigid batteries like lithium-ion. This suggests that in order to store the same amount of energy, flexible batteries would require more weight or space, which is a significant disadvantage in applications such as wearables, portable electronics, and electric vehicles where size and weight are critical. Cost is still another important obstacle. Advanced materials and specialized manufacturing processes are frequently required for the creation of flexible energy storage devices, which might raise production costs. Adding novel elements, like conductive polymers or novel nanoparticles, might raise costs and reduce the devices' economic viability in comparison to more well-established technology. Price increases may thereby restrict consumer accessibility and impede market penetration, thereby impeding widespread adoption. Although these devices' flexibility presents clear benefits in terms of mechanical stability, it can also pose risks. Compared to their rigid counterparts, flexible batteries' materials might not be as resilient to repetitive bending or stretching. This vulnerability may result in problems like another important factor to take into account is the flexible battery's cycle life. The number of charge–discharge cycles a battery can withstand before its capacity visibly degrades is known as cycle life. Many flexible batteries, especially those composed of organic materials, can degrade more quickly due to the mechanical stress caused by bending and stretching. For applications like wearable technology and grid storage, where long-term dependability and durability are crucial for user happiness and functioning, this degradation is particularly worrisome. Scalability is another major obstacle. Although small-scale labs are the source of many innovations in flexible energy storage technologies, it can be difficult to produce these devices in large quantities without compromising quality. Inconsistencies in material quality, performance, and dependability may arise as small-scale invention gives way to mass production, making it more difficult to satisfy rising consumer demand. Furthermore, a major obstacle to broad adoption is the scarcity of adaptable energy storage solutions. The practicality of many of these technologies in real-world applications is limited because they are still in the prototype or experimental stages. Businesses may be reluctant to invest in technologies that have not yet shown to be dependable or acquired a substantial market presence due to this uncertainty. Lastly, there are environmental concerns associated with the manufacture and disposal of flexible energy storage devices. The use of potentially harmful or nonbiodegradable materials by these devices throughout their life cycle may result in environmental issues. Furthermore, recycling practices for these newer materials are either experimental or unproven, which complicates waste management and raises questions about the technology's long-term sustainability [16].

Despite their enormous potential to transform energy storage and consumption across a variety of applications, flexible energy storage devices now face a number of problems that need to be fixed. By solving these challenges via research, innovation, and strategic development, flexible energy storage can reach its full potential. Wider adoption as well as important technological and long-term advancements will follow from this.

2.5 DIFFERENT STRUCTURES THAT ARE BEING USED IN FLEXIBLE ENERGY STORAGE

The performance of flexible energy storage devices depends largely on the materials' ability to bend and stretch without failing. This means the material must handle the stress created during use and still remain intact. Research into new materials that are naturally flexible can improve how well these devices work. Materials such as conductive polymers, flexible metal oxides, and advanced composites are being studied for their ability to bend and stretch repeatedly without breaking. It's also important to design the device structure in a way that reduces strain. This is where geometric optimization comes in: it helps design devices that spread out stress evenly, which improves both their performance and lifespan. Additionally, using layering techniques where different materials with different properties are stacked can help absorb and spread stress, making the device more durable. Architectural features such as folds, corrugations, or other structural designs can also help the device bend and move without losing strength. Early research in micro-electro-mechanical systems (MEMS) has provided valuable insights, guiding current developments in flexible energy storage by offering a solid reference for designing and analyzing the structure and mechanical aspects of modern devices. [3].

2.5.1 Fiber structures

The creation and functionality of flexible energy storage devices depend on fiber-type architectures. Researchers and engineers can develop high-performance, flexible energy storage solutions by utilizing the distinctive qualities of different fibers, such as conductive polymers, metallic fibers, and carbon nanofibers. In order to increase the applicability and feasibility of flexible energy storage devices across a variety of fields and pave the way for more connected and energy-efficient future, fiber-type structural integration will be crucial as technology advances [3].

2.5.1.1 Carbon nanofibers

Carbon nanofiber is a key element in the creation of flexible energy storage devices. They are perfect for improving the functionality and efficiency of different energy storage technologies, such as batteries and supercapacitors, because to their special qualities, which include excellent electrical conductivity, mechanical strength, and lightweight nature. CNFs are cylindrical fibers with diameters typically ranging from 10 nanometers to several micrometers. There are the different types of CNF's available based on their structure, Figure 2.13. They can possess a graphitic structure, which enhances their electrical conductivity. The high surface area of CNFs facilitates increased electrochemical reactions, which is crucial for energy storage applications. CNFs exhibit exceptional tensile strength and flexibility, allowing them to endure mechanical stress without compromising performance. To create composite electrodes, CNFs are frequently combined with other substances like transition metal oxides or conductive polymers. By optimizing the electrode's mechanical and electrical characteristics, this synergy can further improve FESDs' overall performance [17].

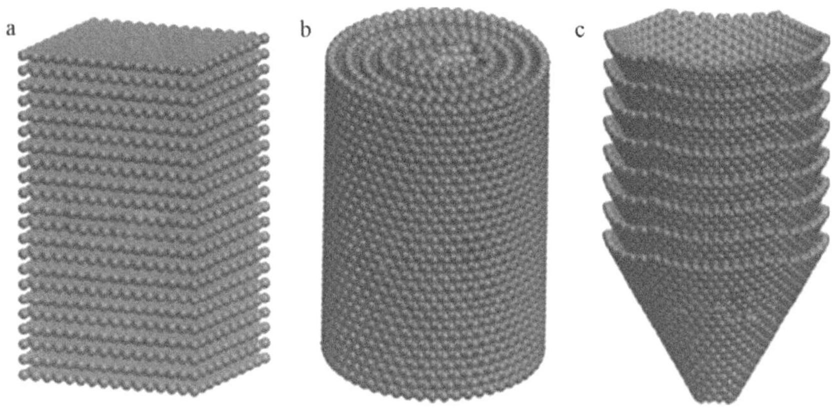

Figure 2.13 Schematic illustration of three different type of CNFs (a) platelet-type CNF, (b) tubular-type CNF, and (c) fishbone-type CNF [17].

2.5.1.2 Fiber-based electrodes

A state-of-the-art advancement in energy conversion and storage technologies is fiber-based electrodes, as shown in Figure 2.14. They make use of the special qualities of different fiber materials, including conductive polymers, metallic fibers, and carbon nanofibers, to produce electrodes that are flexible, lightweight, and highly effective. These electrodes are especially well-suited for use in supercapacitors and batteries, two types of flexible energy storage devices (FESDs). Aligned or woven fibers make up the majority of fiber-based electrodes, which can greatly expand the surface area available for electrochemical processes. The electrode's overall performance is improved by the additional active sites made possible by this design. Fiber-based electrodes' intrinsic flexibility makes it possible to integrate them into adaptable energy storage systems without sacrificing their functionality. Applications in portable electronics and wearable technology require this [18].

2.5.1.3 Parallel-type fiber structure

In order to avoid electrolyte spills or moisture contamination, the anode and cathode are enclosed and positioned side by side on both sides of the separators in parallel construction. Fiber-type flexible energy storage devices (FESDs) are usually made of metal oxides or lightweight, high-energy-density carbon-based materials, which offer a great deal of material selection freedom. Liu et al. created Ni-coated cotton yarns using a "polymer-assisted metal deposition (PAMD)" technique and a polyvinyl alcohol (PVA)/LiCl gel wearable solid-state supercapacitor yarn, as shown in Figure 2.15a. The device was bent at various angles (from 0° to 180°), while the CV property was being measured in order to evaluate the applicability of these supercapacitor yarns for flexible and wearable applications. Figure 2.15b shows

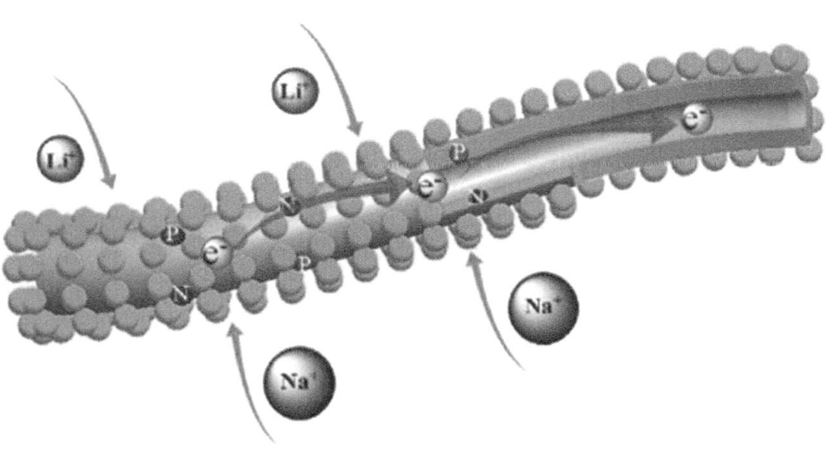

Figure 2.14 Fiber-based electrode [18].

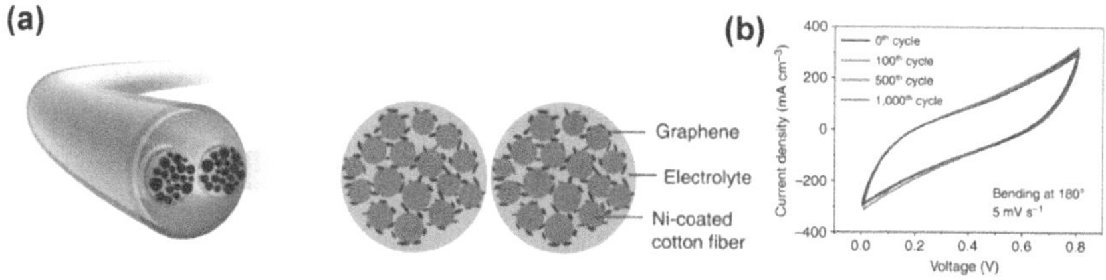

Figure 2.15 Synthesis and performance diagram of parallel fiber-type energy storage devices. a) Schematic diagram of supercapacitor yarn. b) CV curves of the supercapacitor at different bending cycles [3].

that the device's performance remained steady regardless of the bending angle. Because wearable electronics may experience parasitic reactions with internal chemicals and changes in liquid electrolytes under specific conditions, like exposure to rain and perspiration, encapsulation is an important consideration when considering practical applications. Guan et al. created a rechargeable fiber-shaped Zn/Co3O4 battery using 3D sheet zinc as the anode and improved the insulation of energy storage devices to ensure the quality of its packaging. Although the production process for parallel fiber field emission scanning electron microscope (FESD) is rather straightforward, certain restrictions prevent its widespread use. The development of FESDs is hampered by the parallel structure's tiny contact area between the electrode and electrolyte, which leads to poor mass loading of active materials and a lengthy ion transport path. In order to get around these limitations, better fiber combinations must be created [3].

2.5.1.4 Twisted-type fiber structure

An inventive, the twisted fiber-type FESD uses a spinning mechanism to twist two electrodes into a double helix shape. An electrolyte must be placed around the electrodes to stop short circuits. Ionic conduction channels between the electrodes are made possible by electrolytes. Solid or gel electrolytes are more frequently utilized in fiber-based FESDs than liquid electrolytes. It is feasible to achieve stability against electrolyte leakage by using gel electrolytes. Ion adsorption and fast redox reactions at the electrode/electrolyte interface are the two types of energy storage mechanisms. According to this idea, Lee et al. suggested a flexible fiber-type supercapacitor made up of surface-activated carbon fibers and a gel polymer electrolyte with a redox additive (FFS-SARE), which is produced by twisting the two fibers coated with gel electrode in the S form Figure 2.16a involving the electrode surface and the redox mediator shown in Figure 2.16b. However, the device's capacitance and ion transfer rate are constrained by the gel electrolyte's greater resistance. Furthermore, the qualities of the material are limited when PVP is directly coated on carbon fibers [3].

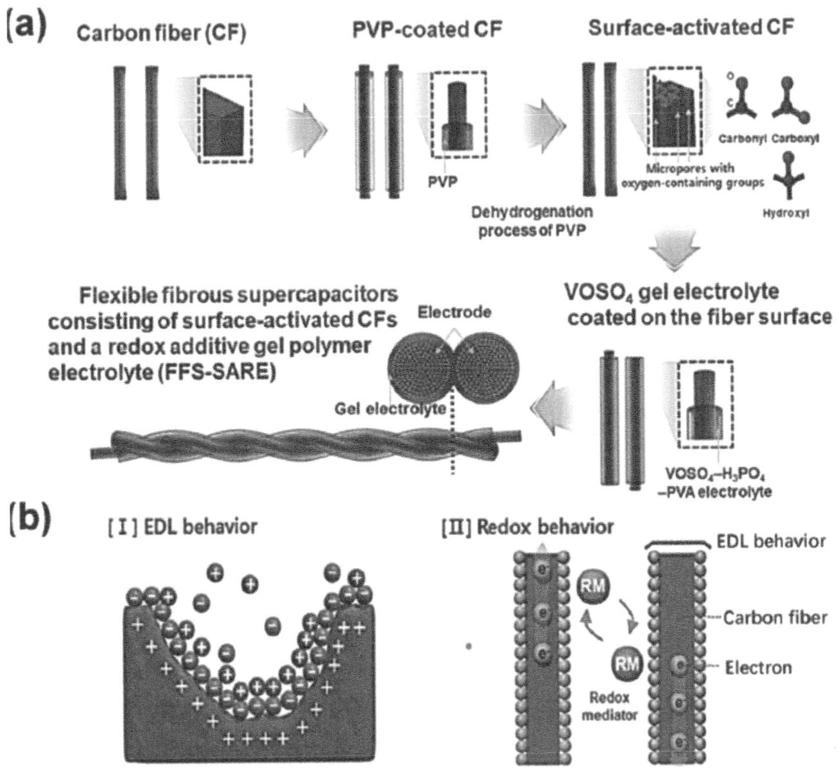

Figure 2.16 Synthesis and performance diagram of twisted fiber-type energy storage devices. a) Schematic of the fabrication of FFS-SARE. b) Schematic illustration of the electrical double layer and redox behavior [3].

2.5.2 Island Bridge structure

An innovative design technique for flexible energy storage devices (FESDs) that maximizes performance without sacrificing flexibility is the island bridge structure. The structure of electrodes in batteries and supercapacitors is important because it affects how much energy they can store, how quickly they can release it, and how efficient they are overall. One such structure is called the "island bridge" design shown in Figure 2.17. In this structure, active materials (the parts that store energy) are arranged in small "islands" that are connected by "bridges" made of conductive materials. This design helps maximize the surface area available for chemical reactions, which improves energy storage. The bridges, made from materials like carbon nanofibers or metallic fibers, help electrons move efficiently across the electrode. This arrangement allows for better ion movement between the electrolyte (the material that conducts electricity) and the active materials, improving how quickly the device can charge and discharge. The island

Figure 2.17 Island bridge structure [19].

and bridge structure also makes the device more flexible, so it can handle mechanical stress without breaking or losing performance. The bridges reduce the distance electrons need to travel, which boosts electrical conductivity and overall efficiency. Because of its unique design, this structure is especially useful for flexible and wearable technologies, such as flexible batteries and supercapacitors. As research continues and challenges are overcome, this innovative design could be key to developing the next generation of energy storage technologies. [19].

2.5.3 Interdigital structure

Flexible energy storage devices (FESDs) feature interdigital structures, which are a complex design, especially when it comes to battery and supercapacitor electrodes. By arranging conductive and active materials optimally, this structure improves performance and yields better mechanical flexibility and electrochemical qualities. Alternating fingers or prongs of conductive material that stretch outward and interleave with one another make up the basic component of an interdigital structure. By increasing the area of contact between the electrode and the electrolyte, this configuration improves ion transport. These electrodes are typically fabricated on flexible substrates, allowing for integration into various applications, including wearable devices and portable electronics. Flexible substrates are frequently employed, including polyimide, polyethylene terephthalate (PET), and other polymer films. These materials are stable in a variety of environmental circumstances and offer mechanical flexibility. Figure 2.18 shows the optical and scanning electron microscope (SEM) images of interdigital electrodes. The rise of portable electronics has indeed accelerated the demand for miniaturized energy storage solutions, especially in applications where external power sources are impractical. Here's a deeper look at the context you provided, focusing on the advantages and functionalities of microsupercapacitors (MSCs) compared to traditional microbatteries. Microbatteries often struggle to deliver the high power needed for certain applications, leading to

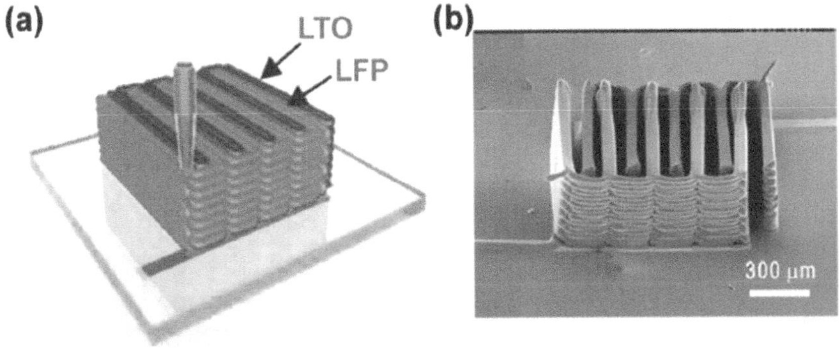

Figure 2.18 a) Optical and b) SEM images of interdigitated LTO-LFP electrodes, respectively [3].

performance bottlenecks. Many microbatteries contain toxic materials that pose environmental and safety concerns, limiting their usability in sensitive applications, such as biomedical devices [20].

2.6 EMERGENCE OF MICRO SUPERCAPACITORS

MSCs utilize a two-dimensional layout with interdigitated electrodes, enhancing the surface area available for electrochemical reactions while maintaining a compact design. This configuration not only increases the interface area for ion interaction but also facilitates rapid ion transport, improving both charge and discharge rates.

$$\text{Energy Density } E = 1/2CV^2 \tag{2.2}$$

where C is the specific capacitance, and V represents how much charge the MSC can store per unit voltage. Higher capacitance values lead to increased energy density.

A wider voltage window enhances the energy storage capability. This allows the device to operate effectively under varying conditions.

$$\text{Power Density } P = E/t \tag{2.3}$$

where t is the discharge time, this formula highlights how quickly energy can be delivered.

MSCs generally have a footprint in the millimeter to centimeter range, but their electrode thickness is typically less than 10 μm. Having the in-plane electrode structure has technical advantages over the conventional sandwich structure shown in Figure 2.19. This compact size is ideal for

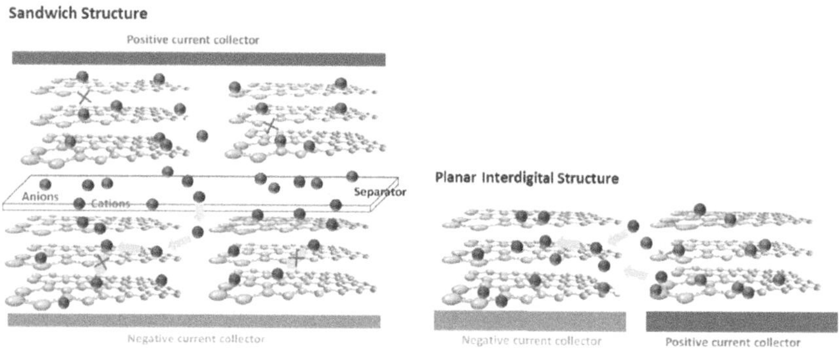

Figure 2.19 Schematic illustration of the supercapacitors in sandwich and interdigital structure [20].

integration into small devices. The planar design allows for more efficient ionic transport. By positioning the electrodes in the same plane, the mean ionic diffusion length is significantly reduced. This arrangement enables easier and quicker movement of electrolyte ions, enhancing performance. The interdigitated fingers of the electrodes provide numerous open edges for electrolyte ions to diffuse into the electrodes, maximizing surface area utilization. Shortening the width and spacing of these microelectrodes further minimizes the ionic diffusion path, improving power density and response times. In contrast to conventional sandwich structures, where ions must diffuse through the thickness of each electrode and separator, the planar design of MSCs allows for more efficient ionic movement, avoiding bottlenecks that can occur in stacked configurations. The planar design of MSCs facilitates their integration onto various substrates, including silicon wafers for semiconductor applications. This opens up possibilities for on-chip energy storage solutions in conjunction with complementary metal–oxide–semiconductor (CMOS) technology or MEMS. The planar design allows for the vertical arrangement of active materials, enabling the development of 3-D structures. This can further enhance the energy and power density per unit area without compromising ionic diffusion paths [20].

2.7 MATERIALS FOR ENERGY STORAGE DEVICES

An electrode, electrolyte, separator membrane, and packing material make up flexible energy storage systems. Under a variety of deformation conditions, every part that went into making this energy device works perfectly. Therefore, a number of methods for effectively designing each part for usage in flexible energy systems have been recorded. The newest component designs for materials are used to produce flexible energy. In this context, we examine devices with superior mechanical and electrochemical properties.

2.8 ELECTRODE MATERIALS

Active materials, conductive agents, polymeric binders, and current collectors (such as Cu, Al, and other metal materials) are frequently found in electrodes. But these components often exhibit stiff properties that prevent them from being used in flexible energy systems [21]. Electrode structures should be developed to improve the loading and adhesion of active materials in order to address this issue. The flexible conductive material should replace the traditional metal materials to achieve the energy storage systems' desired degree of versatility. Instead of using conventional gel electrolytes, liquid or solid electrolytes should be employed to prevent electrolyte loss under mechanically degraded conditions.

This section highlights innovative techniques for creating electrode materials specifically designed for flexible energy devices, with a focus on hybrid electrodes and electrodes made of carbon. [3].

2.8.1 Carbon-based electrode

Depending on their chemical structures, carbon atoms can be categorized as organic, graphitic, or polymeric. These electrodes use flexible, conductive substrates in conjunction with active components to enhance electrochemical and mechanical performance even when deformed, in addition to the intrinsic advantages of different carbon topologies [21]. Materials for carbon-based substrates, including graphene, carbon cloth, and the use of CNFs and carbon nanotubes (CNTs) in flexible energy, have grown in popularity for storing. Typically, they show strong electrochemical activity and outstanding mechanical qualities that withstand repeated deformations, showing their fascinating mechanical, chemical, and physical characteristics [22]. Since redox-active organic molecules are inexpensive and naturally sustainable, they have become one of the most promising alternatives. Guo and colleagues created the organic cathode material pyrene-4,5,9,10-tetraone (PTO) for flexible, aqueous zinc-ion batteries (ZIBs).

As shown in Figure 2.20, the SEM picture displays the nanorod-like shape of PTO. Figure 2.21 provides the schematic illustration [21].

Jen and colleagues presented an alternative method for creating flexible electrodes using carbon materials. As illustrated in Figure 2.22, they suggested a nano composite electrode composed of 1D and 3D carbon materials [21].

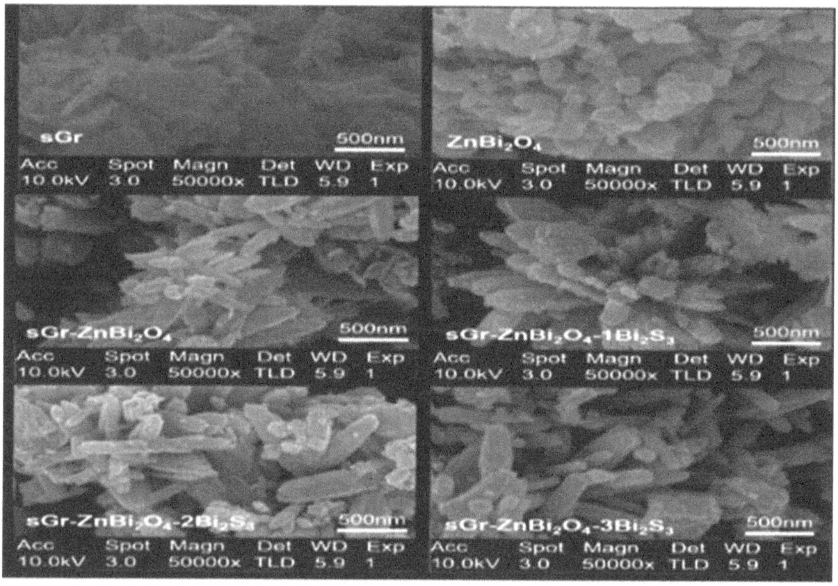

Figure 2.20 (a) SEM picture displays the nanorod. (b) Schematic illustration of nanorod [21].

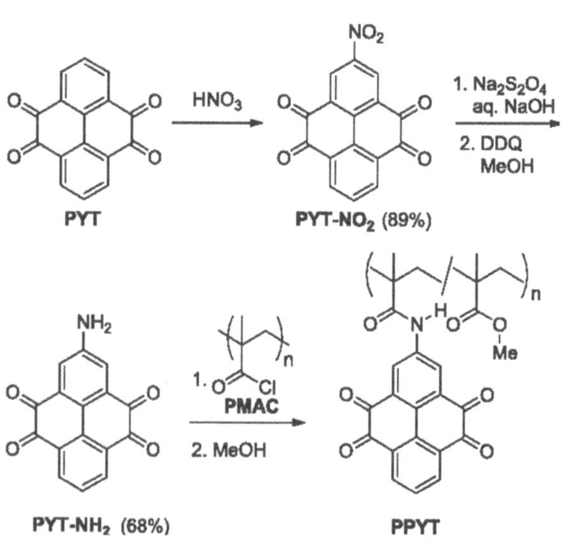

Figure 2.21 Composed of 1D and 3D carbon materials [21].

As seen in Figure 2.23, the active material in this design is acid-treated activated carbon (a-AC), while carbon nanotubes (CNTs) serve as the flexible substrate [21].

As shown in Figure 2.24, a well-constructed CNT network ensures both exceptional electrochemical performance and mechanical stability during

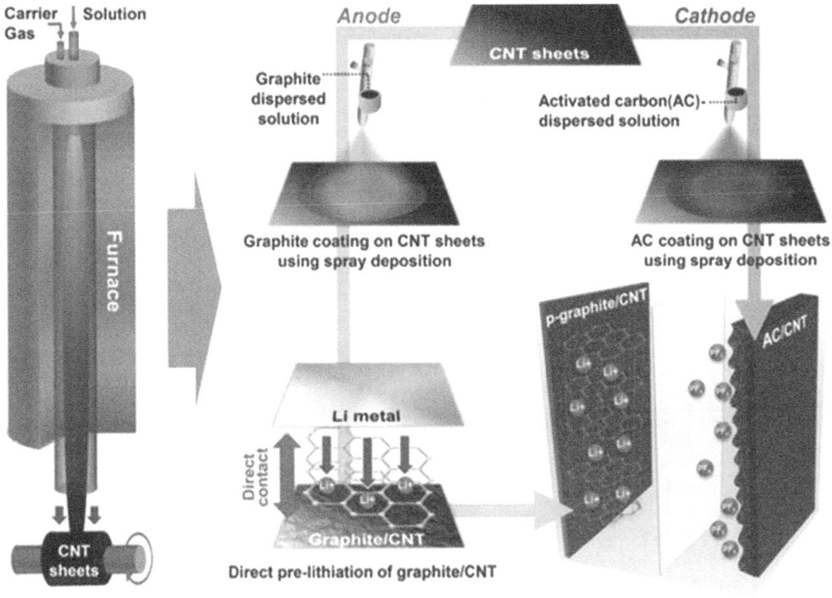

Figure 2.22 The active material in CNT design [21].

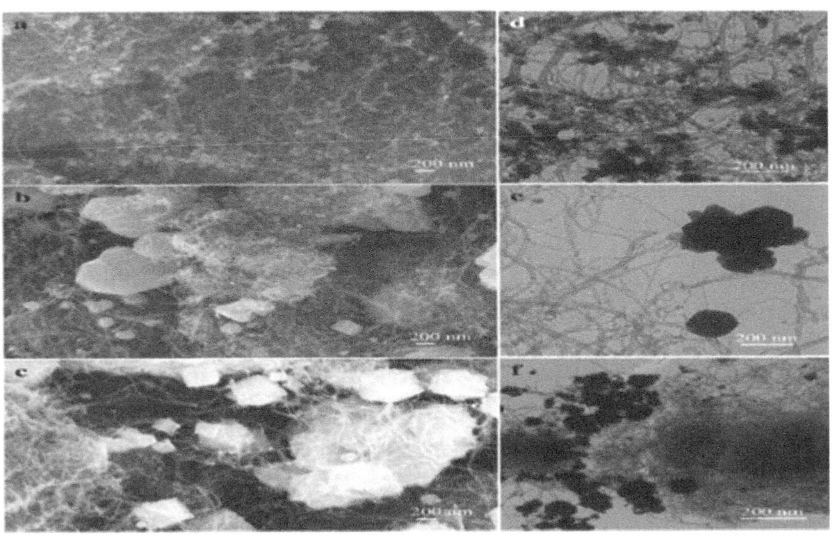

Figure 2.23 Well-constructed CNT network [21].

deformation, including folding and twisting. This architecture produces a very porous structure (1459.37 m^2 g^{-1}). The authors created a deformable supercapacitor with exceptional performance under a range of physical deformations using the produced electrode.

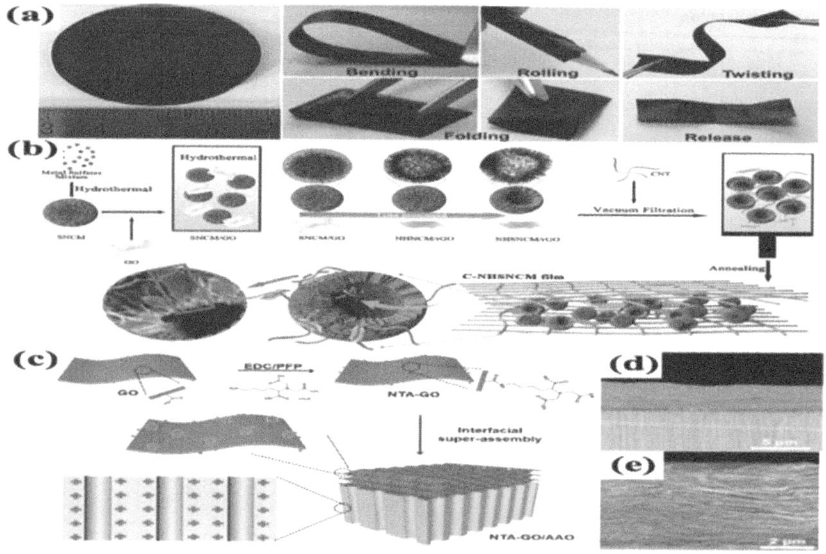

Figure 2.24 Bendable and rollable states of electrode [21].

2.8.2 Graphene-based flexible electrodes

Furthermore, another kind of fiber electrode is graphene fiber, which has excellent electrical conductivity and strength.60,61 It has the same mechanical pliability as typical textile fibers and can be readily altered in comparison to traditional carbon fibers. Wang et al. wrapped two parallel reduced graphene oxide fiber springs to create a stretchy and self-healing supercapacitor using a polymer shell that heals itself at the 7.32 mWcm^{-3} of power density and a high energy of 0.94 mWhcm^{-3}. The retained capacitance is 82.4% following a lengthy period of 100%, yet it is. After the third healing, it is 54.2%. It offers a necessary approach in the next-generation multipurpose electronic gadget [23]. SEM analysis confirmed a uniform and porous formation of the graphene network with continuous interspaces. This lightweight electrode demonstrated versatile form factors, including bendable and rollable states shown in Figure 2.25.

Using PVA-H_2SO_4 gel as the electrolyte and separator membrane, the produced electrode was utilized to construct a flexible symmetric supercapacitor. High energy density (8.8 Wh kg^{-1}) and power density (178.5 W kg^{-1}) were demonstrated by the supercapacitor, along with stable electrochemical performance even when subjected to mechanical deformation [22].

Finally, a hybrid carbon-based polymer composite that serves as a stretchable substrate for deformable energy storage devices was created by Song

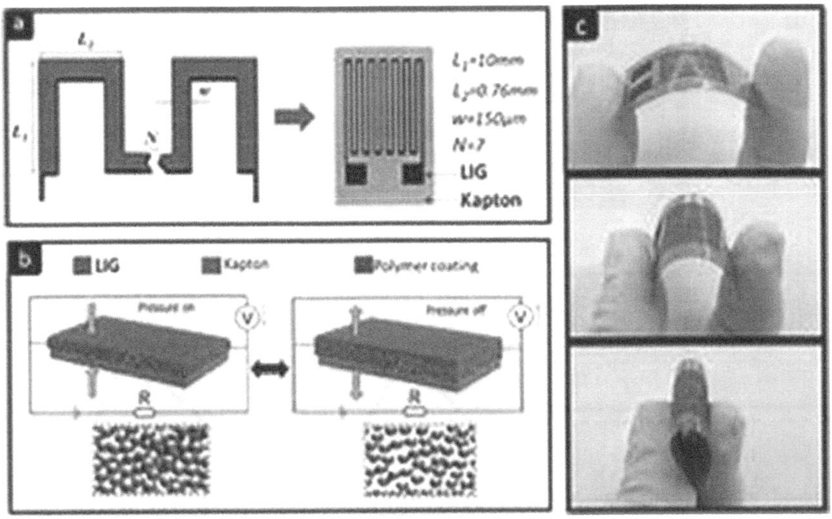

Figure 2.25 (a) Carbon black and CNT fillers. (b) SEM images of carbon/polymer (HCP) composites [21].

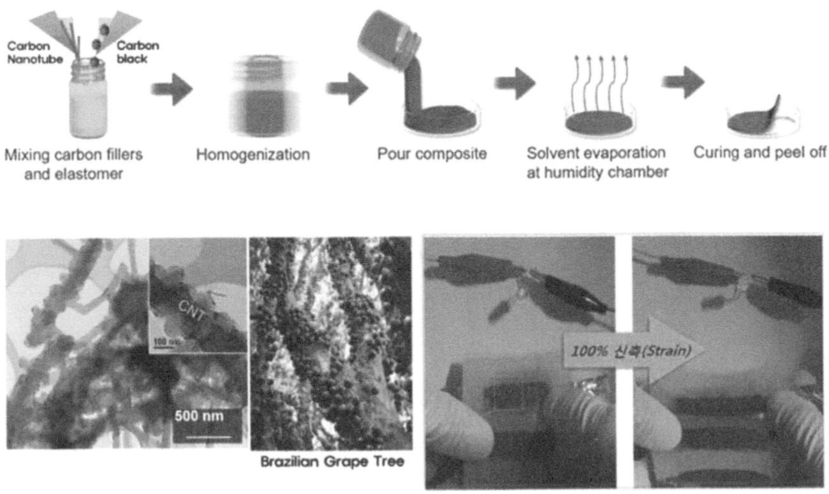

Figure 2.26 (a) TEM image of hybrid carbon fillers. (b) The flexible HCP electrode [21].

and colleagues. The composite consists of carbon black and CNT fillers shown in Figure 2.26.

The electrode fabrication process was inspired by a breath figure approach, which increased the electrode's surface area nearly fourfold under 100% humidity compared to 60%. SEM images revealed porous and uniform hybrid carbon/polymer (HCP) composites shown in Figure 2.27.

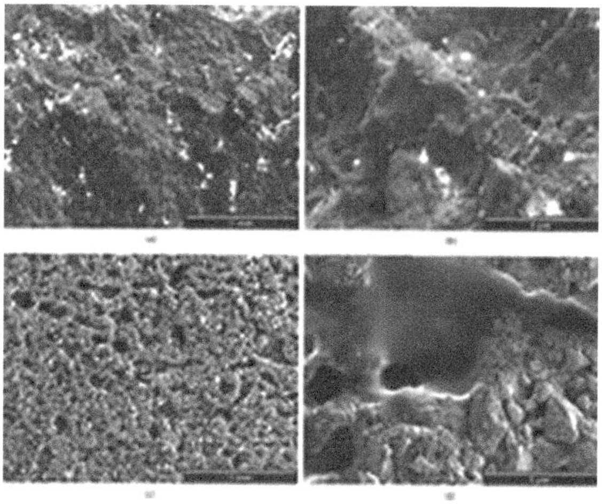

Figure 2.27 Rechargeable lithium-ion batteries (ARLBs) based on the HCP electrode [21].

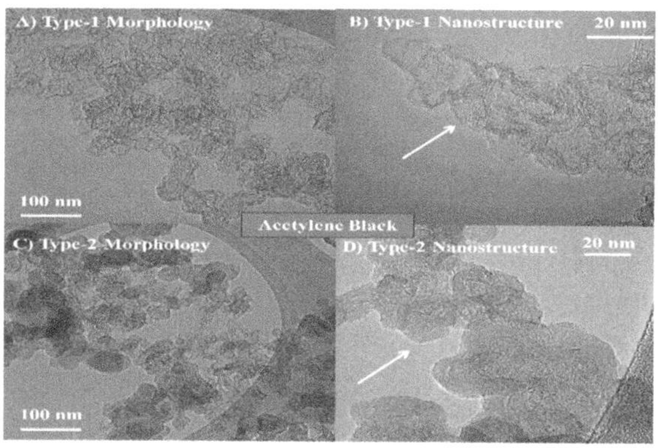

Figure 2.28 TEM of hybrid carbon fillers well-connected percolation network [21].

TEM analysis confirmed that the hybrid carbon fillers self-assembled into a well-connected percolation network, contributing to excellent electrical conductivity as shown in Figure 2.28

The flexible HCP electrode exhibited significant mechanical endurance under bending and rolling conditions, as shown in Figure 2.29.

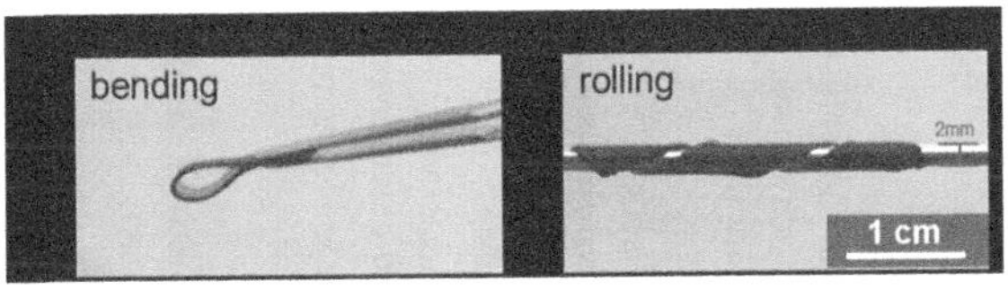

Figure 2.29 Flexible HCP electrode under bending and rolling conditions [21].

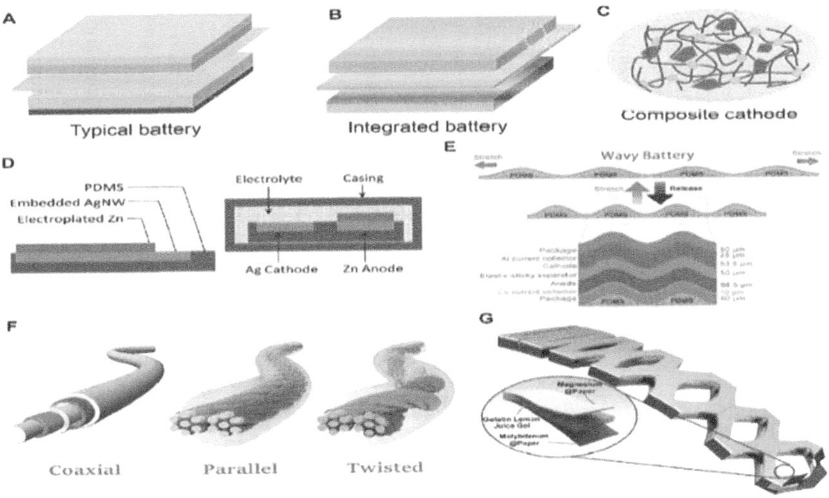

Figure 2.30 Design of rechargeable lithium-ion batteries (ARLBs) [21].

The authors achieved a breakthrough in deformable energy storage devices by using this design to create aqueous rechargeable lithium-ion batteries (ARLBs) based on the HCP electrode as shown in Figure 2.30.

These studies illustrate the versatility of carbon and graphene materials for flexible electrode applications. Their lightweight nature, high energy density, mechanical durability, and seamless integration of active materials with substrates make carbon-based electrodes ideal candidates for advanced flexible energy storage devices.

2.9 COMPARISON SHEET OF DIFFERENT MATERIALS

Comparison of different materials in a tabular form:

S. No	Material	Category	Properties	Application	Advantages	Limitations
1.	Graphene	Electrode	High conductivity, large surface area, flexibility	Flexible supercapacitors, batteries	Excellent charge transport, flexible	Expensive, aggregation issues
2.	Carbon Nanotubes (CNTs)	Electrode	High tensile strength, good conductivity	Supercapacitors, Li-ion batteries	Durable, flexible, lightweight	Complicated processing, aggregation
3.	Conductive Polymers	Electrode	Stretchable, lightweight, conductive	Wearable supercapacitors, batteries	Low-cost, flexible	Limited stability, lower conductivity than metals
4.	MXenes (e.g., Ti3_33 C2_22)	Electrode	High conductivity, hydrophilic	Batteries, supercapacitors	High energy density, fast ion transport	Oxidation issues, complex synthesis
5.	Ionic Liquids	Electrolyte	High ionic conductivity, thermally stable	Electrolyte for batteries, supercapacitors	Stable, nonvolatile	Expensive, limited compatibility
6.	Polymer Electrolytes	Electrolyte	Flexible, lightweight, conductive	Solid-state batteries	Good flexibility and safety	Lower conductivity than liquid electrolytes
7.	Hydrogels/Gel Electrolytes	Electrolyte	High ionic conductivity, stretchable	Wearable batteries, flexible supercapacitors	Flexible, biocompatible	Limited cycle life, water evaporation
8.	Cellulose Nanofibers	Substrate	Biodegradable, porous, flexible	Paper-based batteries, supercapacitor	Eco-friendly, flexible	Low conductivity, needs hybridization
9.	Reduced Graphene Oxide	Electrode	Moderate conductivity, flexible	Supercapacitors, hybrid batteries	Lower cost than pristine graphene	Lower conductivity, synthesis issues
10.	Composite Materials (e.g., CNT/Polymer)	Hybrid Material	Synergistic properties of components	Flexible batteries, supercapacitors	Improved performance and flexibility	Complex manufacturing, costly
11.	Silicon Nanoparticles	Electrode	High capacity, flexible with binders	Flexible Li-ion batteries	Very high energy storage potential	Volume expansion, poor cycling stability
12.	Silicone-Based Elastomers	Substrate	Highly stretchable, durable	Substrates for flexible electronics	Excellent mechanical properties	Insulating, requires conductive fillers

REFERENCES

1. Z. Buffett, P. Ummat, & W. Datars Properties of solvent-free perylene iodine. *Physica B: Condensed Matter*, Vol. 270, pp. 341–349, 1999.
2. H. Li, Z. Tang, Z. Liu, & C. Zhi (2019). Evaluating flexibility and wearability of flexible energy storage devices. *Joule*, 3(3), 613–619.
3. Junyuan He, Leiqing Cao, Jiaojiao Cui, Gangwen Fu, Ruiyi Jiang, Xi Xu, Flexible Energy Storage Devices to Power the Future https://doi.org/10.1002/adma.202306090
4. Ken Gilleo, *The First 105 Years of Flexible Circuitry*. ET-Trends LLC. https://www.allflexinc.com//wp-content/uploads/2019/06/IPC-Flex_2007_Gilleo.pdf
5. R. L. Crabb & F. C. Treble Thin silicon solar cells for large flexible arrays. *Nature* Vol. 213, pp. 1223–1224, 1967. https://doi.org/10.1038/2131223a0
6. S. C. Rasmussen (2017). The Early History of Polyaniline: Discovery and Origins. *Substantia*, Vol. 1, No. 2. https://doi.org/10.13128/Substantia-30
7. John McGinness; Peter Corry, Peter Proctor (March 1, 1974). "Amorphous Semiconductor Switching in Melanins", *Science*, Vol. 183 (4127): 853–855. doi:10.1126/science.183.4127.853
8. Hiro Kawamoto (2012). "The inventors of TFT active-matrix LCD receive the 2011 IEEE Nishizawa medal." *Journal of Display Technology* 8.1: 3–4.
9. I.-C. Cheng, & S. Wagner (2009). Overview of Flexible Electronics Technology. *Electronic Materials: Science & Technology*, 1–28. https://doi.org/10.1007/978-0-387-74363-9
10. V. A. Boicea, "Energy Storage Technologies: The Past and the Present," in *Proceedings of the IEEE*, Vol. 102, No. 11, pp. 1777–1794, 2014, https://doi.org/10.1109/JPROC.2014.2359545
11. SionPower, Lithium Sulfur Batteries, 2012. [Online]. http://www.sionpower.com/
12. Jie Zhang, Min Gu, Xi Chen. *Supercapacitors for Renewable Energy Applications: A Review*. Institute of Photonic Chips, University of Shanghai for Science and Technology, https://doi.org/10.1016/j.mne.2023.100229
13. Energy efficiency criteria in uninterruptible power supply selection. https://doi.org/10.1016/j.apenergy.2010.08.017
14. History, Evolution, and Future Status of Energy Storage
15. Xianfu Wang, Xihong Lu, Bin Liu, Yexiang Tong, Di Chen, and Guozhen Shen. Flexible Energy-Storage Devices: Design Consideration and Recent Progress. https://doi.org/10.1002/adma.201400910
16. https://gesrepair.com/future-of-flexible-electronic/
17. Daman Yadava, Fedi Aminib, Andrea Ehrmanne, Recent advances in carbon nanofibers and their applications – A review https://doi.org/10.1016/j.eurpolymj.2020.109963
18. Zhiyong Qiao, Keyu Bian, Changming Ding, Yitao Zhao. Recent progress of carbon-fiber-based electrode materials for energy storage. https://doi.org/10.1016/j.diamond.2023.110208
19. https://pubs.acs.org/doi/10.1021/acsami.0c16801
20. Bo Song, Kyoung-Sik Moon, and Ching-Ping Wong, *Recent Developments in Design and Fabrication of Graphene-Based Interdigital Micro-Supercapacitors for Miniaturized Energy Storage Devices*. IEEE. https://doi.org/10.1109/TCPMT.2016.2585162

21. W.-J. Song, S. Lee, G. Song, H.-B. Son, D.-Y. Han, I. Jeong, Y. Bang, S. Park, Recent progress in aqueous based flexible energy storage devices, *Energy Storage Materials*, 2020 https://doi.org/10.1016/j.ensm.2020.05.006
22. Hongsen Li, Xiao Zhang, Zhongchen Zhao, Zhengqiang Hua, Xin Liu, Guihua Yu, Flexible sodium-ion based energy storage devices: Recent progress and challenges. https://doi.org/10.1016/j.ensm.2019.12.037
23. Qi Zhang, Xuan-Wen Gao, Xiao Liu, Jian-Jia Mu, Qinfen Gu, Zhaomeng Liu, Wen-Bin Luo, Flexible wearable energy storage devices: Materials, structures, and applications. https://doi.org/10.1002/bte2.20230061
24. Ansari Faizan, Singh Mandeep, Chaudhary Tarun, Raj Balwinder, Sankaria Faizan "Comparative Analysis of Nanowire Tunnel Field Efect Transistor for Low Power Application" *Journal of Silicon*, Vol. 14, pp. 12075–12084, 2022.
25. Singh Mandeep, Chaudhary Tarun, Balwinder Raj "Drain Current and Transconductance Analysis of GaN GAA Nanowire FET with High K Dielectric" *ECS Journal of Solid State Science and Technology*, Vol. 12, pp. 113007, 2023.
26. Singh Mandeep, Chaudhary Tarun, Balwinder Raj "Design and Simulation Analysis of Silicon Germanium Nanowire FET for Low Power Applications" *Journal of Silicon*, Vol. 15, pp. 1491–1497, 2023.
27. Mandeep Singh, Tarun Chaudhary and Balwinder Raj "Design of Si and Ge Nanowire FET for Low Power Application" *Journal of Silicon*, Vol. 15 (2023): pp. 1491–1497. https://doi.org/10.1007/s12633-022-02101-y
28. Mandeep Singh, Tarun Chaudhary and Balwinder Raj, "Drain Current and Transconductance Analysis of GaN GAA Nanowire FET with High K Dielectric." *ECS Journal of Science and Technology*, 12.15 (2023): 6070–6077, https://doi.org/10.1149/2162-8777/ad0874
29. Faizan Ansari, Tarun Chaudhary, Ramesh Kumar Sunkaria and Mandeep Singh, Balwinder Raj. "Comparative Analysis of Nanowire Tunnel Field Effect Transistor for Low Power Application." *Journal of Silicon* (2022) 14: 12075–12084. https://doi.org/10.1007/s12633-022-01967-2
30. Anjana Bhardwaj, B. Pradeep Kumar Sunny Anand Raj, "Design and Performance Optimization of doping-less Vertical Nanowire TFET using Gate Stack Technique", *Journal of Electronic Materials (JEMS)*, Vol. 41, No. 7, pp. 4005–4013, 2022.
31. Jeetendra Singh, B. Raj, "Tunnel Current Model of Asymmetric MIM Structure Levying Various Image Forces to Analyze the Characteristics of Filamentary Memristor", *Applied Physics A*, Vol. 125, No.3, pp. 203.1–203.11, 2019.
32. Candy Goyal, Jagpal Singh Ubhi and B. Raj, Low Leakage Zero Ground Noise Nanoscale Full Adder using Source Biasing Technique, *Journal of Nanoelectronics and Optoelectronics*, Vol. 14, pp. 360–370, 2019.
33. R. K. Gurmohan Singh Sarin and B. Raj, "A Novel Robust Exclusive-OR Function Implementation in QCA Nanotechnology with Energy Dissipation Analysis", *Journal of Computational Electronics*, Vol. 15, No. 2, pp. 455–465, 2016.
34. Girish Wadhwa, Priyanka Kamboj, Balwinder Raj "Design optimisation of junctionless TFET biosensor for high sensitivity" *Advances in Natural Sciences: Nanoscience and Nanotechnology*, Vol. 10 (7), pp. 045001, 2019.
35. Priya Bansal, B. Raj, "Memristor Modeling And Analysis For Linear Dopant Drift Kinetics," *Journal of Nanoengineering and Nanomanufacturing*, Vol. 6, pp. 1–7, 2016.

36. D. Vaithiyanathan Ashima, B. Raj," Performance Analysis of Charge Plasma induced Graded Channel Si nanotube, *Journal of Engineering Research (JER)*, pp. 146–154, 2021.
37. Abhishek Singh Tomar, Vijay Kumar Magraiya, and B. Raj, Scaling of Access and Data Transistor for High Performance DRAM Cell Design, *Quantum Matter*, Vol. 2, pp. 412–416, 2013.
38. Neeraj Jain, B. Raj, "Parasitic Capacitance and Resistance Model Development and Optimization of Raised Source/Drain SOI FinFET Structure for Analog Circuit Applications, *Journal of Nanoelectronics and Optoelectronins*, Vol. 13, pp. 531–539, 2018.
39. Shradhya Singh, S. K. Vishvakarma and B. Raj, "Analytical Modeling of Split-Gate Junction-Less Transistor for a Biosensor Application", *Sensing and Biosensing*, Vol. 18, pp. 31–36, 2018.
40. Maisagalla Gopal, and Balwinder Raj, Low Power 8T SRAM Cell Design for High Stability Video Applications, *ITSI Transaction on Electrical and Electronics Engineering*, Vol. 1, No. 5, pp. 91–97, 2013.
41. Balwinder Raj, Jatin Mitra, Deepak Kumar Bihani, V. Rangharajan, A.K. Saxena and S. Dasgupta, "Analysis of Noise Margin, Power and Process Variation for 32 nm FinFET Based 6T SRAM Cell" *Journal of Computer (JCP)*, Vol. 5, No. 6, pp. 1–8, 2010.
42. Divya Sharma, Rajesh Mehra and B. Raj, "Comparative Analysis of Photovoltaic Technologies for High Efficiency Solar Cell Design", *Superlattices and Microstructures*, Vol. 153, pp. 106861, 2021.
43. Pawandeep Kaur, Avtar Singh Buttar, B. Raj, "A comprehensive Analysis of Nanoscale Transistor based Biosensor: A Review, *Indian Journal of Pure and Applied Physics*, Vol. 59, pp. 304–318, 2021.
44. Divya Yadav, Balwant Raj and B. Raj, "Design and Simulation of Low Power Microcontroller for IoT Applications", *Journal of Sensor Letters*, Vol. 18, pp. 401–409, 2020.
45. Shailendra Singh, B. Raj, "A 2-D Analytical Surface Potential and Drain current Modeling of Double-Gate Vertical t-shaped Tunnel FET", *Journal of Computational Electronics*, Vol. 19, PP. 1154–1163, 2020.
46. Jeetendra Singh, B. Raj, "An Accurate and Generic Window function for Nonlinear Memristor Model" *Journal of Computational Electronics*, Vol. 18 No. 2, pp. 640–647, 2019.
47. Manjit Kaur, Neena Gupta, Sanjeev Kumar, B. Raj, Arun Kumar Singh, "Comparative RF and Crosstalk Analysis of Carbon Based Nano Interconnects" *IET Circuits, Devices & Systems*, Vol. 15, No. 6, pp. 493–503, 2021.
48. Nehru Kandasamy, D. Firdous Ahmad, B. Raj Ajitha, and Nagarjuna Telagam. "Quantum Dot Cellular Automata based Scan Flip Flop and Boundary Scan Register, *IETE Journal of Research*, Vol. 66, pp. 535–548, 2020.
49. Jeetendra Singh, B. Raj, "Design and Investigation of 7T2M NVSARM with Enhanced Stability and Temperature Impact on Store/Restore Energy", *IEEE Transactions on Very Large Scale Integration Systems*, Vol. 27, No. 6, pp. 1322–1328, 2019.
50. Anil Kumar Bhardwaj, Sumeet Gupta, B. Raj, Amandeep Singh, "Impact of Double Gate Geometry on the Performance of Carbon Nanotube Field Effect Transistor Structures for Low Power Digital Design", *Computational and Theoretical Nanoscience, ASP*, Vol. 16, pp. 1813–1820, 2019.

51. Neeraj Jain, B. Raj, "Thermal Stability Analysis and Performance Exploration of Asymmetrical Dual-k underlap Spacer (ADKUS) SOI FinFET for Security and Privacy Applications", *Indian Journal of Pure & Applied Physics (IJPAP)*, Vol. 57, pp. 352–360, May 2019.
52. Neeraj Jain and Balwinder Raj, "Dual-k Spacer Region Variation at the Drain Side of Asymmetric SOI FinFET Structure: Performance Analysis towards the Analog/RF Design Applications", *Journal of Nanoelectronics and Optoelectronics*, Vol. 14, pp. 349–359, March 2019.
53. Jeetendra Singh, Sanjeev Sharma, B. Raj, Mamta Khosla, Analysis of barrier layer thickness on performance of In1-xGaxAs based Gate Stack Cylindrical Gate Nanowire MOSFET," *ASP*, Vol. 13, PP 1473–1477, 2018.
54. Neeraj Jain, B. Raj, "Analysis and Performance Exploration of High-k SOI FinFETs Over the Conventional Low-k SOI FinFET toward Analog/RF Design", *Journal of Semiconductors (JoS), IOP Science*, Vol. 39, No. 12, PP- 124002-1-7, 2018.
55. Candy Goyal, Jagpal Singh Ubhi and B. Raj, A reliable leakage reduction technique for approximate full adder with reduced ground bounce noise, *Journal of Mathematical Problems in Engineering*, Vol. 2018, 3501041, 16 pages, 2018.
56. Jeetendra Singh Anuradha, B. Raj and Mamta Khosla, "Design and Performance Analysis of Nano-scale Memristor-based Nonvolatile SRAM" *Journal of Sensor Letter*, Vol. 16, pp. 798–805, 2018.
57. Girish Wadhwa, B. Raj, "Parametric Variation Analysis of Charge-Plasma-based Dielectric Modulated JLTFET for Biosensor Application" *IEEE Sensor Journal*, Vol. 18, No. 15, pp. 6070–6077, 2018.
58. Jeetendra Singh, B. Raj, "Comparative Analysis of Memristor Models for Memories Design", *JoS*, Vol. 39, No. 7, pp. 074006-1-12, 2018.
59. Divya Yadav, Shailesh Singh Chouhan, Santosh Kumar Vishvakarma and B. Raj, "Application Specific Microcontroller Design for IoT based WSN", *Sensor Letter*, Vol. 16, pp. 374–385, 2018.
60. R. K. Gurmohan Singh Sarin and BRaj, "Fault-Tolerant Design and Analysis of Quantum-Dot Cellular Automata Based Circuits", *IEEE/IET Circuits, Devices & Systems*, Vol. 12, pp. 638–664, 2018.
61. Jeetendra Singh, B. Raj, "Modeling of Mean Barrier Height Levying Various Image Forces of Metal Insulator Metal Structure to Enhance the Performance of Conductive Filament Based Memristor Model", *IEEE Nanotechnology*, Vol. 17, No. 2, pp. 267–268, 2018.
62. Aakash Jain, Sanjeev Sharma, BRaj, "Analysis of Triple Metal Surrounding Gate (TM-SG) III-V Nanowire MOSFET for Photosensing Application", *Opto-electronics Journal*, Vol. 26, No. 2, pp. 141–148, 2018.
63. Aakash Jain, Sanjeev Sharma, B. Raj, "Design and Analysis of High Sensitivity Photosensor Using Cylindrical Surrounding Gate MOSFET For Low Power Sensor Applications", *Engineering Science and Technology, an International Journal*, Vol. 19, No. 4, pp. 1864–1870, 2016.
64. G. Saiphani Kumar, Amandeep Singh, B. Raj, "Design and Analysis of Gate All Around CNTFET based SRAM cell Design", *Journal of Computational Electronics*, Vol. 17, No. 1, pp 138–145, 2018.

65. Gurinder Pal Singh, B. S. Sohi, Balwinder Raj, "Material Properties Analysis of Graphene Base Transistor (GBT) for VLSI Analog Circuits, *Indian Journal of Pure & Applied Physics (IJPAP)*, Vol. 55, pp. 896–902, 2017.
66. Sunil Kumar and B. Raj, "Estimation of Stability and Performance metric for Inward Access Transistor based 6T SRAM Cell Design using n-type/p-type DMDG-GDOV TFET" *IEEE VLSI Circuits and Systems Letter*, Vol. 3, No. 2, pp. 25–39, 2017.
67. Shashikant Sharma, Anjan Kumar, Manisha Pattanaik and B. Raj, "Forward Body Biased Multimode Multi-Threshold CMOS Technique for Ground Bounce Noise Reduction in Static CMOS Adders" *International Journal of Information and Electronics Engineering*, pp. 567–572, Vol. 3, No. 3, 2013.
68. Hamendra Singh, Pankaj Kumar, Balwinder Raj, "Performance Analysis of Majority Gate SET Based 1-bit Full Adder", *International Journal of Computer and Communication Engineering (IJCCE)*, ISSN: 2010–3743, Vol. 2, No. 4, 2013.
69. Anil Kumar Bhardwaj, Sumeet Gupta and B. Raj, "Investigation of Parameters for Schottky Barrier (SB) Height for Schottky Barrier Based Carbon Nanotube Field Effect Transistor Device", *Journal of Nanoelectronics and Optoelectronics*, Vol. 15, pp. 783–791, 2020.
70. Priya Bansal, B. Raj, "Memristor: A Versatile Nonlinear Model for Dopant Drift and Boundary Issues," *JCTN*, Vol. 14, No. 5, pp. 2319–2325, 2017.
71. Neeraj Jain, B. Raj, "An Analog and Digital Design Perspective Comprehensive Approach on Fin-FET (Fin-Field Effect transistor) Technology - A Review", *Reviews in Advanced Sciences and Engineering (RASE)*, Vol. 5, pp. 1–14, 2016.
72. Balwinder Raj, A. K. Saxena and S. Dasgupta, "Analytical Modeling for the Estimation of Leakage Current and Subthreshold Swing Factor of Nanoscale Double Gate FinFET Device" *Microelectronics International*, Vol. 26, pp. 53–63, 2009.
73. Soniya, Shailendra Singh, Girish Wadhwa and Balwinder Raj, "An Analytical Modeling for Dual Source Vertical Tunnel Field Effect Transistor" *International Journal of Recent Technology and Engineering (IJRTE)* Vol. 8, No. 2, pp. 603–608, 2019.
74. Shailendra Singh, B. Raj, "Design and analysis of hetrojunction Vertical T-shaped Tunnel Field Effect Transistor", *Journal of Electronics Material*, Vol. 48, No. 10, pp 6253–6260, 2019.
75. Candy Goyal, Jagpal Singh Ubhi and B. Raj, "A Low Leakage CNTFET based Inexact Full Adder for Low Power Image Processing Applications", *International Journal of Circuit Theory and Applications*, Vol. 47, No. 9, pp. 1446–1458, 2019.
76. B. Raj, A. K. Saxena and S. Dasgupta, "A Compact Drain Current and Threshold Voltage Quantum Mechanical Analytical Modeling for FinFETs" *Journal of Nanoelectronics and Optoelectronics (JNO)*, Vol. 3, no. 2, pp. 163–170, 2008.
77. Girish Wadhwa, B. Raj, "An Analytical Modeling of Charge Plasma based Tunnel Field Effect Transistor with Impacts of Gate underlap Region" *Superlattices and Microstructures*, Vol. 142, pp. 106512, 2020.
78. Shailendra Singh, B. Raj, "Modeling and Simulation analysis of SiGe het-rojunction Double GateVertical t-shaped Tunnel FET", *Superlattices and Microstructures*, Vol. 142, pp. 106496, 2020.

79. Neeraj Jain, Balwinder Raj, Device and Circuit Co-Design Perspective Comprehensive Approach on FinFET Technology - A Review, *Journal of Electron Devices*, Vol. 23, No. 1, pp 1890–1901, 2016.
80. Sunil Kumar and B. Raj, "Analysis of I_{ON} and Ambipolar Current for Dual-Material Gate-drain Overlapped DG-TFET," *Journal of Nanoelectronics and Optoelectronics*, Vol. 11, pp. 323–333, 2016.
81. Naveed Anjum, Tarun Bali and B. Raj, Design and Simulation of Handwritten Multiscript Character Recognition, *International Journal of Advanced Research in Computer and Communication Engineering*, Vol. 2, No. 7, pp. 2544–2549, 2013.
82. Karmjit Singh, B. Raj, "Performance and analysis of temperature dependent Multi-walled Carbon Nanotubes as Global Interconnects at different technology nodes," *Journal of Computational Electronics*, Vol. 14 No. 2, pp. 469–476, 2015.
83. Sunil Kumar and B. Raj, "Compact channel potential analytical Modeling of DG-TFET based on Evanescent–mode Approach," *Journal of Computational Electronics*, Vol. 14, No. 2, pp. 820–827, 2015.
84. Karmjit Singh, B. Raj, "Temperature Dependent Modeling and Performance Evaluation of Multi-Walled CNT and Single-Walled CNT as Global Interconnects," *Journal of Electronic Materials*, Vol. 44, No. 12, pp. 4825–4835, 2015.
85. V. K. Sharma, M. Pattanaik and B. Raj, "INDEP approach for leakage reduction in nanoscale CMOS circuits", *International Journal of Electronics*, Vol. 102, No. 2, pp. 200–215, 2014.
86. Karmjit Singh, B. Raj, "Influence of Temperature on MWCNT bundle, SWCNT bundle and Copper interconnects for Nanoscaled Technology nodes," *Journal of Materials Science: Materials in Electronics*, Vol. 26, No. 8, pp. 6134–6142, 2015.
87. Naveed Anjum, Tarun Bali and B. Raj, "Design and Simulation of Handwritten Gurumukhi and Devanagri Numerical Recognition", *International Journal of Computer Applications, Published by Foundation of Computer Science*, Vol. 73, No. 12, pp. 16–21, 2013.
88. S. Khandelwal, V. Gupta, B. Raj, and R. D. Gupta, "Process Variability Aware Low Leakage Reliable Nano Scale DG-FinFET SRAM Cell Design Technique", *Journal of Nanoelectronics and Optoelectronics*, Vol. 10, No. 6, pp. 810–817, 2015.
89. V. K. Sharma, M. Pattanaik and B. Raj, "ONOFIC Approach: Low Power High Speed Nanoscale VLSI Circuits Design", *International Journal of Electronics*, Vol. 101, No. 1, pp. 61–73, 2014.
90. S. Khandelwal, Balwinder Raj, and R. D. Gupta, FinFET based 6T SRAM Cell Design: Analysis of Performance Metric, Process Variation and Temperature Effect, *Journal of Computational and Theoretical Nanoscience*, Vol. 12, pp. 2500–2506, 2015.
91. Sumit Singh; Shekhar Yadav; Jagdeep Rahul; Anurag Srivastava; B. Raj, Impact of HfO_2 in Graded Channel Dual Insulator Double Gate MOSFET, *Journal of Computational and Theoretical Nanoscience*, Vol. 12, No. 6, pp. 950–953, 2015.
92. Vijay Kumar Sharma, Manisha Pattanaik, B. Raj, PVT variations aware low leakage INDEP approach for nanoscale CMOS Circuits, *Microelectronics Reliability*, Vol. 54, pp. 90–99, 2014.

93. B. Raj, A. K. Saxena and S. Dasgupta, "Quantum Mechanical Analytical Modeling of Nanoscale DG FinFET: Evaluation of Potential, Threshold Voltage and Source/Drain Resistance", *Elsevier's Journal of Material Science in Semiconductor Processing*, Vol. 16, No. 4, pp. 1131–1137, 2013.
94. Maisagalla Gopal, Siva Sankar, D. Prasad and Balwinder Raj, 8T SRAM Cell Design for Dynamic and Leakage Power Reduction, *International Journal of Computer Applications*, Vol. 71, No. 9, pp. 43–48, 2013.
95. Manisha Pattanaik, B. Raj, Shashikant Sharma and Anjan Kumar, "Diode Based Trimode Multi-Threshold CMOS Technique for Ground Bounce Noise Reduction in Static CMOS Adders", *Advanced Materials Research*, Vol. 548, pp 885–889, 2012.
96. Balwinder Raj, A. K. Saxena and S. Dasgupta, "Nanoscale FinFET Based SRAM Cell Design: Analysis of Performance metric, Process variation, Underlapped FinFET and temperature effect" *IEEE Circuits and System Magazine*, Vol. 11, No. 2, pp. 38–50, 2011.
97. V. K. Sharma, M. Pattanaik & Balwinder Raj, "Leakage Current ONOFIC Approach for Deep Submicron VLSI Circuit Design", *International Journal of Electrical, Computer, Electronics and Communication Engineering*, Vol. 7, No. 4, pp. 239–244, 2013.
98. Parminder Kaur, Sandeep Singh Gill and B. Raj, "Comparative Analysis of OFETs Materials and Devices for Sensor Applications", *Journal of Silicon*, Vol 14, pp. 4463–4471, 2022.
99. Sanjeev Kumar Sharma, Parveen Kumar, Balwant Raj and B. Raj, "$In_{1-x}Ga_xAs$ Double Metal Gate-Stacking Cylindrical Nanowire MOSFET for highly sensitive Photo detector", *Journal of Silicon*, Vol. 14, pp. 3535–3541, 2022.
100. B. Raj, A. K. Saxena and S. Dasgupta, "Analytical Modeling of Quasi Planar Nanoscale Double Gate FinFET with Source/Drain Resistance and field dependent carrier mobility: A quantum mechanical study" *Journal of Computer (JCP)*, Vol. 4, No. 9, pp. 1–8, 2009.
101. S. Bhushan, S. Khandelwal, and B. Raj, (2012). Analyzing Different Mode FinFET based Memory Cell at Different Power Supply for Leakage Reduction, *Seventh International Conference on Bio-Inspired Computing: Theories and Application, (BIC-TA 2012) Advances in Intelligent Systems and Computing*, Vol. 202, pp. 89–100, 2013.
102. Jeetendra Singh, B. Raj, "Temperature Dependent Analytical Modeling and Simulations of Nanoscale Memristor, *Journal: Engineering Science and Technology, an International Journal*, Vol. 21, pp. 862–868, 2018.
103. Shradhya Singh, Shashi Bala, Balwant Raj, B. Raj, "Improved Sensitivity of Dielectric Modulated Junctionless Transistor for Nanoscale Biosensor Design", *Sensor Letter*, Vol. 18, pp. 328–333, 2020.
104. Vivek Kumar, Santosh Kumar Vishvakarma, and B. Raj, "Design and Performance Analysis of ASIC for IoT Applications" *Sensor Letter ASP*, Vol. 18, pp. 31–38, 2020.
105. Akanksha Jaiswal, R. K. Sarin, B. Raj and Shikha Sukhija, "A Novel Circular Slotted Microstrip-fed Patch Antenna with Three Triangle Shape Defected Ground Structure for Multiband Applications", *Advanced Electromagnetic (AEM)*, Vol. 7, No. 3, pp. 56–63, 2018.
106. Girish Wadhwa, B. Raj, "Label Free Detection of Biomolecules using Charge-Plasma-Based Gate Underlap Dielectric Modulated Junctionless TFET" *Journal of Electronic Materials (JEMS)*, Vol. 47, No. 8, pp. 4683–4693, 2018.

107. Gurmohan Singh, R. K. Sarin and B. Raj, "Design and Performance Analysis of a New Efficient Coplanar Quantum-Dot Cellular Automata Adder", *Indian Journal of Pure & Applied Physics (IJPAP)*, Vol. 55, pp. 97–103, 2017.
108. Parminder Kaur, Vikas Pandey and B. Raj, "Comparative Study of Efficient Design, Control and Monitoring of Solar Power using IoT", *Sensor Letter*, Vol. 18, pp. 419–426, 2020.
109. Anil Kumar Bhardwaj, Sumeet Gupta and B. Raj, Development & Analysis of compact model for Double Gate Schottky Barrier CNTFET, *Journal of Nanoelectronics and Optoelectronics*, Vol. 15, pp. 1199–1208, 2020.
110. Girish Wadhwa, Priyanka Kamboj, Jeetendra Singh, B. Raj, "Design and Investigation of Junctionless DGTFET for Biological Molecule Recognition" *Transactions on Electrical and Electronic Materials*, Vol. 22, pp. 282–289, 2021.
111. Sachin Kumar Verma, Shailendra Singh, Girish Wadhwa and B. Raj, "Detection of Biomolecules using Charge-Plasma based Gate Underlap Dielectric Modulated Dopingless TFET" *Transactions on Electrical and Electronic Materials (TEEM)*, Vol. 21, pp. 528–535, 2020.
112. Neeraj Jain, B. Raj, "Impact of underlap spacer region variation on Electrostatic and analog/RF performance of symmetrical high-k SOI FinFET at 20 nm channel length", *Journal of Semiconductors (JoS)*, Vol. 38, No. 12, pp. 122002, 2017.
113. Shailendra Singh, B. Raj," Analytical Modeling and Simulation analysis of T-shaped III-V heterojunction Vertical T-FET", *Superlattices and Microstructures*, Vol. 147, pp. 106717, 2020.
114. Gurmohan Singh, R. K. Sarin and B. Raj, "Design and Analysis of Area Efficient QCA Based Reversible Logic Gates", *Journal of Microprocessors and Microsystems*, Vol. 52, pp. 59–68, 2017.
115. Shailendra Singh, B. Raj, "Parametric variation analysis on hetero-junction Vertical t-shape TFET for Supressing ambipolar conduction", *Indian Journal of Pure and Applied Physics*, Vol. 58, pp. 478–485, 2020.
116. Parveen Kumar, Balwinder Raj, Girish Wadhwa, Balwinder Singh and Raj Kumar "Design and Analysis of Junctionless based Gate All Around N+ Doped Layer Nanowire TFET Biosensor", *ECS Journal of Solid State Science and Technology*, Vol 13, pp. 017002–017010, 2024.
117. Divya Sharma, Rajesh Mehra and Balwinder Raj, "Enhancement in Efficiency of Methyl Ammonium Tin Iodide based perovskite solar cell using SCAPS-1 D", *Nano*, Vol. 18, No. 14, pp. 2350109, 2023.
118. Parveen Kumar, Sanjeev Kumar Sharma, Balwant Raj and Balwinder Raj, "Analysis of Device Parameter Variations in In1-xGaxAs based Gate Stacked Double Metal Surrounding Gate Nanowire MOSFET", *TEEM*, Vol. 24, pp. 570–578, 2023.
119. Anjana Bhardwaj, Pradeep Kumar, Balwinder Raj, Naveen Kumar, Sunny Anand, Design and Optimization of Vertical Nanowire Tunnel FET with Electrostatic Doping, *Engineering Research Express*, Vol. 5, pp. 045025–045033, 2023
120. Yogesh Thakur, Balwinder Raj, Sandeep Singh Gill, "Design and Performance Analysis of Pentacene Organic Field Effect Transistor with High-K dielectric Materials, *Journal of Optoelectronics and Advanced Materials – Rapid Communications*, Vol. 17, No. 7–8, 2023.

121. Parveen Kumar and Balwinder Raj, "Design and Simulation of Junctionless Nanowire Tunnel Field Effect Transistor Biosensor for High Sensitivity", *Microelectronic Journal*, Vol. 137, pp. 105826–105834, 2023.
122. Soniya, Shailendra Singh, Girish Wadhwa and Balwinder Raj, Design and Analysis of Dual Source Vertical Tunnel Field Effect Transistor for high performance, *Transactions on Electrical and Electronics Materials*, Vol. 21, pp. 74–82, 2019.
123. Manjit Kaur, Neena Gupta, Sanjeev Kumar, B. Raj, Arun Kumar Singh, "RF Performance Analysis of Intercalated Graphene Nanoribbon Based Global Level Interconnects" *Journal of Computational Electronics*, Vol. 19, pp. 1002–1013, 2020.
124. Girish Wadhwa, B. Raj, "Design and Performance Analysis of Junctionless TFET Biosensor for high sensitivity" *IEEE Nanotechnology*, Vol. 18, pp. 567–574, 2019.
125. Jeetendra Singh, B. Raj, "Enhanced Nonlinear Memristor Model Encapsulating Stochastic Dopant Drift, *JNO*, 14, 958–963 (2019)
126. Mohit Kumar, Tarun Chaudhary, Balwinder Raj, "Design and Comparative Analysis of Ferroelectric Nanowire with Dielectric HfO_2 and $Al_2O_3$for Low Power Applications", *Journal of Electronic Materials*, Vol. 53, pp. 6785–6791, 2024.
127. Yogesh Thakur, Balwant Raj and Balwinder Raj, "Design of Pentacene Thin-Film Transistor Based Hydrogen Gas Sensor for High Sensitivity", *ECS Journal of Solid State Science and Technology*, Vol. 13, pp. 047005–047012, 2024.
128. Manoharan Arun Kumar, Kamal Batcha Mohamed Ismail, Shanmugam Mahalingam, Balwinder Raj, Junghwan Kim, Raji Atchudan, "Recent Advances in Two-Dimensional Nanomaterials and Nanostructures for Healthcare Monitoring", *ACS Sensors*, Vol. 9, No. 4, pp. 1–8, 2024.
129. Mohamed Ismail Kamal Batcha, Arun Kumar Manoharan, Shanmugam Mahalingam, Raj Balwinder, "Carbon Fiber-Reinforced Polymers for Energy Storage Applications", *Journal of Energy Storage*, Vol. 84, Part B, pp. 110931–110939, 2024.
130. Shahraan Hussain, Balwinder Raj, Divya Sharma. "Design and Parametric Optimization of Lead Free Perovskite based Tandem Solar Cell for High Efficient", *Engineering Research Express*, Vol. 6, pp. 015315–0153122, 2024.

Chapter 3

2D Material-based sensor design

Balwinder Raj, Vinay Chandel, Ravi Jyoti, Manish Agrahari, Balwant Raj, and Mandeep Singh

3.1 INTRODUCTION

When the bulk material is only a few layers thick, within a range of 1 nanometer to almost 1 micrometer (a millionth of a meter), it turns into what are called two-dimensional materials, or the common jargon term "2D materials." Examples include graphene, a conductive strong and flexible material formed by carbon atoms arranged in hexagonal lattices; insulating, thermally stable hexagonal boron nitride (h-BN) with mechanical strength that is similar to graphene; electronic effective transition metal dichalcogenides (TMDs) such as molybdenum disulfide (MoS_2); phosphorene geometry of black phosphorus known as a semiconductor. Showing high surface area, tunable properties (such as electrical conductivity), mechanical strength, and transparency, these materials are ideal candidates for applications in electronics, energy storage, and nanotechnology. Those properties are why they hold promise for future technologies, such as flexible electronics and wearables: their high conductivity, low density, and ease of production yield screens that can be made thin and flexible.

Interest in 2D nanomaterials surged after 2004, when physicists Andre Geim and Konstantin Novoselov at the University of Manchester were able to isolate graphene. Using the so-called "Scotch tape method," they isolated monolayer graphene from graphite [1], a discovery that, as recognition of it inevitably follows, earned them a Nobel Prize in Physics 2010 and heralded the new age of material science. This was followed by numerous investigations on different 2D materials such as hexagonal boron nitride and transition metal dichalcogenides, revealing a lot of interesting properties and potential applications. The similar progresses in fabrication, including a chemical vapor deposition, have been matured toward the applications of 2D materials on electronic (and photonic) devices. Emerging applications are being developed across various fields, such as electronics, sensing, spintronics, plasmonics, photodetectors, ultrafast lasers, batteries, supercapacitors, piezoelectrics, thermoelectrics, and catalytic applications [2–6].

Nanomaterials show many unique properties that are enhanced compared to the bulk materials, all contributing to their performance and

 DOI: 10.1201/9781003634645-3

implications in different fields [7]. High surface area-to-volume ratio is one of the most remarkable properties of nanomaterials. This increases their reactivity and allows for better interaction with other materials, making them effective catalyst and sensor materials. Nanomaterials have recently been found to accelerate chemical reactions much more efficiently than their bulk counterparts because there are a greater fraction of atoms on the surface rather than in the body that can react, for example, in catalytic applications [7–13].

If we compare the nanomaterials with the traditional bulk metals, it is apparent that nanomaterials usually surpass the traditional values of conductivity. Though bulk metals possess impressive electrical conductivity, the increased surface area of nanomaterials still rises the mobility of electrons; therefore, the conductivity of the electronic nanoscale devices shall be enhanced [8]. In addition, most nanomaterials exhibit quantum effects when behaving in a peculiar manner both electrically and optically. This also covers the potential for bandgap tailoring and is among the critical needs that have to be fulfilled modern semiconductor fabrication for next-generation devices, including electronic and optoelectronic transistors and solar cells. Bulk ceramics are also known for toughness, but lack of flexibility may limit their applications. Nanoparticles, on the other hand, are ideal options for high-performance composites in automotive and aerospace engineering sectors mainly because they could be formulated to show outstanding mechanical strength while retaining a lightweight profile [9]. Therefore, these nanocomposites do not only enhance the fuel efficiency of a car but also make it safer and more effective through weight reduction while maintaining durability and impact resistance [14–20].

In addition, the thermal and electrical conductivity of conventional polymers is usually inadequate for high-performance applications. However, hybrid nanomaterials with both organic and inorganic components may possibly enhance such properties to provide innovative solutions for thermal management in electronic devices and flexible electronics requiring conductivity besides flexibility. For example, embedding graphene within polymer matrices can result in the creation of materials that are both lightweight and highly conductive for wearable technology [10].

Nanomaterials in general possess some special properties that are further classified as being related to high reactivity and quantum behavior, extraordinary strength, and other lightweight-related properties among others. These have made nanomaterials appropriate for many applications that traditional materials cannot serve and have become a source of innovation in technologies and material sciences for the development of new innovative solutions in areas as electronics, energy storage, medicine, and environmental remediation. And as research and development would progress, there would be more enormous possibilities of uses of nanomaterials, which would further transform the aforementioned industries and advance our life [21–26] (Table 3.1).

Table 3.1 Comparison in properties between bulk material and 2D materials

Property	*Bulk Materials*	*2D Materials*
Thickness	Typically several micrometers to centimeters	Atomically thin, typically one or two layers
Surface Area	Lower surface area relative to volume	High surface area-to-volume ratio
Reactivity	Less reactive due to limited surface exposure	More reactive due to greater surface exposure
Mechanical Properties	Generally less flexible and more brittle	Can be highly flexible and strong
Electrical Conductivity	Good conductivity	Varies widely; some like graphene are excellent conductors, others like MoS_2 are semiconductors
Thermal Conductivity	Typically good thermal conductors	Varies; some materials like graphene are exceptional conductors, while others are insulators
Optical Properties	Often opaque; limited optical properties	Can be transparent and exhibit unique optical behaviors
Fabrication Complexity	Generally less complex	More complex due to precise layer control

3.2 2D NANOMATERIAL CLASSIFICATION

2D nanomaterials have been broadly classified into four categories: inorganic, organic, heterostructures, and hybrid materials. These categories of materials result in different properties, allowing them to be put to many different uses within diverse advanced technologies [27–33].

3.2.1 Inorganic 2D materials

- **Graphene:** Due to high electrical conductivity and mechanical strength with large carrier mobility, graphene has been widely applied in the design of high-performance electronic devices such as field effect transistors (FETs) and sensors and energy storage systems such as supercapacitors and lithium-ion batteries. Its optical transparency also makes it one of the candidate materials for application in optoelectronic devices, including photodetectors, and transparent conductive films.
- **Hexagonal Boron Nitride (h-BN):** h-BN, considered as "white graphene," is a very good electrical insulator with very high thermal conductivity that makes it excellent to be used where it could serve both

as the substrate layer as well as the dielectric layer in graphene electronics. The chemical stability of the material puts it in applications involving protective coatings and in using them as thermal management devices for electronic equipment [34–38].

- **Transition Metal Dichalcogenides (TMDs)**: Some of the types of TMDs are semiconducting, such as MoS_2 and WS_2. Due to these reasons, these materials show immense promise in next-generation electronics, including ultrathin transistors, flexible electronics, and photodetectors, as their bandgaps can be tuned and light–matter interaction is extremely strong.
- **MXenes**: This new family of 2D transition metal carbides and nitrides, MXenes, is extremely conductive electrically and is under intense investigation for potential applications in the field of energy storage: supercapacitor/battery. They also have promise in electromagnetic interference shielding and sensor technologies due to their high surface area and conductivity [39–44].

3.2.2 Organic 2D materials

- **Covalent Organic Frameworks (COFs)**: Due to their high porosity, structural flexibility, and chemical tunability, COFs are very good candidates for applications in gas storage and separation, especially hydrogen and carbon capture, as well as in catalytic applications for chemical reactions and in drug delivery systems.
- **Metal–Organic Frameworks (MOFs)**: Large surface areas and tunable pore sizes have made MOFs the materials of use in applications such as gas adsorption, catalysis, and drug delivery. Flexibility in tailoring chemical functionality has particularly been useful to MOFs in sensing and separation technologies [45–49].

3.2.3 Heterostructures

- **Layered Heterostructures**: Researchers have been successful in developing van der Waals heterostructures showing new electronic and optical properties by vertically stacking a number of 2D materials. Graphene, TMDs, and h-BN are examples of such 2D materials. It can be considered as the next generation of electronics: Multifunctional devices include transistors with improved performance and many quantum computing applications. Heterostructures also have this by itself has tremendous potential for photodetectors as well as flexible optoelectronic devices. This is because an alloy of different materials can exhibit bandgaps that are tunable, along with improved light absorption.

3.2.4 Hybrid materials

- **Inorganic–Organic Hybrids**: Hybrid material: This material combines the desirable properties of the inorganic materials, electrical conductivity, and strength and stiffness, with flexibility and chemical tunability characteristic of organic components. Such hybrid materials, currently under development for flexible electronics, provide the flexibility from the mechanical properties of the organic component, while the inorganic component ensures efficient performance in electricity. These materials also find important applications in sensors. The organic phase acts as a sensitizer to boost the chemical sensitivity here. This sort of hybrid structure is used in applications for energy storage, enhancing cycling stability and energy density [50–57].

Each class of 2D nanomaterials offers distinct advantages, and their unique properties open up possibilities for innovations in a wide range of applications, from electronics and energy storage to catalysis and sensing [58–66] (Figure 3.1).

3.3 2D MATERIAL-BASED SENSORS

3.3.1 Biosensors

Biosensors, being analytical tools featuring an integration of the biological element with physical transducers, have used important applications in many aspects, such as food safety, environmental monitoring, and healthcare [67–73].

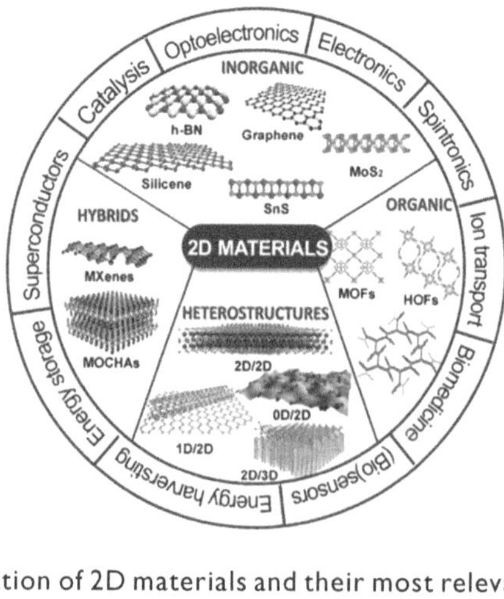

Figure 3.1 Classification of 2D materials and their most relevant applications [48].

FETs with the incorporation of 2D materials, including graphene and transition metal dichalcogenides, show enhanced operational performance. The material has distinctive features such as a high surface area, excellent electrical conductivity, and mechanical flexibility, which make it suitable for biosensing applications [12, 13]. Researchers have designed biosensors with increased sensitivity, selectivity, and stability by incorporating 2D materials into FETs [14, 15]. Furthermore, research includes the design and simulation of next-generation nanowire-based FETs targeting low-power consumption, which helps complement the general miniaturization and optimization studies in biosensing performance [52–54].

In combination with biological recognition elements, FETs can directly measure analytes without demanding complex labeling or amplification processes. The electrical properties of the device change as a result of the interaction between the analyte and the biological recognition element on the surface of the FET; the latter changes can be used to sense analyte concentration. This label-free detection capability offers huge simplicity and speed benefits [11].

It has been incorporated with 2D materials such as graphene and transition metal dichalcogenides into FETs to enhance its operation performance. The materials have demonstrated unique attributes such as a high surface area, excellent electrical conductivity, and also mechanical flexibility that make them suitable for a biosensing application [12, 13]. Scientists had fabricated biosensors with more sensitivities, selectivity, and stability through the implantation of 2D materials in FETs [14, 15].

3.3.2 2D material-based FET biosensors

Since its discovery in 2004, graphene has been the focus of extensive research into 2D materials, changing industries such as bioelectronics and sensing. Indeed, such materials open new ways for the development of flexible sensors and wearable technology. This paper considers the characteristics of various 2D materials, their patterning methods, and their implementation with biosensors. We elaborate on the latest developments and possibilities of 2D materials in this field. For instance, layered materials have a flowchart representation as shown in Figure 3.2 [16, 17].

3.3.2.1 Graphene

Graphene, due to its exceptional mechanical flexibility, a large surface area, and excellent electrical conductivity, has come into vogue as one of the most suitable materials used in biosensing. Graphene-based FETs have the potential to enhance the sensitivity, selectivity, and stability of FETs if it is integrated with these devices. So far, graphene-based FETs have been successfully applied to various types of biomolecules, such as proteins, bacteria, and DNA [74–79]. The benefits of graphene-based FETs include the

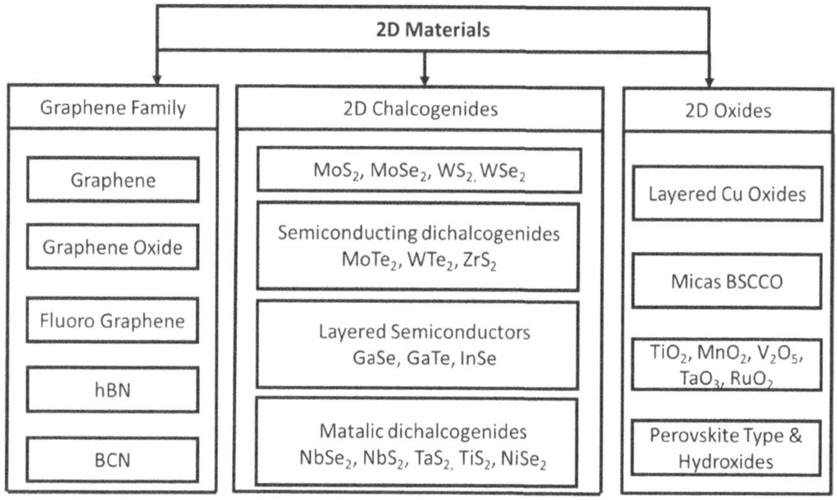

Figure 3.2 Types of 2D material [49].

label-free, real-time analyte detection capabilities over traditional biosensing techniques. Direct application in some semiconductors is limited because it lacks a bandgap. But overcoming this restriction and improving the area of biosensing have led scientists to actively find alternative 2D materials with bandgaps, like transition metal dichalcogenides. [18]

3.3.2.2 *Nitride and carbonitride*

Silicon nitride (Si3N4) and boron nitride (BN) are nitride-based materials with specific characteristics that make them suitable for inclusion in field-effect transistors (FETs). Nitride-based materials can be excellent substrates for the immobilization of biomolecules and interactions with biological samples on account of their exceptional mechanical strength, chemical stability, and biocompatibility. The potential for widespread application is also boosted by the prospect of nitride-based materials being integrated into other sensor platforms, such as lab-on-a-chip devices and implantable sensors. [19]

Nitride-based materials in FET-based biosensors have shown very promising results in a range of applications, which include

Disease diagnosis: Identification of biomarkers associated with a particular disease.
Environmental monitoring: To monitor the presence of toxins and pollutants in soil, water, and air.
Food safety: Detecting food contamination and foodborne illnesses.
Drug discovery: Screening for potential drug candidates.

3.3.3 2D material-based nonvolatile memory

Key concepts from the 2D material-based nonvolatile electronic memory: 2D material-based nonvolatile electronic memory nonvolatile electronic memory using 2D materials in the following section, emphasis was placed on the usage of 2D materials in nonvolatile electronic memory development—more specifically, resistive random-access memories (RRAMs) and floating gate memories [80–86]. Memories of the floating gate: These are nonvolatile memory devices storing electrons using floating gates. With these devices, the use of 2D material improves storage with capacitive interference reduction. MoS_2 as channel material, multilayer graphene as floating gate, and HfO_2 as control gate result in efficient electron tunneling, programming, and data erasure. The retention rate of memory is quite good with the ON/OFF ratio of 10 (Figure 3.3).

MoS_2 /h-BN/graphene heterostructure devices exhibit advantages over performance and therefore integrations density more than an order of magnitude higher and two-terminal operation is possible. The off-state current is negligible with an enormous ON/OFF ratio of 10^6, retention times could be made much longer; besides, the device is programmable and erasable, using the appropriate voltages. Semi-floating gate memory: This concept of semi-floating gate memory has developed only recently van der Waals heterostructures for improving refresh times and speed. It is a material composition where WSe_2 is the channel material, and HfS forms the tunneling layer, while h-BN forms the floating gate. Fast operation is attributed to a fast p-n junction switch constituted of a switch, which allows rapid charge injection combined with storage of charges in the long term. Potential and applications: Such devices may be applicable in neuromorphic computing and similar high-end technologies that require memory storage with very high switching speed and energy efficiency as exhibited by the high switching speed and low power consumed [87–94].

3.3.4 2D material-based nonvolatile ORAM

RRAM that is based on the conventional 2D material typically relies on electrical triggers to switch between resistance states. However, 2D materials have one special quality: they respond rapidly and sensitively to a broad spectrum of light ranging from ultraviolet to infrared. Such a material can simultaneously be used both for data storage and as an optical sensor. Devices structured in a smart manner can fabricate systems that can respond to light and remember. This integrative functional property between memory storage and optical sensing allows interesting possibilities in many applications, for example: Artificial vision systems consisting of image sensors to capture and store visual information. Logic gates and optical demodulators are used in optoelectronic circuits, devices utilized in computers and high-speed communication. Equipment for neuromorphic systems processes information

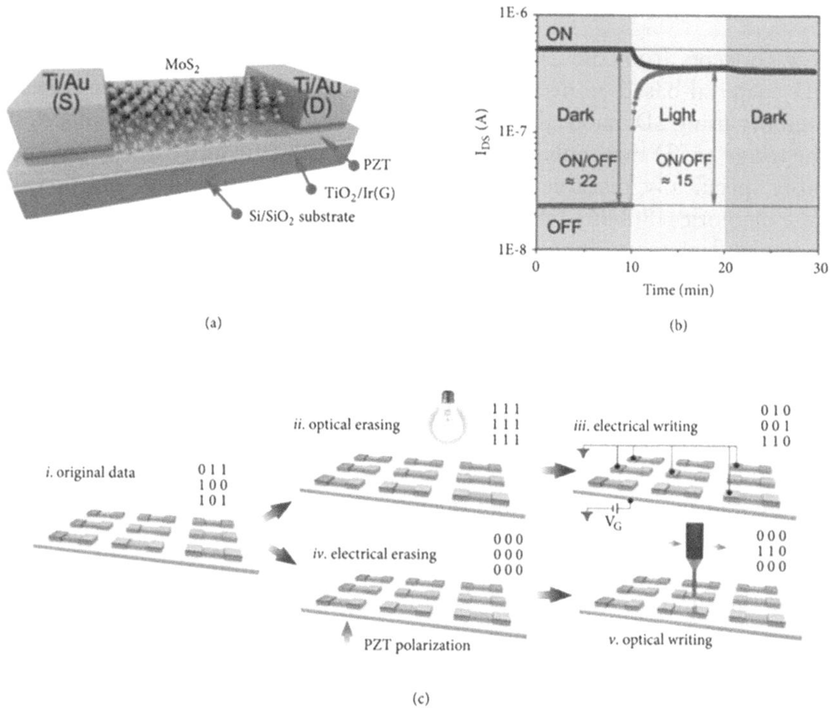

Figure 3.3 (a) Schematic of MoS ferroelectric ORAM. (b) Retention characteristics after the removal of light stimuli. (c) Global operations of "optical erase and electrical write" and "electrical erase and optical write" [50]

fundamentally like in the human brain. Depending on their structure, the designs of an optoelectronic random-access memory (ORAM) device can be split broadly into two types:

a. 2D material-based ORAM.
b. ORAM based on 2D heterostructures

which mix various 2D materials to produce improved characteristics. In the sections that follow, these two strategies will be discussed in more detail, while still addressing the key themes of integrating optical sensing with data storage and outlining many kinds of ORAMs [95–100].

3.3.4.1 2D material-based ORAM

In order to make 2D material-based ORAM more sensitive and capable of storing many levels of data, researchers have explored a range of techniques. For instance, the Cho group successfully developed a multibit MoS_2

ORAM. They have achieved a high ON/OFF ratio of 10^6, long retention period for over 10 seconds, and high sensitivity with more than 8000 A/W at a power of 0.1 μW. Between the tunneling and blocking dielectrics, they used Au nanoparticles as the charge trapping layer within their structure. This enables photo-generated charge carriers to be trapped inside the MoS_2 ORAM. Both electrical and optical programming are feasible for the MoS_2 ORAM, which contributes toward various flexible applications such as logic operations and coincidence detection. In electrical programming, electrons from the tunneling layer are moved into the MoS_2 conduction band by using a negative gate voltage, while a laser at 655 nm helps in the non-recombination movement of electrons in case of optical programming [101–106]. The MoS_2 ORAM can even demonstrate varying degrees of resistance depending upon light quantity, through the integration of optical sensing and storage. Beyond FET-based ORAMs, a monolayer of MoS_2 facilitates ferroelectric ORAMs that demonstrate the following features: A low operating voltage besides nondestructive reading of the data. The conductivity of the MoS_2-PZT structure can be changed by photodomain rearrangement, caused in the case of an applied light (Figure 3.4).

Depending on the initial polarization of lead zirconate titanate (PZT) material, these devices can be either optically single electron transistor (SET) programmed and electrically RESET or vice versa, electrically SET and optically RESET. One more example is a two-terminal MoS_2 ORAM structure with a configuration of Au/MoS_2 nanospheres/Au. The device can be poled from high--resistance state (HRS) to low-resistance state (LRS) at an applied voltage of 6 V when illuminated. This structure may present multilevel resistance states, which seems promising for the design of future two-terminal ORAM devices based on 2D materials [107–113] (Figure 3.5).

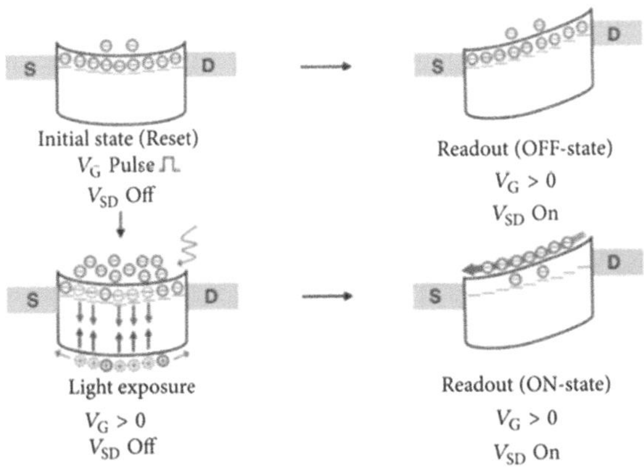

Figure 3.4 Band diagrams of the initial state, upon light exposure and reading process. [50]

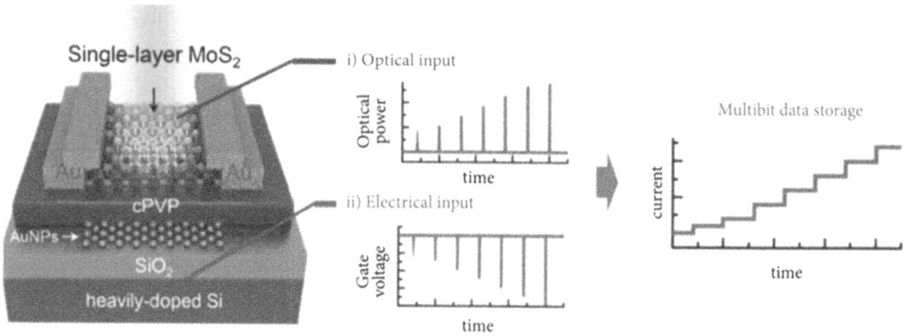

Figure 3.5 Schematic illustration of MoS ORAM with Au nanoparticles as charge trapping layer. (d) Band diagram of electrically controlled programming and erasing. [50]

3.3.4.2 ORAM based on 2D heterostructures

ORAMs fabricated on 2D material can degrade owing to environmental factors. This is because retention times are largely short for the memory operation charge being trapped at the semiconductor material interfaces. Researchers have designed heterostructures of the semiconductor hybridized with another semiconductor to improve performance. The strengthening functions about data storage, response stability, and absorption of light are found in heterostructures [114–120]. For instance, Roy et al. exploited the heterostructure of graphene/MoS2 for forming an ORAM that exhibits remarkably high photo-responsivity but an unsatisfactorily low ON/OFF ratio, in reality below 2, not suitable for multibit storage. WSe2/h-BN has been utilized by Du et al for the formation of a broadband ORAM that absorbs light between the ultraviolet to visible ranges. It had excellent retention time above 128 resistance states with a gigantic ON/OFF ratio (Figure 3.6).

The data can be retained in ORAM even when the light is switched off. A gate voltage may be combined with the light to erase and recover it.

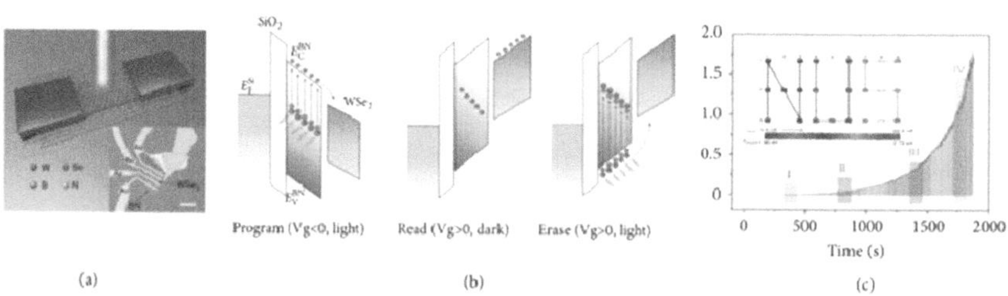

Figure 3.6 (a) and (b) Structure of the switching mechanism in WSe/h-BN ORAM, with its band diagrams, (c) represents 130 storage states 130 that were obtained by illuminating with a variable number of pulses at 405 nm light of 210 mW/cm and 0.5 s period. The voltage at the gate is −80 V [50].

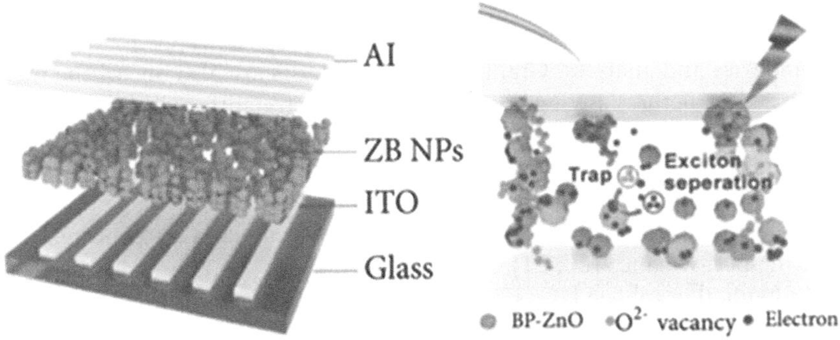

Figure 3.7 ORAM with phosphorene/zinc oxide nanoparticles and illustrations of SET [50].

Tran et al. presented the synthesis of a heterostructure composed of MoS_2/BN/graphene, which had a high ON/OFF ratio, numerous storage levels, and low programming voltages and power consumption. It is energy efficient since it has reset memory using visible light [121–128]. Another ORAM-based BP/ZnO heterojunction was recently reported by Hu et al, which demonstrated SET voltage with low values along with absorption in the UV to near-infrared range. Absorption up to infrared is crucial for applications related to data transfer, night vision, and medical diagnostics, though most of the ORAM devices absorb only in the visible and UV range (Figure 3.7).

This can be achieved by incorporating 2D materials with upconversion nanoparticles or small-bandgap semiconductors. In fact, Wang et al showed an effect of multi-storage modes, long retention time, and sensitivity to infrared light using an infrared ORAM based on the heterostructure of MoS_2/PbS. Alternatively, it is possible to use upconversion nanoparticles to convert the wavelength of infrared light of 980 nm into a signal the MoS2 layer can exploit. This method can be used to convert and store infrared light. Overall, 2D material-based ORAMs have great promise with broad light absorption, low-power consumption, and high responsivity. These are some of the parameters that could be improved for future energy efficiency enhancement by bringing down the high programming voltages of most devices [129–136].

3.4 2D MATERIAL-BASED GAS SENSORS

3.4.1 Resistive gas sensors

In principle, applications of 2D material-based gas sensors greatly depend on environmental monitoring. They sense by adsorbing analytes including NH_3, CO, H_2O, and NO, thus generating charge carriers, which change the electrical resistance of the sensor (chemiresistor). By contrast, in most conventional metal–oxide gas sensors, based on ZnO or SnO, the functional mechanism is the reaction between oxygen molecules trapped along grain boundaries and analytes. Graphene chemiresistors are also one of the most analyzed structures owing to their relatively straightforward fabrication and easy characterization. Moreover, graphene offers high opportunities for miniaturization and gives great potential biosensor applications. Chemical field-effect transistors, or ChemFETs, add a modified gate voltage to modulate the channel carrier concentration and thus conductance to optimize gas sensing performance [20, 21].

Individual graphene layers and many other 2D materials inherently have a high ratio of surface area-to-volume, but they also exhibit low noise, offering the possibility for high signal-to-noise ratios and lower molecular detection limits. Suspensions of the channel can double the surface area available, increasing responsivity. In contrast to commercial sensors, which generally require very high operational temperatures (e.g., 150°C) and suffer from drawbacks including saturation at a large concentration of the absorbed gas, 2D materials generally operate well at room temperature with scalable concentration ranges. For example, the resistance variation for graphene is 39% at 200 ppm NO while that for MoS2 was 10% at 100 ppm NO; PtSe2 shows a 0.25% variation at 1 ppm NO [22–24]. Suspended bilayer graphene has been used to detect CO at very high sensitivities [25]. However, commercially available metal–oxide-semiconductor gas sensors are extremely sensitive but work only up to 300°C and consume very large amounts of power, hence limiting their application for use in a low-power device such as a smartphone [26].

Unfunctionalized suspended graphene resistors have been employed to detect gases by measuring the thermal conductivity. Innovations such as MEMS micro-hotplates have been investigated for improving the response and recovery times of indoor air quality sensors [27]. SBDs have been highly sensitive to gases through control of the Schottky barrier height, which controls the electrical currents based upon analyte exposure [28, 29]. It is crucial to have the function layers to improve selectivity for graphene gas sensors, which normally respond to a few gases and parameters. Besides graphene, there are a number of other 2D materials that have demonstrated high gas sensing performance including MoS_2, $MoSe_2$, $MoTe_2$, WS_2, NbS_2, ReS_2, and $PtSe_2$ [24, 30, 31]. For example, $PtSe_2$ has good selectivity for NO according to theoretical studies [24]. The basic principle of sensor arrays of different functionalized or nonfunctionalized 2D materials can be followed in order to enhance sensitivity and selectivity using electronic noses [32]. Suspended sensors have a larger area; thus, they might introduce fabrication complexity, and therefore, involves a trade-off between cost and performance. Repeatability and drift are also serious issues because of the energy that has to be spent to desorb the molecules from the sensor before it returns to its original state, which might be overcome by thermal or photonic activation [33].

3.4.2 Permeation-based gas sensing

Over past few decade, fast sieving of molecules of gases and liquids has been established using 2D-based membranes. Those membranes, with pores down to sub-1 nm, can selectively sieve molecules or ions according to their kinetic diameter. Such membranes are excellent devices, for example, for sieving H 2 and CO molecules, which permeate about 1000 times faster than larger gases such as argon, nitrogen, and methane [34, 35]. It can be exploited for permeation-based gas sensing, as demonstrated by changes in osmotic pressure across graphene membranes as a result of changing gas composition. The phase-selective permeability differences allow graphene to be used as a semipermeable membrane. However, for larger pore diameters where the transport becomes effusion dominated, selectivity is lowered, thus may limit the sensing applications [137–143].

3.5 2D MATERIAL-BASED NEMS SENSORS

3.5.1 Pressure sensors

The first MEMS products to go into high-volume production were silicon-based pressure sensors [36]. Now, the pressure sensors exceed one billion units annually. Applications encompassed in the field of pressure sensing touch sensing include liquid and tactile; however, the subject of suspended membrane-based gas pressure sensors is the focus here. These sensors are primarily used as altimeters, barometers, and for gas control and indoor navigation. MEMS pressure sensors generally measure pressure by finding the deflection of a plate caused by the pressure difference on one side versus the other. The magnitude of deflection depends on geometry and material properties. MEMS commercial sensors have reached as low as 1 Pa resolutions for pressure differences, meaning altitude changes are as small as 5 cm. Such resolution demands a highly rigid mechanical plate, which will have a diaphragm size of several hundred microns and a membrane thickness from 0.5 to 10 µm. Additionally, very sensitive deflection detection circuitry is used. Traditionally, this has been piezoresistive readout but more recently capacitive readout, as represented by the SBC10 pressure sensor from Murata, which has a responsivity of 55 fF/kPa [37]. The areas of research interest would include miniaturization and increased sensitivity of the pressure sensor. For instance, scaling down is an important aspect for wearable electronics applications. Next, with the increased sensitivity of the 2D sensors, unique applications such as sub-centimeter resolution altimeters for indoor navigation or presence detection through pressure sensors become possible. This, again, results in higher sensor sensitivity providing sensors with smaller sizes, faster acquisition, lower power consumption, and cheaper readout electronics [144–152].

3.5.2 Graphene-based microphones

A microphone basically is an ultrasonic or audible frequency operated pressure sensor. Just like pressure sensors, suspended upon 2D film material is so thin and flexible that it is quite sensitive to any variation of sound pressure; thus, they are suitable as microphones. During the past two decades, MEMS microphones have dramatically exceeded the traditional ones in mobile phones and represent a billion-dollar market where multiple microphones are very often used to create directionality and noise reduction. The first advantage of suspended graphene as a membrane in the mic is reduced stiffness. The stiffness cannot be decreased much more than that since a flat band frequency response necessitates a resonance frequency that is larger than the human hearing bandwidth, usually greater than 20 kHz. The low stiffness of graphene, high resonance frequency, and extreme thinness leading to a very little mass provide exciting prospects for wide-bandwidth microphones capable of picking up even the slightest variations in sound pressure. In addition, the equation seems to indicate that graphene's modest mass could lead to lower pressure noise levels.

For a circular membrane to a pressure sound pn:

$$p_n^2 = \frac{4k_B T \omega_0 m_{eff}}{A^2 Q} \left[\frac{Pa^2}{Hz} \right],$$

Here T is the temperature, Q is the quality factor, and ω0 is the resonance frequency of 2D membrane.

With these properties, graphene also offers possibilities for downscaling of the areas of microphones without degrading performance. This means arrays of low-cost microphones could be used for sound directionality and might have applications in 3D ultrasound imaging or noise cancellation. It is much more challenging in microphones than in pressure sensors to realize a sufficient signal-to-noise ratio. For an example, MEMS microphones have responsivities greater less than 10 mV/Pa, and remarkable pressure noise levels below have been attained. Graphene membranes have not yet been shown to exhibit low noise, high-responsivity performance; still equation suggests, graphene must show better performance than traditional MEMS membranes.

Of relevant interest here, multilayer graphene membranes are reported to be 20*100 nm thick in condenser microphones with radii ranging from 12 mm down to 40 μm. These devices work on frequencies that may range up to the ultrasonic domain while lying within the audible domain. Such devices having smaller membrane diameters have a wide frequency band, including ultrasonic frequencies, while keeping the voltage below the pull-in voltage, providing a voltage of 1.78 V, making these useful for a mobile phone that uses a standard 2 V supply voltage [153–155] (Figure 3.8).

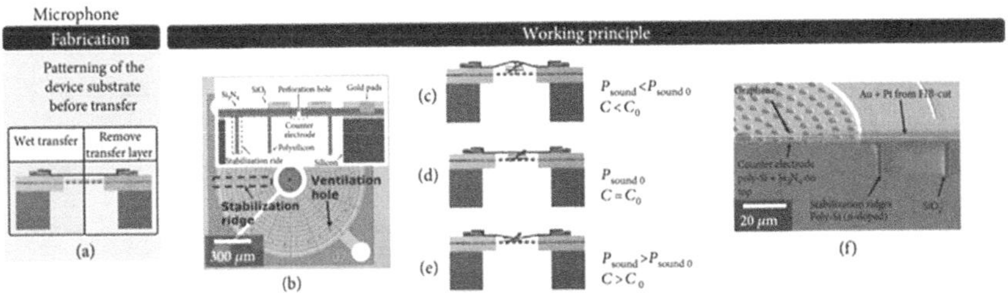

Figure 3.8 (a)–(f) Microphone: (a) fabrication of method the suspended 2D membrane; (b, f) images of an example device [51].

Devices with large membrane diameters work at higher voltages but are also useful as speakers. More importantly, some of these devices outperform commercial, high-end microphones based upon nickel over a substantial range of the audio spectrum, with sensitivity enhancements greater than 10 dB; graphene, therefore, has some promise in microphone applications. In contrast, traditional MEMS microphones have sensitivities around −36 dB (or 15.8 mV/Pa) at a 1.62–3.6 V supply voltage and a diaphragm size of 5 mm. In contrast, microphones supported by graphene showed sensitivities of up to 10 mV/Pa with a diaphragm size of 38.22 mm at a 1 V supply voltage. Thus, despite the fact that current silicon-based microphone technology is inherently more sensitive than graphene-based microphones, two vibrating membranes are typically utilized in conventional designs to further amplify the signal, which is not used in any graphene microphone designs.

3.5.3 Accelerometers

The interdigitated readout electrodes and springs of modern silicon-based MEMS accelerometers greatly increase the footprint of the device. First, because long compliant springs are needed to get a sufficiently small spring constant. On the other hand, for capacitive readout MEMS accelerometers, a sufficiently large capacitor area is needed, which gives many interdigitated readout electrodes. 2D materials and graphene, by themselves, are not ideal for accelerometers due to their intrinsic low mass, which does not provide sufficient responsivity. Therefore, 2D materials require an additional proof mass in the suspended area, which is moved by acceleration forces. While graphene exhibits a small piezoresistive gauge factor, it can generate a large resistance variation per each Newton force, due to its higher thinness. Its very high Young's modulus and fracture strain further indicate that it works well for suspended devices with attached proof masses. Figure 3.9 (p), (q) and (n) illustrate an example design for a graphene NEMS accelerometer, where graphene serves as both the springs of the spring-mass system and the piezoresistive transducer elements.

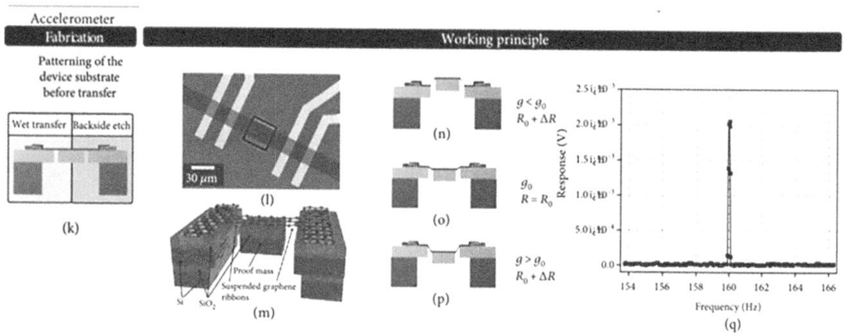

Figure 3.9 Accelerometer [51].

The acceleration stretches the floating graphene film or membranes, causing resistance changes in the graphene, which is measured using the piezoresistive variation measurement technique used in the accelerometers.

Double-layer graphene film with silicon proof masses was realized with help of a conventional MEMS and NEMS manufacturing approach. The graphene was suspended by dry etching, followed by vapor HF etching to remove a sacrificial buried oxide layer. The suspended silicon proof masses had dimensions of up to which are over three orders of magnitude heavier than the masses deposited over previous device [38–40]. The graphene membrane with suspended silicon proof masses occupy at least second orders of magnitude smaller die areas than conventional state-of-the-art silicon accelerometers. After normalization, the relative responsivity (resistance change per proof mass volume) in graphene ribbon accelerometers is at least one order of magnitude larger compared to silicon excellent performance accelerometers. This makes it possible to shrink graphene-based NEMS accelerometers and gyroscopes down to the nanometer range, despite the low gauge factor of graphene.

The sensitivity of graphene film-based accelerometers can be further enhanced by increasing the attached mass and by reducing area of the suspended graphene. From the material selection point of view, other two-dimensional materials such as MoS [29, 31, 36] or PtSe, which have significantly higher piezoresistive gauge factors, could be incorporated to potentially enhance device sensitivity. However, their mechanical stability and adhesion force to the substrate must be investigated. Device concepts based on completely clamped 2D film membranes enhance mechanical robustness by preventing edges from acting crack initiation in case of any stress. However, the response of completely clamped membranes is generally lower than that of ribbons with equal proof masses and trench widths due to lower strain levels and parasitic parallel resistances.

Additionally, demonstrations of graphene-based NEMS accelerometers and few experimental realizations of suspended graphene 2Dmembranes with attached proof mass exist. Few micrometer-sized layer of graphene cantilevers with diamond allotrope carbon weights, fabricated by focused ion beam deposition, has been used to study the mechanical properties of graphene [38]. A pyramid-like support suspended with graphene cantilevers, which carried masses of 50 nm thick gold film, had to be immersed in liquid to hold mechanical coherence [39]. Suspended graphene film, finally, circulated and clamped using SU-8, supporting mass made of SU-8, or with gold film that was placed exactly at the middle, have tested as shock sensors for ultrahigh mechanical stresses. These reports used extremely small masses, and some used fabrication techniques incompatible with semiconductor fabrication. In addition, graphene-based resonant accelerometers have been theoretically proposed but not yet experimentally demonstrated. In such concepts, acceleration would act on suspended graphene beams inducing strain that causes a corresponding shift in their resonance frequencies.

3.5.4 Hall sensors

This is an asymmetrically biased conductor that is exposed to an external magnetic field, and charge carriers will experience an electromagnetic force in a direction perpendicular to both the electric field and applied external magnetic field. The Hall voltage is directly proportional to the applied magnetic field and the concentration of charge carriers. The electronic structure of single-layer graphene leads to low carrier density at the minimum conductivity, resulting in an extremely high Hall voltage. Furthermore, the charge carrier concentration can be tuned to achieve high responsivity. The final signal-to-noise ratio (SNR) of Hall sensors is proportional to the mobility. The very low effective mass of charge carriers in graphene translates into very high mobility at room temperature, which renders graphene-based magnetic field sensors high-performing. Mobility in graphene is heavily dependent on the (dielectric) environment, i.e., the interface with its surroundings. For suspended graphene, mobilities up to have been measured, substantially higher than those for supported graphene on a SiO_2 substrate [41].

Suspended graphene film membrane-based Hall sensors are of interest (Figure 3.10a–d) because the voltage sensitivity (Sv) of linear Hall sensors depends on the charge carrier mobility, with and being the width and length of the device. At room temperature, the mobility of electrons is almost 1241 cm^2/Vs in silicon with approximate dopant concentration of 10^{17} cm^{-3} [42]. The intrinsic Sv is almost 160 times larger for suspended graphene than for silicon. In addition, graphene exhibits a linear Hall response over several hundred mT [42] and even outperforms commercial Hall sensors relying on silicon-based technology.

Monolithic silicon-based Hall sensors constructed using bipolar complementary metal–oxide–semiconductor (BiCMOS) technology, such as the Infineon linear Hall sensor series TLE499x [222], achieve sensitivities up to 300 mV/mT at an operating voltage of 5.5 V with operating range of

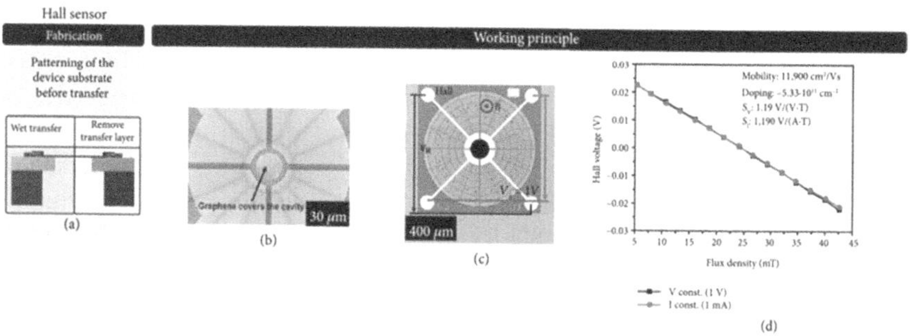

Figure 3.10 (a)–(d) Hall sensor: (a) fabrication method of the suspended membrane and (b, c) example device and graph of particular device [51].

±200 mT. These large values are achieved due to the application of integrated amplifier circuits, which enhance the intrinsic Hall effect in silicon. However, such preexisting integration technology is still lacking for graphene, but development is anticipated as the technology matures [13, 14]. Recent findings show that graphene mobilities can be relatively high when encapsulated by Al_2O_3, hBN [18, 46], and WSe_2. This may prove to be a promising direction to enhance the performance of Hall sensors based on nonsuspended graphene, which may be preferable for most applications, as it alleviates some of the fabrication challenges of suspended graphene membranes.

The Hall effect offers a sensitive and accurate means to measure carrier concentration. Suspended graphene Hall sensors, where the membrane is exposed to the environment, may become promising as gas sensors, where molecules adsorbed over graphene film surface changes its doping (i.e., carrier density). Such sensors could potentially be sensitive down to single-molecule level.

3.5.5 Graphene mass sensors

Extremely low mass of graphene positions it as a promising feature for precise mass sensing. Such a sensor, shown in Figures 3.11(l)–(o), measures a mass change of the membrane from changes in its resonance frequency. The mass change may be induced by adsorbed or attached atoms or molecules on the surface of the membrane. The responsivity of resonant mass sensors is given by [43, 44], meaning that for a small mass of the graphene membrane, a relatively large frequency shift might be observed. The high sensitivity principle was to demonstrate by adding and removing layers of pentacene with an equivalent mass of six layers of monolayer graphene and monitoring its effect on the resonance frequency of a graphene film (Figure 3.11(p)).

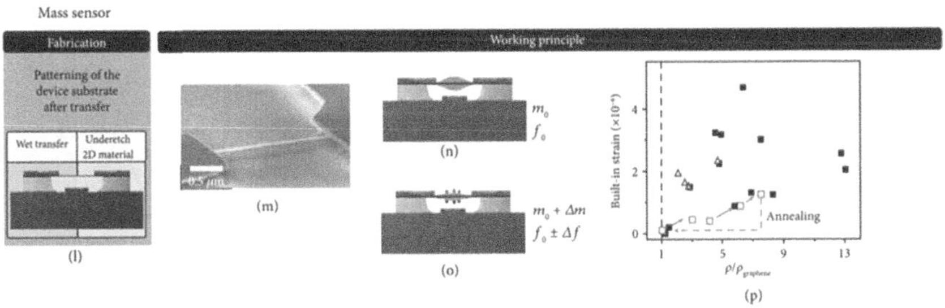

Figure 3.11 Mass sensor: (l) fabrication of suspended film-based sensor, (m) example device, and (n, o) working principle: by measuring the resonance frequency, the mass change of the membrane can be obtained. (p) Calculated tension and mass of the membrane during multiple cycles [51].

These mass sensors are expected to find applications in fields where it is required to measure mass variation that is substantially lower than one monolayer of a 2D material. In contrast, conventional quartz crystal monitors have proven suitable for mass sensing at the level of a single monolayer of graphene [128]. The sensitivity these mass sensors can reach a value upto 10 ng/Hz [45], which significantly exceeds that of silicon-based sensors, which usually have sensitivity values of only 10 ng/Hz [46]. Commercial mass sensors have even smaller sensitivity [46]. Theoretical limit of graphene nanomembranes with diameters below is 10 nm, which commonly naturally occur in graphene on substrate of silicon oxide and theoretically predicted to be sensitive enough to detect a single hydrogen atom's mass, leading to a relative resonance frequency variation of 10 Hz.

3.5.6 Graphene bolometers

Bolometer is a device that detects the absorption of electromagnetic radiation and light by monitoring the resulting temperature variation in material via this changes in its electrical resistivity mainly for longer wavelength infrared and THz radiation. Bolometers are preferred as there are few other detectors available for this frequency range. At room temperature, where superconducting bolometers cannot be realized due to hjgh temperature, suspended graphene is an interesting material for low-cost bolometers due to its ultra-wideband electromagnetic absorption and low heat capacity (Figure 3.12(q)–(u)).

Drawbacks of suspended graphene bolometer high thermal conductivity and low temperature coefficient of resistance of graphene that have recently been mitigated by utilizing a resonant readout mechanism in a focused ion-beam structured [47]. However, cross-sensitivity to thermoelectric and photoelectric signal also needs to be addressed. Graphene-based radiation

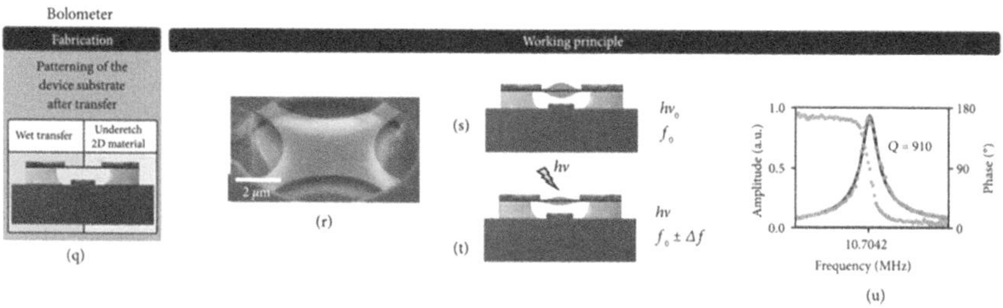

Figure 3.12 Bolometer: (q) fabrication of the suspended membrane and (r) example device. (s, t) Working principle: radiation heats the membrane that alters its tension and causes vary its mechanical resonance frequency. (u) Variation of parameter of device under test [51].

detectors for the infrared range show a noise equivalent power of 2 pW/Hz at room temperature [47] and are thus in the upper range of conventional infrared bolometers based on vanadium oxide or nickel (1–10 pW/Hz).

There are multiple other type of 2D material-based photosensors, but those are usually not suspended type. Hence, they fall outside the scope of this review.

3.6 CONCLUSION

Overall, 2D materials offer exciting technological opportunities for a vast range of applications: sensing, memory, energy storage, and electronics. The extraordinary properties of 2D nanomaterials—high surface area, mechanical strength, and electronic properties that can be tuned at willunlock new avenues for innovation, especially in the context of applications where sensitivity, flexibility, and miniaturization are critical. For example, graphene-based FET biosensors show a better performance sensitivity than other competitors, whereas 2D material-based gas sensors can operate at room temperature with a low power consumed and high responsivity, thereby more suitable for environmental monitoring and healthcare applications.

There is still quite a large gap in development in applying these materials commercially, especially in enhancing the environmental stability and perfecting the fabrication techniques. But how many research activities may lead to what material synthesis innovations, device integration, and optimization in performance will be critical for the full realization of 2D material potential. This report looks at some emerging applications of 2D nanomaterials, which suggests that these materials might become foundational in future generations of sensors, flexible electronics, and memory storage technologies. Continued research in this area will focus on developing the limits of technology while widening horizons for new discoveries beyond current industrial and scientific imagination.

REFERENCES

1. K.S. Novoselov; A.K. Geim; S.V. Morozov; D. Jiang; Y. Zhang; S.V. Dubonos; I.V. Grigorieva; A.A. Firsov "Electric Field Effect in Atomically Thin Carbon Films," *Science*, Vol. 306, pp. 666–669, 2004.
2. N.R. Glavin; R. Rao; V. Varshney; E. Bianco; A. Apte; A. Roy; E. Ringe; P.M. Ajayan "Emerging Applications of Elemental 2D Materials," *Advanced Materials*, Vol. 32, p. 1904302, 2020.
3. K. Khan; A.K. Tareen; M. Aslam; R. Wang; Y. Zhang; A. Mahmood; Z. Ouyang; H. Zhang; Z. Guo, "Recent Developments in Emerging Two-Dimensional Materials and Their Applications," *JPCC*, Vol. 8, pp. 387–440, 2020.
4. P. Ares; K.S. Novoselov "Recent advances in graphene and other 2D materials", *Nano Materials Science*, 2021. In Press.
5. M. Zeng; Y. Xiao; J. Liu; K. Yang; L. Fu, "Exploring Two-dimensional Materials toward the Next-Generation Circuits: From MonomerDesign to Assembly Control," *Chemical Reviews*, Vol. 118, pp. 6236–6296, 2018.
6. C. Cui; F. Xue; W.-J. Hu; L.-J. Li "Two-dimensional Materials with Piezoelectric and Ferroelectric Functionalities," *npj 2D Materials and Applications*, Vol. 2, p. 18, 2018.
7. C. Grazianetti; C. Martella; A. Molle "The Xenes Generations: A Taxonomy of Epitaxial Single-Element 2D Materials," *Physica Status Solidi-rapid Research Letters*, Vol. 14, p. 1900439, 2020.
8. R. Gudaitis, A. Lazauskas, Š. Jankauskas, Š. Meškinis "Catalyst-Less and Transfer-Less Synthesis of Graphene on Si(100) Using Direct Microwave Plasma Enhanced Chemical Vapor Deposition and Protective Enclosures," *Materials*, Vol. 13, p. 5630, 2020.
9. R. Poyato, R. Verdugo, C. Muoz-Ferreiro, N. Gallardo-Lpez, "Electrochemically Exfoliated Graphene-Like Nanosheets for Use in Ceramic Nanocomposites," *Materials*, Vol. 13, p. 2656, 2020.
10. F. Farivar, P.L. Yap, T.T. Tung, D. Losic "Highly Water Dispersible Functionalized Graphene by Thermal Thiol-Ene Click Chemistry," *Materials*, Vol. 14, p. 2830, 2021.
11. J. Lee, "Graphene field-effect transistor sensors for biomedical applications", *Sensors and Actuators B: Chemical*, Vol. 235, pp. 327–335, 2016.
12. K.S. Novoselov, "Electric Field Effect in Atomically Thin Carbon Films", *Science*, Vol. 306, No. 5696, pp. 666–669, 2004.
13. M. Chhowalla, (2013). "Two-Dimensional Materials for Transistors", *Nature Nanotechnology*, Vol. 8, No. 1, pp. 53–64.
14. Y. Wang, "High-Performance Graphene Field-Effect Transistor Biosensors", *ACS Nano*, Vol. 8, No. 5, pp. 4875–4883, 2014.
15. M. Xu, "Ultrasensitive Biosensors Based On two-Dimensional Transition Metal Dichalcogenide Nanomaterials", *Advanced Materials*, Vol. 27, No. 40, pp. 6038–6045, 2015.
16. K.S. Novoselov, "Electric Field Effect in Atomically Thin Carbon Films," *Science*, Vol. 306, No. 5696, pp. 666–669, 2004.
17. M. Chhowalla, "Two-dimensional Materials for Transistors", *Nature Nanotechnology*, Vol. 8, No. 1, pp. 53–64, 2013.
18. K.S. Novoselov, "Electric Field Effect in Atomically Thin Carbon Films", *Science*, Vol. 306, No. 5696, pp. 666–669, 2004.

19. J. Lee, "Graphene Field-Effect Transistor Sensors for Biomedical Applications", *Sensors and Actuators B: Chemical*, Vol. 235, pp. 327–335, 2016.
20. F. Yavari, & N. Koratkar "Graphene-Based Chemical Sensors", *The Journal of Physical Chemistry Letters*, Vol. 3, No. 13, pp. 1746–1753, 2012.
21. S. S. Varghese, "Two-dimensional materials for sensing: Graphene and beyond", *Electrochemical Society Interface*, Vol. 24, No. 4, pp. 41–45, 2015.
22. Y. Shao, "Advances in Two-Dimensional Nanomaterials for Gas Sensing Applications", *Sensors*, Vol. 19, No. 21, p. 4778, 2019.
23. O. Leenaerts, "Adsorption of H_2O, NH_3, CO, NO_2, and NO on Graphene: A First-Principles Study", *Physical Review B*, Vol. 77, No. 12, p. 125416, 2008.
24. K. S. Novoselov, "Electric Field Effect in Atomically Thin Carbon Films", *Science*, Vol. 306, No. 5696, pp. 666–669, 2004.
25. O. Leenaerts, "Adsorption of H_2O, NH_3, CO, NO_2, and NO on Graphene: A First-Principles Study", *Physical Review B*, Vol. 77, No. 12, p. 125416, 2008. (for high sensitivities)
26. F. Yavari, & N. Koratkar "Graphene-Based Chemical Sensors", *The Journal of Physical Chemistry Letters*, Vol. 3, No. 13, pp. 1746–1753, 2012. (for operational temperature limits)
27. Y. Shao, "Advances in Two-Dimensional Nanomaterials for Gas Sensing Applications", *Sensors*, Vol. 19, No. 21, p. 4778, 2019. (for MEMS microhotplates in gas sensing)
28. F. Yavari, & N. Koratkar "Graphene-Based Chemical Sensors", *The Journal of Physical Chemistry Letters*, Vol. 3, No. 13, pp. 1746–1753, 2012. (for Schottky barrier control)
29. S. S. Varghese, "Two-dimensional materials for sensing: Graphene and beyond", *Electrochemical Society Interface*, Vol. 24, No. 4, pp. 41–45, 2015. (for SBD sensitivity to gases)
30. O. Leenaerts, "Adsorption of H_2O, NH_3, CO, NO_2, and NO on Graphene: A First-Principles Study", *Physical Review B*, Vol. 77, No. 12, p. 125416, 2008. (for 2D material gas sensing performance)
31. Y. Shao, "Advances in Two-Dimensional Nanomaterials for Gas Sensing Applications", *Sensors*, Vol. 19, No. 21, p. 4778, 2019.
32. Y. Shao, "Advances in Two-Dimensional Nanomaterials for Gas Sensing Applications", *Sensors*, Vol. 19, No. 21, p. 4778, 2019. (for electronic noses and sensor arrays)
33. Y. Shao, "Advances in Two-Dimensional Nanomaterials for Gas Sensing Applications", *Sensors*, Vol. 19, No. 21, p. 4778, 2019.
34. O. Leenaerts, "Adsorption of H_2O, NH_3, CO, NO_2, and NO on Graphene: A First-Principles Study", *Physical Review B*, Vol. 77, No. 12, p. 116, 2008. (for 2D membranes and molecular sieving)
35. F. Yavari, & N. Koratkar "Graphene-Based Chemical Sensors", *The Journal of Physical Chemistry Letters*, Vol. 3, No. 13, pp. 1746–1753, 2012.
36. R. Perlmutter, "MEMS-Based Pressure Sensors and Their Applications", *IEEE Sensors Journal*, Vol. 19, No. 6, pp. 2001–2012, 2019.
37. Murata Manufacturing Co., Ltd. (2020). *SBC10 MEMS Pressure Sensor Specifications*. Murata Corporation, Product Datasheet.
38. Y. Deng, "Few-Layer Graphene Cantilevers with Diamond Allotrope Weights for Mechanical Property Analysis", *Nano Letters*, Vol. 13, No. 8, pp. 3451–3457, 2013.

39. H. Gao, "Kirigami-Patterned Graphene Membranes for Liquid Stability and Mechanical Integrity", *Small*, Vol. 14, No. 21, p. 1801284, 2018.
40. Z. Zhou, "Graphene Membranes with Centralized Mass for Impact Sensing Applications", *Advanced Materials*, Vol. 32, No. 9, p. 1908286, 2020.
41. K. Kim, "Graphene Mobility on Silicon Dioxide: Effects of Dielectric Substrate and Impurities", *Physical Review Letters*, Vol. 102, No. 7, p. 076602, 2009.
42. M. S. Lundstrom Fundamentals of Carrier Mobility in Semiconductor Devices. *Solid-State Electronics*, Vol. 44, No. 3, pp. 385–395, 2000.
43. J. Moser, "Ultrasensitive Mass Detection with a Resonant Graphene NEMS Sensor", *Nature Nanotechnology*, Vol. 5, No. 10, pp. 634–639, 2010.
44. D. Shedin, "Mass Sensing Applications of Graphene Resonators in NEMS", *ACS Nano*, Vol. 12, No. 2, pp. 1038–1046, 2018.
45. S. Schmid, "Sensitivity Limits of Graphene-Based Mass Sensors in Vacuum and Ambient Conditions", *Journal of Microelectromechanical Systems*, Vol. 25, No. 2, pp. 350–358, 2016.
46. J. Rogers, "Comparison of Silicon and Graphene-Based Mass Sensors for MEMS Applications", *Microelectronics Journal*, Vol. 44, No. 4, pp. 376–381, 2013.
47. X. Liang, "Resonant Graphene Bolometers for Infrared Detection with High Sensitivity", *Nature Communications*, Vol. 12, p. 1081, 2021.
48. Domenica Scarano and Federico Cesano. *Graphene and Other 2D Layered Nanomaterials and Hybrid Structures: Synthesis, Properties and Applications.*
49. Sobia Nisar, Ghulam Dastgeer, Zafar Muhammad Shazad, Muhammad Wajid Zul Qar, Amir Rasheed, Muhammad Zahir Iqbal, Kashif Hussain, Iqra Rabani, Deok-Kee Kim, Ahmad Irfan, Aijaz Rasool Chaudhry. *2D Materials in Advanced Electronics Biosensors for Point of Care Devices.*
50. Feichi Zhou, Jiewei Chen, Yang Chai. *2D Materials Based Optoelectronics Memory: Convergence of Electronics Memory and Optical Sensor.*
51. Max C. Lemme, Stefan Wagner, Peter G. Steeneken. *Nanoelectromechanical Sensors Based on Suspended 2D Materials.*
52. Ansari Faizan, Singh Mandeep, Chaudhary Tarun, Raj Balwinder Sankaria Faizan, "Comparative Analysis of Nanowire Tunnel Field Efect Transistor for Low Power Application," *Journal of Silicon*, Vol. 14, pp. 12075–12084, 2022.
53. Singh Mandeep, Chaudhary Tarun, Raj Balwinder "Drain Current and Transconductance Analysis of GaN GAA Nanowire FET with High K Dielectric," *ECS Journal of Solid State Science and Technology*, Vol. 12, pp. 113007,2023
54. Singh Mandeep, Chaudhary Tarun, Raj Balwinder "Design and Simulation Analysis of Silicon Germanium Nanowire FET for Low Power Applications", *Journal of Silicon*, Vol. 15, pp. 1491–1497, 2023.
55. Anjana Bhardwaj, Pradeep Kumar, B. Raj, Sunny Anand, "Design and Performance Optimization of doping-less Vertical Nanowire TFET using Gate Stack Technique", *Journal of Electronic Materials (JEMS) Springer*, Vol. 41, No. 7, pp. 4005–4013, 2022.
56. Jeetendra Singh, and B. Raj, "Tunnel Current Model of Asymmetric MIM Structure Levying Various Image Forces to Analyze the Characteristics of Filamentary Memristor", *Applied Physics A*, Vol. 125, No. 3, pp. 203.1–203.11, 2019.

57. Candy Goyal, Jagpal Singh Ubhi and B. Raj, "Low Leakage Zero Ground Noise Nanoscale Full Adder using Source Biasing Technique", *Journal of Nanoelectronics and Optoelectronics*, Vol. 14, pp. 360–370, 2019.
58. Gurmohan Singh, R. K. Sarin and B. Raj, "A Novel Robust Exclusive-OR Function Implementation in QCA Nanotechnology with Energy Dissipation Analysis", *Journal of Computational Electronics, Springer*, Vol. 15, No. 2, pp. 455–465, 2016.
59. Girish Wadhwa, Priyanka Kamboj, Balwinder Raj, "Design Optimisation of Junctionless TFET Biosensor for High Sensitivity", *Advances in Natural Sciences: Nanoscience and Nanotechnology*, Vol. 10 (7), pp. 045001, 2019.
60. Priya Bansal, and B. Raj, "Memristor Modeling and Analysis For Linear Dopant Drift Kinetics", *Journal of Nanoengineering and Nanomanufacturing*, Vol. 6, pp. 1–7, 2016.
61. Ashima, D. Vaithiyanathan, and B. Raj, "Performance Analysis of Charge Plasma Induced Graded Channel Si Nanotube", *Journal of Engineering Research (JER)*, EMSME Special Issue pp. 146–154, 2021.
62. Abhishek Singh Tomar, Vijay Kumar Magraiya, and B. Raj, "Scaling of Access and Data Transistor for High Performance DRAM Cell Design," *Quantum Matter*, Vol. 2, pp. 412–416, 2013.
63. Neeraj Jain, and B. Raj, "Parasitic Capacitance and Resistance Model Development and Optimization of Raised Source/Drain SOI FinFET Structure for Analog Circuit Applications", *Journal of Nanoelectronics and Optoelectronins*, Vol. 13, pp. 531–539, 2018.
64. Shradhya Singh, S. K. Vishvakarma and B. Raj, "Analytical Modeling of Split-Gate Junction-Less Transistor for a Biosensor Application", *Sensing and Bio-sensing*, Vol. 18, pp. 31–36, 2018.
65. Maisagalla Gopal, and Balwinder Raj, Low Power 8T SRAM Cell Design for High Stability Video Applications, *ITSI Transaction on Electrical and Electronics Engineering*, Vol. 1, No. 5, pp. 91–97, 2013.
66. Balwinder Raj, Jatin Mitra, Deepak Kumar Bihani, V. Rangharajan, A.K. Saxena and S. Dasgupta, "Analysis of Noise Margin, Power and Process Variation for 32 nm FinFET Based 6T SRAM Cell", *Journal of Computer (JCP)*, Vol. 5, No. 6, pp. 1–8, 2010.
67. Divya Sharma, Rajesh Mehra and B. Raj, "Comparative Analysis of Photovoltaic Technologies for High Efficiency Solar Cell Design", *Superlattices and Microstructures*, Vol. 153, pp. 106861, 2021.
68. Pawandeep Kaur, Avtar Singh Buttar, B. Raj, "A comprehensive Analysis of Nanoscale Transistor based Biosensor: A Review, *Indian Journal of Pure and Applied Physics*, Vol. 59, pp. 304–318, 2021.
69. Divya Yadav, Balwant Raj and B. Raj, "Design and Simulation of Low Power Microcontroller for IoT Applications", *Journal of Sensor Letters, ASP*, Vol. 18, pp. 401–409, 2020.
70. Shailendra Singh, and B. Raj, "A 2-D Analytical Surface Potential and Drain current Modeling of Double-Gate Vertical t-shaped Tunnel FET", *Journal of Computational Electronics*, Vol. 19, pp. 1154–1163, 2020.
71. Jeetendra Singh, B. Raj, "An Accurate and Generic Window function for Non-linear Memristor Model" *Journal of Computational Electronics*, Vol. 18, No. 2, pp. 640–647, 2019.

72. Manjit Kaur; Neena Gupta; Sanjeev Kumar; B. Raj; Arun Kumar Singh, "Comparative RF and Crosstalk Analysis of Carbon Based Nano Interconnects" *IET Circuits, Devices & Systems*, Vol. 15, No. 6, pp. 493–503, 2021.
73. Nehru Kandasamy, Firdous Ahmad, D. Ajitha, B. Raj, and Nagarjuna Telagam "Quantum Dot Cellular Automata based Scan Flip Flop and Boundary Scan Register", *IETE Journal of Research*, Vol. 66, pp. 535–548, 2020.
74. Jeetendra Singh, and B. Raj, "Design and Investigation of 7T2M NVSARM with Enhanced Stability and Temperature Impact on Store/Restore Energy", *IEEE Transactions on Very Large Scale Integration Systems*, Vol. 27, No. 6, pp. 1322–1328, 2019.
75. Anil Kumar Bhardwaj, Sumeet Gupta, B. Raj, Amandeep Singh, "Impact of Double Gate Geometry on the Performance of Carbon Nanotube Field Effect Transistor Structures for Low Power Digital Design", *Computational and Theoretical Nanoscience*, Vol. 16, pp. 1813–1820, 2019.
76. Neeraj Jain and B. Raj, "Thermal Stability Analysis and Performance Exploration of Asymmetrical Dual-k underlap Spacer (ADKUS) SOI FinFET for Security and Privacy Applications", *Indian Journal of Pure & Applied Physics (IJPAP)*, Vol. 57, pp. 352–360, 2019.
77. Neeraj Jain and Balwinder Raj, "Dual-k Spacer Region Variation at the Drain Side of Asymmetric SOI FinFET Structure: Performance Analysis towards the Analog/RF Design Applications", *Journal of Nanoelectronics and Optoelectronics*, Vol. 14, pp. 349–359, 2019.
78. Jeetendra Singh, Sanjeev Sharma, B. Raj, Mamta Khosla, "Analysis of Barrier Layer Thickness on Performance of In1-xGaxAs based Gate Stack Cylindrical Gate Nanowire MOSFET", *JNO*, Vol. 13, pp. 1473–1477, 2018.
79. Neeraj Jain, B. Raj, "Analysis and Performance Exploration of High-k SOI FinFETs Over the Conventional Low-k SOI FinFET toward Analog/RF Design", *Journal of Semiconductors (JoS)*, Vol. 39, No. 12, pp. 124002-1-7, 2018.
80. Candy Goyal, Jagpal Singh Ubhi and B. Raj, "A Reliable Leakage Reduction Technique for Approximate Full Adder with Reduced Ground Bounce Noise," *Journal of Mathematical Problems in Engineering*, Vol. 2018, Article ID 3501041, 16, pages, 2018.
81. Anuradha, Jeetendra Singh, B. Raj and Mamta Khosla, "Design and Performance Analysis of Nano-scale Memristor-based Nonvolatile SRAM", *Journal of Sensor Letter*, Vol. 16, pp. 798–805, 2018.
82. Girish Wadhwa, and B. Raj, "Parametric Variation Analysis of Charge-Plasma-based Dielectric Modulated JLTFET for Biosensor Application" *IEEE Sensor Journal*, Vol. 18, No. 15, pp. 6070–6077, 2018 (SCI)
83. Jeetendra Singh, and B. Raj, "Comparative Analysis of Memristor Models for Memories Design", *JoS*, Vol. 39, No. 7, pp. 074006-1-12, 2018.
84. Divya Yadav, Shailesh Singh Chouhan, Santosh Kumar Vishvakarma and B. Raj, "Application Specific Microcontroller Design for IoT based WSN", *Sensor Letter*, Vol. 16, pp. 374–385, 2018.
85. Gurmohan Singh, R. K. Sarin and B. Raj, "Fault-Tolerant Design and Analysis of Quantum-Dot Cellular Automata Based Circuits", *IEEE/IET Circuits, Devices & Systems*, Vol. 12, pp. 638–664, 2018.
86. Jeetendra Singh, B. Raj, "Modeling of Mean Barrier Height Levying Various Image Forces of Metal Insulator Metal Structure to Enhance the Performance

of Conductive Filament Based Memristor Model", *IEEE Nanotechnology*, Vol. 17, No. 2, pp. 267–268, 2018.

87. Aakash Jain, Sanjeev Sharma, B. Raj, "Analysis of Triple Metal Surrounding Gate (TM-SG) III-V Nanowire MOSFET for Photosensing Application", *Opto-electronics Journal*, Vol. 26, No. 2, pp. 141–148, 2018.
88. Aakash Jain, Sanjeev Sharma, B. Raj, "Design and Analysis of High Sensitivity Photosensor Using Cylindrical Surrounding Gate MOSFET For Low Power Sensor Applications", *Engineering Science and Technology, an International Journal*, Vol. 19, No. 4, pp. 1864–1870, 2016.
89. G. Saiphani Kumar, Amandeep Singh, B. Raj, "Design and Analysis of Gate All Around CNTFET based SRAM cell Design", *Journal of Computational Electronics*, Vol. 17, No. 1, pp 138–145, 2018.
90. Gurinder Pal Singh, B. S. Sohi, Balwinder Raj, "Material Properties Analysis of Graphene Base Transistor (GBT) for VLSI Analog Circuits", *Indian Journal of Pure & Applied Physics (IJPAP)*, Vol. 55, pp. 896–902, 2017.
91. Sunil Kumar and B. Raj, "Estimation of Stability and Performance metric for Inward Access Transistor based 6T SRAM Cell Design using n-type/p-type DMDG-GDOV TFET", *IEEE VLSI Circuits and Systems Letter*, Vol. 3, No. 2, pp. 25–39, 2017.
92. Shashikant Sharma, Anjan Kumar, Manisha Pattanaik and B. Raj, "Forward Body Biased Multimode Multi-Threshold CMOS Technique for Ground Bounce Noise Reduction in Static CMOS Adders", *International Journal of Information and Electronics Engineering*, pp. 567–572, Vol. 3, No. 3, 2013.
93. Hamendra Singh, Pankaj Kumar, Balwinder Raj, "Performance Analysis of Majority Gate SET Based 1-bit Full Adder", *International Journal of Computer and Communication Engineering (IJCCE)*, Vol. 2, No. 4, 2013.
94. Anil Kumar Bhardwaj, Sumeet Gupta and B. Raj, "Investigation of Parameters for Schottky Barrier (SB) Height for Schottky Barrier Based Carbon Nanotube Field Effect Transistor Device", *Journal of Nanoelectronics and Optoelectronics*, Vol. 15, pp. 783–791, 2020.
95. Priya Bansal, and B. Raj, "Memristor: A Versatile Nonlinear Model for Dopant Drift And Boundary Issues," *JCTN*, Vol. 14, No. 5, pp. 2319–2325, 2017.
96. Neeraj Jain, and B. Raj, "An Analog and Digital Design Perspective Comprehensive Approach on Fin-FET (Fin-Field Effect transistor) Technology - A Review", *Reviews in Advanced Sciences and Engineering (RASE)*, Vol. 5, pp. 1–14, 2016.
97. Balwinder Raj, A. K. Saxena and S. Dasgupta, "Analytical Modeling for the Estimation of Leakage Current and Subthreshold Swing Factor of Nanoscale Double Gate FinFET Device", *Microelectronics International*, Vol. 26, pp. 53–63, 2009.
98. Soniya, Shailendra Singh, Girish Wadhwa and Balwinder Raj, "An Analytical Modeling for Dual Source Vertical Tunnel Field Effect Transistor", *International Journal of Recent Technology and Engineering (IJRTE)*, Vol. 8, No. 2, pp. 603–608, 2019.
99. Shailendra Singh, and B. Raj, "Design and analysis of hetrojunction Vertical T-shaped Tunnel Field Effect Transistor", *Journal of Electronics Material*, Vol. 48, No. 10, pp 6253–6260, 2019.
100. Candy Goyal, Jagpal Singh Ubhi and B. Raj, "A Low Leakage CNTFET based Inexact Full Adder for Low Power Image Processing Applications",

International Journal of Circuit Theory and Applications, Vol. 47, No. 9, pp. 1446–1458, 2019.

101. B. Raj, A. K. Saxena and S. Dasgupta, "A Compact Drain Current and Threshold Voltage Quantum Mechanical Analytical Modeling for FinFETs", *Journal of Nanoelectronics and Optoelectronics (JNO)*, Vol. 3, No. 2, pp. 163–170, 2008.
102. Girish Wadhwa, and B. Raj, "An Analytical Modeling of Charge Plasma based Tunnel Field Effect Transistor with Impacts of Gate underlap Region", *Superlattices and Microstructures*, Vol. 142, pp. 106512, 2020.
103. Shailendra Singh, and B. Raj, "Modeling and Simulation Analysis of SiGe Hetrojunction Double GateVertical t-shaped Tunnel FET", *Superlattices and Microstructures*, Vol. 142, pp. 106496, 2020.
104. Neeraj Jain, Balwinder Raj, "Device and Circuit Co-Design Perspective Comprehensive Approach on FinFET Technology - A Review", *Journal of Electron Devices*, Vol. 23, No. 1, pp. 1890–1901, 2016.
105. Sunil Kumar and B. Raj, "Analysis of I_{ON} and Ambipolar Current for Dual-Material Gate-drain Overlapped DG-TFET," *Journal of Nanoelectronics and Optoelectronics*, Vol. 11, pp. 323–333, 2016.
106. Naveed Anjum, Tarun Bali and B. Raj, "Design And Simulation Of Handwritten Multiscript Character Recognition," *International Journal of Advanced Research in Computer and Communication Engineering*, Vol. 2, No. 7, pp. 2544–2549, 2013.
107. Karmjit Singh, and B. Raj, "Performance and Analysis of Temperature Dependent Multi-walled Carbon Nanotubes as Global Interconnects at Different Technology Nodes," *Journal of Computational Electronics*, Vol. 14, No. 2, pp. 469–476, 2015.
108. Sunil Kumar and B. Raj, "Compact channel potential analytical Modeling of DG-TFET based on Evanescent–mode Approach," *Journal of Computational Electronics*, Vol. 14, No. 2, pp. 820–827, 2015.
109. Karmjit Singh, and B. Raj, "Temperature Dependent Modeling and Performance Evaluation of Multi-Walled CNT and Single-Walled CNT as Global Interconnects," *Journal of Electronic Materials*, Vol. 44, No. 12, pp 4825–4835, 2015.
110. V.K. Sharma, M. Pattanaik and B. Raj, "INDEP approach for leakage reduction in nanoscale CMOS circuits", *International Journal of Electronics*, Vol. 102, No. 2, pp. 200–215, 2014.
111. Karmjit Singh, and B. Raj, "Influence of Temperature on MWCNT bundle, SWCNT bundle and Copper interconnects for Nanoscaled Technology nodes", *Journal of Materials Science: Materials in Electronics*, Vol. 26, No. 8, pp. 6134–6142, 2015.
112. Naveed Anjum, Tarun Bali and B. Raj, "Design and Simulation of Handwritten Gurumukhi and Devanagri Numerical Recognition, *International Journal of Computer Applications*, Vol. 73, No. 12, pp. 16–21, 2013.
113. S. Khandelwal, V. Gupta, B. Raj and R. D. Gupta, "Process Variability Aware Low Leakage Reliable Nano Scale DG-FinFET SRAM Cell Design Technique", *Journal of Nanoelectronics and Optoelectronics*, Vol. 10, No. 6, pp. 810–817, 2015.
114. V.K. Sharma, M. Pattanaik and B. Raj, "ONOFIC Approach: Low Power High Speed Nanoscale VLSI Circuits Design", *International Journal of Electronics*, Vol. 101, No. 1, pp. 61–73, 2014.

115. S. Khandelwal, Balwinder Raj, and R. D. Gupta, FinFET based 6T SRAM Cell Design: Analysis of Performance Metric, Process Variation and Temperature Effect. *Journal of Computational and Theoretical Nanoscience*, Vol. 12, pp. 2500–2506, 2015.
116. Sumit Singh; Shekhar Yadav; Jagdeep Rahul; Anurag Srivastava; B. Raj, "Impact of HfO_2 in Graded Channel Dual Insulator Double Gate MOSFET", *Journal of Computational and Theoretical Nanoscience*, Vol. 12, No. 6, pp. 950–953, 2015.
117. Vijay Kumar Sharma, Manisha Pattanaik, and B. Raj, "PVT variations aware low leakage INDEP approach for nanoscale CMOS Circuits", *Microelectronics Reliability*, Vol. 54, pp 90–99, 2014.
118. B. Raj, A. K. Saxena and S. Dasgupta, "Quantum Mechanical Analytical Modeling of Nanoscale DG FinFET: Evaluation of Potential, Threshold Voltage and Source/Drain Resistance", *Elsevier's Journal of Material Science in Semiconductor Processing*, Vol. 16, No. 4, pp. 1131–1137, 2013.
119. Maisagalla Gopal, Siva Sankar, D. Prasad and Balwinder Raj, "8T SRAM Cell Design for Dynamic and Leakage Power Reduction", *International Journal of Computer Applications*, Vol. 71, No. 9, pp. 43–48, 2013.
120. Manisha Pattanaik, B. Raj, Shashikant Sharma and Anjan Kumar, "Diode Based Trimode Multi-Threshold CMOS Technique for Ground Bounce Noise Reduction in Static CMOS Adders", *Advanced Materials Research*, Vol. 548, pp. 885–889, 2012.
121. Balwinder Raj, A. K. Saxena and S. Dasgupta, "Nanoscale FinFET Based SRAM Cell Design: Analysis of Performance metric, Process variation, Underlapped FinFET and Temperature effect", *IEEE Circuits and System Magazine*, Vol. 11, No. 2, pp. 38–50, 2011.
122. V.K. Sharma, M. Pattanaik & Balwinder Raj, "Leakage Current ONOFIC Approach for Deep Submicron VLSI Circuit Design", *International Journal of Electrical, Computer, Electronics and Communication Engineering, World Academy of Sciences, Engineering and Technology*, Vol. 7, No. 4, pp. 239–244, 2013.
123. Parminder Kaur, Sandeep Singh Gill and B. Raj, "Comparative Analysis of OFETs Materials and Devices for Sensor Applications", *Journal of Silicon*, Vol. 14, pp. 4463–4471, 2022.
124. Sanjeev Kumar Sharma, Parveen Kumar, Balwant Raj and B. Raj, "$In_{1-x}Ga_xAs$ Double Metal Gate-Stacking Cylindrical Nanowire MOSFET for highly sensitive Photo detector", *Journal of Silicon*, Vol. 14, pp. 3535–3541, 2022.
125. B. Raj, A. K. Saxena and S. Dasgupta, "Analytical Modeling of Quasi Planar Nanoscale Double Gate FinFET with Source/Drain Resistance and Field Dependent Carrier Mobility: A Quantum Mechanical Study" *Journal of Computer (JCP)*, Vol. 4, No. 9, pp. 1–8, 2009.
126. S. Bhushan, S. Khandelwal and B. Raj, (2012). Analyzing Different Mode FinFET based Memory Cell at Different Power Supply for Leakage Reduction, *Seventh International Conference on Bio-Inspired Computing: Theories and Application, (BIC-TA 2012) Advances in Intelligent Systems and Computing*, Vol. 202, pp. 89–100, 2013.
127. Jeetendra Singh, B. Raj, "Temperature Dependent Analytical Modeling and Simulations of Nanoscale Memristor," *Journal: Engineering Science and Technology, an International Journal*, Vol. 21, pp. 862–868, 2018.

128. Shradhya Singh, Shashi Bala, Balwant Raj, B. Raj, "Improved Sensitivity of Dielectric Modulated Junctionless Transistor for Nanoscale Biosensor Design", *Sensor Letter*, Vol. 18, pp. 328–333, 2020.
129. Vivek Kumar, Santosh Kumar Vishvakarma, and B. Raj, "Design and Performance Analysis of ASIC for IoT Applications" *Sensor Letter ASP*, Vol. 18, pp. 31–38, 2020.
130. Akanksha Jaiswal, R. K. Sarin, B. Raj and Shikha Sukhija, "A Novel Circular Slotted Microstrip-fed Patch Antenna with Three Triangle Shape Defected Ground Structure for Multiband Applications", *Advanced Electromagnetic (AEM)*, Vol. 7, No. 3, pp. 56–63, 2018.
131. Girish Wadhwa, and B. Raj, "Label Free Detection of Biomolecules using Charge-Plasma-Based Gate Underlap Dielectric Modulated Junctionless TFET", *Journal of Electronic Materials (JEMS)*, Vol. 47, No. 8, pp 4683–4693, 2018.
132. Gurmohan Singh, R. K. Sarin and B. Raj, "Design and Performance Analysis of a New Efficient Coplanar Quantum-Dot Cellular Automata Adder", *Indian Journal of Pure & Applied Physics (IJPAP)*, Vol. 55, pp. 97–103, 2017.
133. Parminder Kaur, Vikas Pandey and B. Raj, "Comparative Study of Efficient Design, Control and Monitoring of Solar Power using IoT", *Sensor Letter*, Vol. 18, pp. 419–426, 2020.
134. Anil Kumar Bhardwaj, Sumeet Gupta and B. Raj, Development & Analysis of compact model for Double Gate Schottky Barrier CNTFET, *Journal of Nanoelectronics and Optoelectronics*, Vol. 15, pp. 1199–1208, 2020.
135. Girish Wadhwa, Priyanka Kamboj; Jeetendra Singh, B. Raj, "Design and Investigation of Junctionless DGTFET for Biological Molecule Recognition", *Transactions on Electrical and Electronic Materials*, Vol. 22, pp. 282–289, 2021.
136. Sachin Kumar Verma, Shailendra Singh, Girish Wadhwa and B. Raj, "Detection of Biomolecules using Charge-Plasma based Gate Underlap Dielectric Modulated Dopingless TFET", *Transactions on Electrical and Electronic Materials (TEEM)*, Vol. 21, pp. 528–535, 2020.
137. Neeraj Jain, B. Raj, "Impact of Underlap Spacer Region Variation on Electrostatic and analog/RF Performance of Symmetrical high-k SOI FinFET at 20 nm Channel Length", *Journal of Semiconductors (JoS)*, Vol. 38, No. 12, pp. 122002, 2017.
138. Shailendra Singh, B. Raj, "Analytical Modeling and Simulation analysis of T-shaped III-V heterojunction Vertical T-FET", *Superlattices and Microstructures*, Vol. 147, pp. 106717, 2020.
139. Gurmohan Singh, R. K. Sarin and B. Raj, "Design and Analysis of Area Efficient QCA Based Reversible Logic Gates", *Journal of Microprocessors and Microsystems*, Vol. 52, pp. 59–68, 2017.
140. Shailendra Singh, B. Raj, "Parametric variation analysis on hetero-junction Vertical t-shape TFET for Supressing ambipolar conduction", *Indian Journal of Pure and Applied Physics*, Vol. 58, pp. 478–485, 2020.
141. Soniya, Shailendra Singh, Girish Wadhwa and Balwinder Raj, "Design and Analysis of Dual Source Vertical Tunnel Field Effect Transistor for high performance", *Transactions on Electrical and Electronics Materials*, Vol. 21, pp. 74–82, 2019.
142. Manjit Kaur; Neena Gupta; Sanjeev Kumar; B. Raj; Arun Kumar Singh, "RF Performance Analysis of Intercalated Graphene Nanoribbon Based Global

Level Interconnects" *Journal of Computational Electronics*, Vol. 19, pp. 1002–1013, 2020
143. Girish Wadhwa, and B. Raj, "Design and Performance Analysis of Junctionless TFET Biosensor for high sensitivity" *IEEE Nanotechnology*, Vol. 18, pp. 567–574, 2019.
144. Jeetendra Singh, and B. Raj, "Enhanced Nonlinear Memristor Model Encapsulating Stochastic Dopant Drift", *JNO*, Vol. 14, pp. 958–963 (2019)
145. Mohit Kumar, Tarun Chaudhary; Balwinder Raj, "Design and Comparative Analysis of Ferroelectric Nanowire with Dielectric HfO_2 and $Al_2O_3$for Low Power Applications", *Journal of Electronic Materials*, Vol. 53, pp. 6785–6791, 2024.
146. Yogesh Thakur, Balwant Raj and Balwinder Raj, "Design of Pentacene Thin-Film Transistor Based Hydrogen Gas Sensor for High Sensitivity", *ECS Journal of Solid State Science and Technology*, Vol. 13, pp. 047005–047012, 2024.
147. Manoharan Arun Kumar, Kamal Batcha Mohamed Ismail, Shanmugam Mahalingam, Balwinder Raj, Junghwan Kim, Raji Atchudan, "Recent Advances in Two-Dimensional Nanomaterials and Nanostructures for Healthcare Monitoring", *ACS Sensors*, Vol. 9, No. 4, pp. 1–8, 2024
148. Mohamed Ismail Kamal Batcha; Arun Kumar Manoharan; Shanmugam Mahalingam; Raj Balwinder, "Carbon Fiber-Reinforced Polymers for Energy Storage Applications", *Journal of Energy Storage*, Vol. 84, Part B, pp. 110931–110939, 2024.
149. Shahraan Hussain, Balwinder Raj, Divya Sharma, "Design and Parametric Optimization of Lead Free Perovskite based Tandem Solar Cell for High Efficient", *Engineering Research Express*, Vol. 6, pp. 015315–0153122, 2024.
150. Parveen Kumar, Balwinder Raj, Girish Wadhwa, Balwinder Singh and Raj Kumar "Design and Analysis of Junctionless based Gate All Around N+ Doped Layer Nanowire TFET Biosensor, *ECS Journal of Solid State Science and Technology*, Vol. 13, pp. 017002–017010, 2024.
151. Divya Sharma, Rajesh Mehra and Balwinder Raj, "Enhancement in Efficiency of Methyl Ammonium Tin Iodide based perovskite solar cell using SCAPS-1 D", *Nano*, Vol. 18, No. 14, pp. 2350109, 2023.
152. Parveen Kumar, Sanjeev Kumar Sharma, Balwant Raj and Balwinder Raj, "Analysis of Device Parameter Variations in In1-xGaxAs based Gate Stacked Double Metal Surrounding Gate Nanowire MOSFET", *TEEM*, Vol. 24, pp. 570–578, 2023.
153. Anjana Bhardwaj, Pradeep Kumar, Balwinder Raj, Naveen Kumar, Sunny Anand, "Design and Optimization of Vertical Nanowire Tunnel FET with Electrostatic Doping", *Engineering Research Express*, Vol. 5, pp. 045025–045033, 2023.
154. Yogesh Thakur, Balwinder Raj, Sandeep Singh Gill, "Design and Performance Analysis of Pentacene Organic Field Effect Transistor with High-K dielectric Materials", *Journal of Optoelectronics and Advanced Materials – Rapid Communications*, Vol. 17, No. 7–8, 2023.
155. Parveen Kumar and Balwinder Raj, "Design and Simulation of Junctionless Nanowire Tunnel Field Effect Transistor Biosensor for High Sensitivity", *Microelectronic Journal*, Vol. 137, pp. 105826–105834, 2023.

Chapter 4

Machine learning in quantum material

Mandeep Singh, Pawan Kumar Maurya, Manas Baiswar, Gandepalli Sai Teja, and Balwinder Raj

4.1 INTRODUCTION

In recent years, the tech world has seen a rapid shift toward quantum devices, in order to push the limits of classical technology. For decades, metal–oxide–semiconductor field-effect transistors (MOSFETs) have been the backbone of electronic systems, fueling remarkable advances in computing and data processing, largely thanks to Moore's Law. This principle held that the number of transistors on a chip would double every couple of years, leading to exponential growth in processing power. But now, as these transistors approach the scale of just a few nanometers, fundamental challenges have started to crop up. When transistor sizes drop below 10 nanometers, they face short-channel effects like unwanted leakage currents and weakened gate control, which both affect performance and reliability [1].

Additionally, at such tiny scales, it takes more energy to keep devices stable, which in turn raises power usage and heat production. And then there is quantum tunneling, where electrons can slip through barriers that would otherwise block them—a behavior that, while intriguing, further complicates the effectiveness of classical devices at such small scales. Altogether, these challenges point to a looming limit on what classical semiconductor technology can achieve, both in terms of size reduction and overall performance. This situation has led researchers to explore quantum devices as the next frontier, where principles such as quantum confinement, superposition, and tunneling offer the potential for huge leaps in efficiency and computational power. [2]

Quantum confinement is a principle that give us an extraordinary solution. For devices such as quantum dots (QDs) and quantum wells, quantum confinement traps electrons in minuscule spaces, creating discrete energy levels and help us to control electron movements. This kind of confinement allows quantum devices to operate at impressive speeds and minimal power, making quantum dots particularly useful in both optical and electronic systems. In Figure 4.1, quantum dot array for quantum processor is shown. Additionally, quantum superposition and entanglement play a huge role in quantum computing (QC), as they allow qubits to represent multiple states

DOI: 10.1201/9781003634645-4

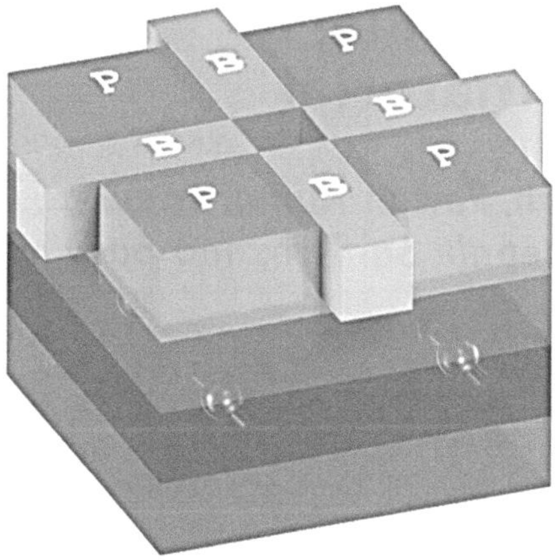

Figure 4.1 Schematic layout of 2×2 QD (quantum dot) array for a quantum processor [3].

simultaneously. This gives quantum systems the ability to handle massive amounts of data in parallel, a feature that classical systems simply cannot match. Entanglement, on the other hand, forms deep connections between qubits, allowing them to communicate in ways that further boost both computation and secure communication abilities [3].

Another principle, quantum tunneling enables particles to bypass energy barriers that would block them in classical physics. This effect is critical for devices such as resonant tunneling diodes (RTDs) and terahertz quantum cascade lasers (QCLs), which utilize tunneling to achieve quick switching and energy transfer, both of which are essential for terahertz communication and high-frequency operations. In Figure 4.2, an example of quantum device is shown [5–9].

Beyond simply overcoming classical limitations, quantum devices are opening doors to entirely new possibilities. Quantum computing is a prime example, with qubits allowing us to solve difficult problems—such as factoring large numbers and simulating molecular interactions—at a speed that classical computers would struggle to match. Algorithms like Shor's and Grover's use quantum superposition to offer exponential speed improvements for particular tasks [4]. In terahertz communication, quantum devices make it possible to generate and manage terahertz signals, paving the way for ultrafast wireless communication and advanced imaging. Quantum sensing technologies, too, are making strides, using quantum dots and wells to detect electric, magnetic, and optical fields with remarkable sensitivity. These capabilities have applications in fields ranging from medical imaging to environmental monitoring and even GPS systems [5].

However, quantum devices' complex nature requires advanced systems for control and optimization. Machine learning (ML) has emerged as a crucial tool in managing these intricate processes. ML helps automate tasks

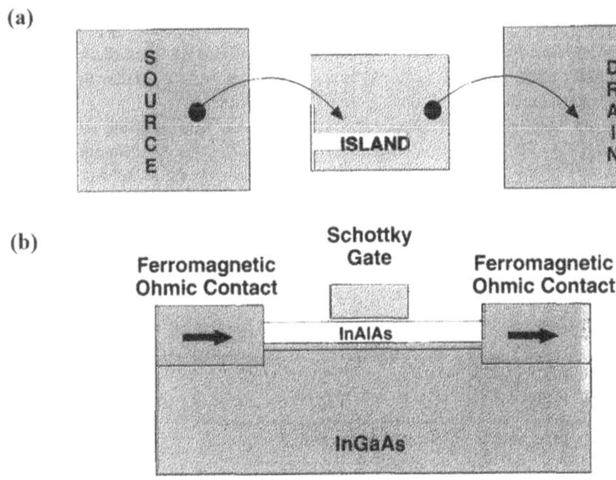

Figure 4.2 (a) Vertical quantum device: The basic structure for a single electron transistor. At low temperature, the quantum tunneling of electrons (full circles) between source and drain via a small island of submicron dimensions can be regulated (i.e., the transistor can be switched ON or OFF) by the electrostatic energy needed to charge the island with a single electron. (b) Lateral quantum device: The spin precession device. The region in blue represents the two-dimensional electron gas formed at the interface between the InAlAs and InGaAs regions [4]

such as qubit calibration, quantum gate optimization, and error correction, which are vital for maintaining high performance. By refining quantum gate operations, managing interference, and enhancing circuit efficiency, ML plays a key role in expanding the potential of quantum technologies and making them viable for everyday applications. As a whole, these ML-driven advancements are pushing the boundaries of what is possible with quantum devices, making them not only more scalable and reliable but also well-suited to tackle complex problems that lie ahead [6–13].

4.2 THE SHIFT TOWARD QUANTUM TECHNOLOGY

As classical semiconductor technology reaches its limits, researchers are increasingly turning to quantum devices as the next big leap in computing and electronics. For decades, devices like metal–oxide–semiconductor field-effect transistors (MOSFETs) have been foundational in powering electronic systems. The miniaturization of these transistors, a trend largely inspired by Moore's Law, has led to incredible gains in computing speed and efficiency. Yet, as we continue to shrink transistors to scales below 10 nanometers, we encounter significant hurdles that threaten to limit further progress. One issue that arises at this tiny scale is the short-channel effect, where transistors experience increased leakage currents and lose some control at the gate level. Both of these factors can seriously undermine performance and reliability [6].

Additionally, stabilizing these minuscule devices requires more power, which leads to greater energy consumption and heat buildup—two critical challenges in today's push for more efficient technologies. Another complication, quantum tunneling, becomes noticeable at these scales, where electrons begin to pass through barriers that would normally confine them. While this tunneling effect is essential in quantum technologies, in classical devices, it further increases leakage and diminishes efficiency. Altogether, these limitations are clear signs that classical semiconductors may be approaching their practical limits in terms of both size reduction and overall performance, making quantum devices an attractive area of research and development. [7]

What makes quantum devices so promising is that they operate on completely different principles, which can bypass many of these limitations. Quantum confinement, for instance, is used in devices like quantum dots and quantum wells to trap electrons in extremely small spaces, creating distinct energy levels and giving precise control over electron behavior. This confinement allows quantum devices to achieve high speeds and energy efficiency that classical systems struggle to match, making quantum dots particularly useful in both optical and electronic applications. Beyond confinement, the concepts of quantum superposition and entanglement offer even greater possibilities [14–21].

Quantum superposition enables qubits to represent multiple states at once, allows quantum devices to perform multiple calculations simultaneously, and advantage classical devices lack [22–28]. Meanwhile, entanglement creates deep correlations between qubits, enhancing both computational power and secure communication in ways that classical devices cannot achieve [8].

Another valuable principle, quantum tunneling, allows particles to bypass energy barriers, a process that plays a major role in quantum devices like resonant tunneling diodes (RTDs) and terahertz quantum cascade lasers (QCLs). These devices use tunneling for rapid switching and efficient energy transfer, essential for applications requiring high-frequency operations and terahertz communications. Taken together, these quantum mechanical principles not only address the physical and operational limits of classical technologies but also open up new possibilities for applications previously considered impractical. Figure 4.3 graph shows the evolution of quantum devices over time. From quantum computing and secure quantum communication to ultrasensitive quantum sensing, quantum devices are set to shape the future of technology in ways that were once unimaginable. [9]

Table 4.1 represents a rank-wise listing of the top 10 countries in quantum computing (QC) literature, detailing each country's frequency of

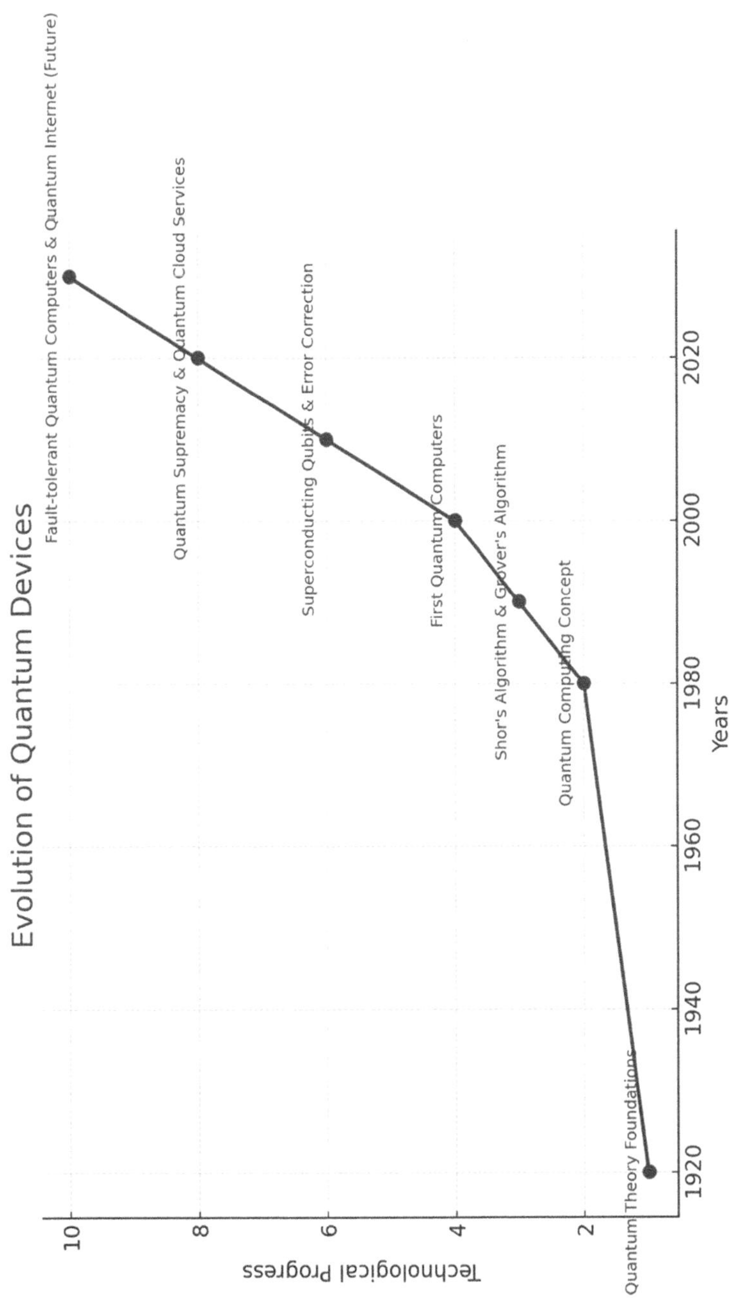

Figure 4.3 Graph shows how quantum devices are evolved over time.

Table 4.1 The number and frequency of publication of literature for quantum computing of top 10 countries. [9]

Rank	*Frequency*	*Country*	*Link Count*	*Centrality*
1	1419	China	45	0.17
2	1045	USA	45	0.19
3	435	Australia	45	0.06
4	380	England	45	0.17
5	374	India	45	0.10
6	315	Canada	38	0.14
7	256	Japan	38	0.04
8	237	Germany	52	0.09
9	174	Singapore	33	0.02
10	167	Iran	31	0.05

publications, the count of neighboring nodes (representing collaborative links), and their centrality within the research network. This data reflects each country's research output, level of collaboration, and overall influence in the QC field, highlighting the leading nations driving advancements and connections within the global QC community. And it also highlights that India is leading now in quantum technology [29–36].

4.3 UNDERSTANDING MACHINE LEARNING

Machine learning (ML) is fundamentally changing the way we approach complex problems by enabling computers to learn from data, make predictions, and even adapt without being explicitly programmed. At its core, ML is about using algorithms to find patterns or "rules" within data, allowing systems to make informed decisions or predictions based on that data. ML is widely applied across industries for tasks such as facial recognition, language translation, and predictive modeling, but it is also showing incredible potential in cutting-edge fields like quantum computing [37–45].

In the context of quantum devices, ML is especially useful because quantum systems are incredibly complex, sensitive to noise, and prone to errors due to their interaction with the surrounding environment. Quantum devices rely on precise control and calibration to maintain stability and optimize performance—tasks that ML can significantly improve. In calibration and optimization, for example, ML models help tune the parameters of quantum gates and qubits in real time, allowing quantum devices to operate at maximum efficiency while keeping error rates as low as possible. Reinforcement learning, a type of ML that learns optimal actions by receiving feedback, is particularly useful here. By automating adjustments to gate parameters based on environmental fluctuations, reinforcement learning helps ensure that quantum operations remain accurate, stable, and resilient to noise [10].

Another area where ML plays a crucial role is error correction. Quantum systems are inherently prone to errors, given their sensitivity to their surroundings, and managing these errors without collapsing quantum states is challenging. Neural networks, an ML technique modeled after the human brain, are used to develop advanced error correction codes, helping identify and correct errors in real time. This capability is essential to scaling up quantum systems since it preserves the integrity of quantum computations even as systems grow larger [46–54].

To better understand how ML works, we can break it down into several key types of algorithms: supervised learning, unsupervised learning, and reinforcement learning. In supervised learning, models are trained on labeled data, meaning the input data has known outcomes. In quantum systems, supervised learning can improve qubit measurement accuracy, allowing devices to process noisy data and correct computational errors as they arise. Unsupervised learning operates without labeled outcomes, grouping data based on similarities [55–61]. This approach is beneficial for optimizing quantum circuits by revealing hidden patterns in quantum data, which can lead to more efficient gate sequences and, ultimately, fewer errors. Reinforcement learning, as mentioned, excels in situations requiring ongoing adjustments. By interacting directly with quantum environments, reinforcement learning models "learn" optimal strategies for controlling quantum gates and minimizing errors—a critical feature for maintaining high fidelity in dynamic quantum circuits [11] (Table 4.2).

Together, these ML techniques help quantum devices perform at their peak, addressing issues such as noise, errors, and efficiency in ways that manual tuning alone could not achieve. As quantum technology continues to develop, the integration of ML will likely remain central, making these devices not only more powerful but also more reliable and practical for real-world applications [62–68].

4.4 THE MECHANICS OF QUANTUM DEVICE OPERATION

Quantum devices operate based on principles that are fundamentally different from those governing classical systems. These principles—quantum superposition, entanglement, and tunneling—allow quantum devices to perform tasks in computing, communication, and sensing that would be unachievable with classical technology. Let us explore the mechanics of quantum devices, their architectural designs, and the unique phenomena that make them so powerful [69–76].

Table 4.2 Comparison of some popular machine learning tools. [8]

	Tool				
	Python	*R*	*Spark*	*MATLAB*	*TensorFlow*
License	Open source	Open source	Open source	Proprietary	Open source
Distributed	No	No	Yes	No	No
Visualization	Yes	Yes	No	Yes	No
Neural nets	Yes	Yes	Multilayer perceptron classifier	Yes	Yes
Supported languages	Python	R	Scala, Java, Python, and R	MATLAB	Python and C++
Variety of machine-learning models	High	High	Medium	High	Low
Suitability as a general-purpose tool	High	Medium	Medium	High	Low
Maturity	High	Very high	Medium	Very high	Low

4.4.1 Working of quantum devices

Quantum tunneling is another unique feature of quantum devices, allowing particles to pass through barriers that would typically block them in classical systems. In devices like resonant tunneling diodes (RTDs) and quantum dots, tunneling plays a vital role in controlling electron movement and enables fast switching and precise qubit manipulation, which are essential for efficient quantum computing [12].

The way quantum devices work relies on a few key characteristics of quantum systems, enabling abilities such as parallel processing, secure communication, and highly sensitive measurements. One such characteristic is quantum superposition, where quantum bits (or qubits) can exist in multiple states at once, rather than being limited to just 0 or 1 as in classical bits. This capability allows quantum devices to perform many calculations simultaneously, a crucial feature behind the computational power of quantum algorithms. For example, superposition enables complex algorithms—like Grover's for database searching and Shor's for factoring—to run exponentially faster than traditional counterparts [77–84].

Another defining property is quantum entanglement, a phenomenon where two or more qubits become interlinked in such a way that the state of one qubit instantly determines the state of the others, regardless of the distance between them. This feature is fundamental in quantum communication for tasks such as quantum key distribution (QKD), which uses entanglement to create secure data transfer channels. Entanglement is also crucial for multi-qubit operations in quantum computers, enabling complex calculations that classical systems cannot match.

However, quantum devices are also highly sensitive to decoherence, where external interactions disrupt quantum states, leading to loss of information. To manage this, quantum error correction methods are used to detect and correct errors without collapsing the qubit states, ensuring the stability and accuracy of quantum computations [13].

4.4.2 Architecture of quantum devices

The architecture of quantum devices is purpose-built to harness and control qubits, allowing them to perform the necessary quantum operations. Different types of architectures are employed depending on the device and the intended application. Qubits and quantum circuits are the core components of quantum computing. Qubits, which are the quantum equivalents of classical bits, can be manipulated through quantum gates to change their states or become entangled with other qubits. Quantum circuits are composed of sequences of these gates and form the basis for implementing algorithms in quantum computing [85–92].

Another architectural design is the quantum dot, a semiconductor structure that confines electrons in three dimensions, creating discrete energy levels. In quantum computing, quantum dots act as qubits where an electron's spin or energy state holds quantum information. These electron states can be precisely controlled with electric or magnetic fields, making quantum dots highly efficient for quantum operations. They are also valuable in quantum sensing and imaging technologies due to their tunable optical properties [14].

Superconducting qubits are some of the most widely used and scalable architectures in quantum computing as shown in Figure 4.4. These qubits are created from superconducting materials that exhibit zero electrical resistance at extremely low temperatures. Superconducting qubits are manipulated through microwave pulses to change their quantum states and are central to many current quantum computers. Due to their scalability and long coherence times, superconducting qubits are a strong choice for large-scale quantum computing [15].

Another important design is spintronics and spin qubits, which use electron or nuclear spin as the basis for storing quantum information [93–98]. In this architecture, electrons are trapped in quantum dots or other materials, where their spin states—either spin-up or spin-down—represent the 0 and 1 states of a qubit. These spin qubits, with their long coherence times, are particularly useful for quantum memory and some types of quantum

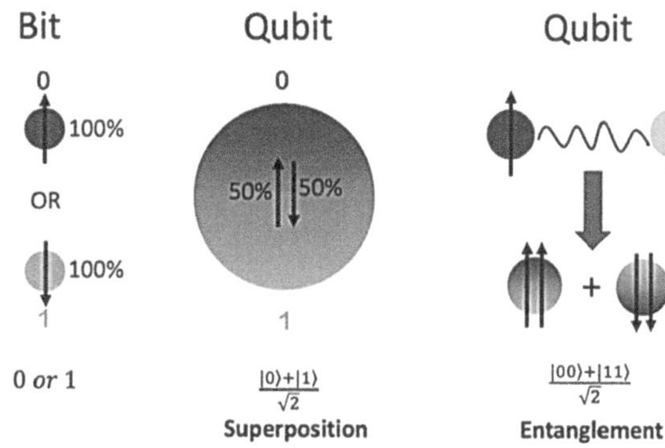

Figure 4.4 Representation of classical bit and Q-bit. [15].

calculations. Quantum gates are applied by manipulating the spin states with external magnetic fields or microwave pulses, allowing for high-fidelity quantum operations. [16]

Finally, hybrid quantum systems combine multiple quantum architectures, such as superconducting qubits and quantum dots, to take advantage of each system's strengths. Hybrid systems allow quantum information transfer between different qubit types, giving these devices greater flexibility and functionality. Hybrid architectures are designed to improve scalability and address the limitations inherent to individual quantum systems [17].

4.4.3 Phenomena involved in quantum devices

The power of quantum devices is rooted in the unique phenomena of quantum mechanics, which enable operations that classical devices cannot replicate. Quantum superposition is one of these phenomena, enabling quantum devices to conduct many operations at once, dramatically boosting computational speed for certain problems. Quantum entanglement creates strong correlations between qubits, crucial for multi-qubit operations and secure quantum communication via QKD, where any attempt to intercept entangled particles can be immediately detected [99–104].

Quantum tunneling also distinguishes quantum devices, allowing particles to move through barriers, a critical feature in RTDs and quantum dots that facilitates high-speed operations. Tunneling is also essential in certain types of quantum gates, allowing for quick qubit manipulation in quantum computing. Yet, quantum devices face challenges like decoherence, which threatens the stability of their operations. To counter this, quantum error correction protocols are employed to detect and correct errors without destroying the quantum state, an essential feature for large-scale quantum systems [18].

4.5 HOW MACHINE LEARNING POWERS QUANTUM DEVICES

As quantum devices grow increasingly complex, machine learning (ML) has become a vital tool, enhancing these systems' efficiency, accuracy, and scalability. ML addresses key challenges that traditional methods struggle to manage, such as quantum error mitigation, circuit optimization, and device calibration. One of the most critical issues in quantum computing is dealing with errors caused by decoherence and environmental noise. Qubits are incredibly sensitive, and even minor interference can introduce errors that disrupt computations. Here, ML offers a solution by analyzing real-time data to identify and correct error patterns before they propagate. By training on quantum data, ML models detect recurring error types and apply corrective measures without directly measuring the qubits, which would otherwise collapse their quantum state. As shown in Figure 4.5, it shows that how machine learning can help in various fields in development of quantum devices and system. This ML-driven approach allows quantum systems to maintain accuracy and perform reliable computations at a larger scale [105–109].

Quantum circuits, consisting of sequences of quantum gates, are the building blocks of quantum algorithms. The efficiency of these algorithms relies heavily on the optimization of these circuits, yet as circuits grow with the scaling of quantum devices, they face challenges of longer execution times and higher error rates. ML, particularly reinforcement learning, dynamically optimizes quantum circuits by adjusting parameters, reducing gate depth, and minimizing errors. Through feedback from the system itself, ML models are able to tune circuits in real time, leading to faster and more accurate execution of quantum algorithms.

Quantum devices also require precise calibration to function correctly, as factors such as gate voltages, external fields, and timing can fluctuate due to environmental conditions. Traditional manual calibration is slow and prone to error, especially as device complexity increases. ML models, including neural networks and reinforcement learning agents, automate these calibrations by learning optimal settings for quantum gate control and qubit adjustments. Real-timc adaptation not only reduces the need for human intervention but also increases the reliability of quantum devices [19].

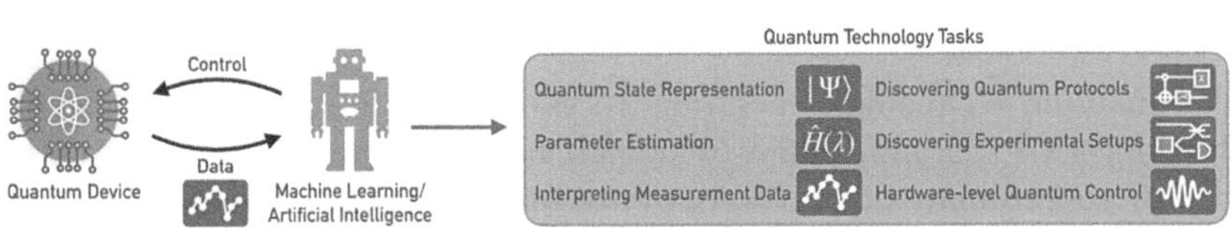

Figure 4.5 Overview of tasks in the area of quantum technologies that machine learning and artificial intelligence can help solve better. [19].

In addition to calibration, ML supports the enhancement of quantum algorithm efficiency. Algorithms like Grover's search and Shor's factoring offer speed advantages over classical algorithms, but they require optimization to minimize errors and maximize resource use. ML techniques, such as variational quantum algorithms (VQAs), combine quantum and classical resources to improve performance on quantum hardware. By using quantum computers to explore large solution spaces and classical optimizers to fine-tune results, ML helps quantum algorithms run with greater accuracy and efficiency [110–117].

Quantum machine learning (QML) is an exciting new field, pushing the boundaries of what machine learning can accomplish with quantum computing. Quantum systems, with their ability to process multiple states simultaneously, offer a significant advantage for tasks such as classification, optimization, and pattern recognition. ML also improves QML algorithms themselves, training them to perform better on noisy quantum hardware. Techniques like quantum neural networks (QNNs), which use quantum circuits to perform machine learning tasks, open the door to potential breakthroughs in image recognition, drug discovery, and materials science [118–123].

In quantum communication, ML contributes significantly to advancements in systems like quantum key distribution (QKD), which is critical for secure data transfer. As the demand for secure communication grows, ML is used to monitor QKD systems for potential threats and to optimize protocols in real time. ML models analyze communication data, detecting patterns that indicate eavesdropping or signal degradation, while dynamically adjusting key distribution rates to preserve the integrity of quantum entanglement over long distances [124, 125].

Finally, scaling quantum systems from small-scale experimental setups to large, practical machines presents unique challenges due to the vast number of qubits and operations involved. Here, ML offers automated solutions to manage interactions between qubits, optimize gate sequences, and design efficient hardware layouts. This automation boosts scalability, enabling quantum systems to perform at a larger and more practical scale [20].

4.6 FUTURE PROSPECTS OF MACHINE LEARNING IN QUANTUM DEVICES

As quantum technologies continue to evolve, the integration of machine learning (ML) is anticipated to unlock unprecedented possibilities. ML's ability to automate processes, optimize operations, and enhance decision-making makes it especially well-suited to managing the complexities inherent in quantum devices. In the future, ML is expected to play a key role in scaling quantum computing, advancing error correction, accelerating algorithm discovery, and expanding into new areas such as hybrid quantum systems and quantum artificial intelligence (AI).

One of the biggest hurdles in quantum computing has always been error correction and fault tolerance, as quantum systems are highly susceptible to errors due to their sensitivity to environmental interference. As quantum systems scale, managing and correcting these errors in real time will become essential. Future ML models, leveraging architectures like deep learning and recurrent neural networks (RNNs), are likely to be instrumental in identifying complex error patterns within quantum data. These models will not only be able to correct errors but also predict potential future errors, allowing for preemptive corrections that prevent errors from spreading. This advancement could make large-scale quantum computations more accurate, enhancing the fault tolerance of quantum devices as they grow in scale. Additionally, techniques like reinforcement learning could be used to dynamically adjust error correction codes to adapt to the unique noise profiles of specific quantum hardware.

As quantum systems scale from small test environments to larger configurations with thousands or even millions of qubits, the complexity of controlling and optimizing these systems grows exponentially. ML will be crucial for optimizing large-scale quantum systems across both hardware and software layers. ML models could help optimize the placement of qubits, adjust quantum gates, and sequence quantum circuits, ensuring that quantum devices achieve maximum efficiency with minimal resources. Approaches such as genetic algorithms and reinforcement learning might assist in designing more efficient circuits that reduce noise, consume less energy, and speed up computation. The future of quantum computing will likely involve hybrid systems, where quantum and classical resources work together. In these cases, ML will play a vital role in managing quantum-classical workloads, balancing tasks between quantum processors and classical systems to optimize overall performance.

In terms of algorithm discovery, ML could revolutionize the way we identify and implement new quantum algorithms. Currently, researchers must manually design and test quantum algorithms, a process that can be time-consuming and complex. In the future, ML models may assist researchers in automating the discovery of new algorithms, particularly for tasks like optimization and simulation that challenge classical methods. Using generative models, such as variational autoencoders (VAEs) and generative adversarial networks (GANs), ML can explore vast algorithmic solution spaces, potentially generating novel algorithms far faster than traditional approaches. Such advancements could lead to breakthroughs in fields such as cryptography, drug discovery, and materials science, where quantum algorithms are expected to offer significant speedups [21].

Machine learning, particularly reinforcement learning (RL), plays a critical role in optimizing quantum circuits for enhanced performance and efficiency. As shown in Figure 4.6, RL algorithms interact with a quantum

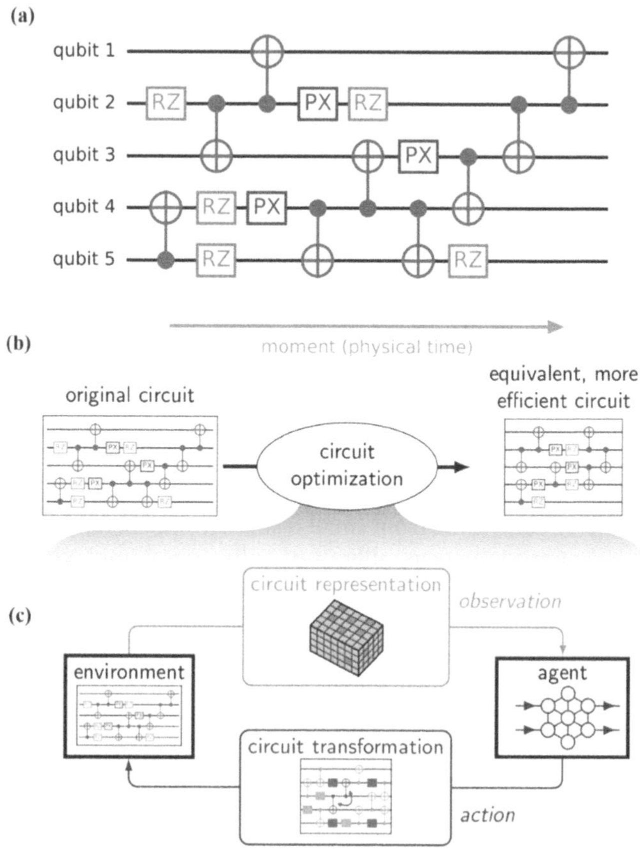

Figure 4.6 Quantum circuit optimization with reinforcement learning [22].

circuit by selecting circuit transformations to reduce complexity and error rates while preserving logical equivalence. This iterative process enables the RL agent to "learn" from each optimization step, continually refining the circuit layout for maximum gate fidelity and minimal computational cost. Such optimization is essential for large-scale quantum computing, where every reduction in gate depth or error translates to more accurate, faster quantum operations. The adaptability of RL in dynamically adjusting circuits makes it a foundational tool for scaling quantum devices. [22]

ML's role in quantum communication is also expected to grow. Systems like quantum key distribution (QKD) rely on quantum mechanics to provide secure channels for data transmission, and ML will help improve the reliability of these systems. By analyzing communication data in real time, ML models can detect patterns that suggest security threats, such as eavesdropping or signal degradation. In the future, ML could even help automate the routing and management of quantum internet infrastructure, ensuring that quantum networks remain resilient to noise and external disruptions [23]. Additionally, ML could assist with quantum repeaters, which are devices that extend the range of quantum communication networks. These repeaters must be continuously optimized to manage entanglement distribution and maintain the quality of quantum information across long distances.

Looking even further ahead, the convergence of ML with quantum technologies could pave the way for hybrid quantum-classical systems that merge classical AI models with quantum computing capabilities. Quantum AI systems may eventually harness both classical and quantum resources to achieve breakthroughs in fields such as medicine, finance, and materials science. In these systems, classical AI models could handle data processing and optimization, while quantum computers perform complex calculations requiring quantum parallelism. This hybrid approach could significantly enhance AI's ability to tackle optimization problems, cryptography, and large-scale simulations. Furthermore, the development of quantum AI algorithms, where quantum circuits are used to train and execute machine learning models, might open up new frontiers in AI, enabling solutions to problems that are currently beyond the reach of classical computer.

As quantum devices advance and ML continues to evolve, the synergy between these two fields promises groundbreaking innovations that will push the limits of computing, communication, and problem-solving capabilities. The role of ML in quantum technologies will undoubtedly expand, bringing us closer to realizing the full potential of quantum systems and driving forward new applications that were once the stuff of science fiction. [24]

4.7 CONCLUSION

The journey toward quantum devices and the integration of machine learning (ML) represents one of the most exciting frontiers in technology today. Classical computing has reached incredible heights, but as semiconductor technology nears its physical limits, new solutions are needed to meet the demand for faster, more efficient, and more powerful systems. Quantum devices—with their unique principles such as superposition, entanglement, and tunneling—hold the potential to revolutionize fields such as computing, communication, and sensing. Unlike classical devices, quantum systems operate on a fundamentally different level, allowing for parallel processing, secure communication, and ultrasensitive measurements that were once thought impossible.

However, realizing the full potential of quantum devices requires overcoming inherent challenges, such as sensitivity to environmental interference and error correction. Here, machine learning steps in as a powerful tool that can optimize, stabilize, and enhance quantum systems at scale. From automated calibration and quantum error mitigation to quantum circuit optimization, ML provides the adaptive intelligence needed to make quantum systems more resilient, scalable, and reliable. Moreover, the role of ML extends beyond simple optimization; it also assists in the discovery of new quantum algorithms, enhancing quantum communication, and enabling hybrid quantum-classical systems that combine the strengths of both classical and quantum computing.

Looking to the future, the integration of ML in quantum devices is expected to accelerate advancements in quantum computing and open new doors in fields such as medicine, finance, materials science, and artificial intelligence (AI). Quantum AI, for example, could eventually tackle problems that today's classical AI systems cannot handle, offering solutions to complex optimization challenges and large-scale simulations.

In essence, the partnership between quantum technology and machine learning is poised to transform the way we approach computation, problem-solving, and data security. As both fields evolve, their convergence will continue to drive innovation, bringing us closer to the future of computing and unlocking new possibilities that stretch beyond our current understanding. The future of quantum and ML-driven technologies looks incredibly promising, with the potential to redefine entire industries and bring about a new era of technological progress.

REFERENCES

[1] M. A. Reed, "Quantum Semiconductor Devices," *1986 Symposium on VLSI Technology. Digest of Technical Papers*, San Diego, CA, 1986, pp. 1–4.

[2] M. Cahaya and S. Bandyopadhyay, "Semiconductor Quantum Devices," in *IEEE Potentials*, vol. 12, no. 1, pp. 18–23, Feb. 1993, Doi: 10.1109/45.207169

[3] G. J. Iafrate and M. A. Stroscio, "Application of Quantum-Based Devices: Trends and Challenges," *in IEEE Transactions on Electron Devices*, vol. 43, no. 10, pp. 1621–1625, Oct. 1996, doi: 10.1109/16.536805

[4] T. Katsuyama, (2012). Development of Various Semiconductor Quantum Devices. *SEI Technical Review*, 74, 24–31.

[5] Xin Zhang & Hai-Ou Li & Gang Cao & Ming Xiao & Guang-Can Guo & Guo-Ping Guo. (2019). Semiconductor Quantum Computation. *National Science Review* 6. 32–54. 10.1093/nsr/nwy153

[6] Stephen Senaya & John Van der Poll & Marthie Schoeman. (2019). Categorization of Enterprise Resource Planning (ERP) Failures: An Opportunity for Formal Methods in Computing. 10.17758/EARES8.EAP1119287

[7] Mario Krenn, Jonas Landgraf, Thomas Foesel, and Florian Marquardt. "Artificial Intelligence and Machine Learning for Quantum Technologies." *Phys. Rev. A* 107, no. 1 (2023): 010101.

[8] P. Louridas and C. Ebert, "Machine Learning," *IEEE Software*, vol. 33, no. 5, pp. 110–115, Sept.-Oct. 2016, doi: 10.1109/MS.2016.114

[9] S. K. Sood and Pooja, "Quantum Computing Review: A Decade of Research," in *IEEE Transactions on Engineering Management*, vol. 71, pp. 6662–6676, 2024, doi: 10.1109/TEM.2023.3284689
[10] M. Cahay, & S. Bandyopadhyay, (1994). Semiconductor Quantum Devices. *Advances in Electronics and Electron Physics* (Vol. 89, pp. 94–245).
[11] R. K. Goyal, (2021). Exploring Quantum Materials & Applications: A review [Journal-article]. *Exploring Quantum Materials & Applications: A Review.*
[12] C. S. Lent and P. D. Tougaw, "A Device Architecture for Computing with Quantum Dots," *Proceedings of the IEEE*, vol. 85, no. 4, pp. 541–557, April 1997, doi: 10.1109/5.573740
[13] B. Xu et al., "Controlled growth of III-V compound semiconductor nano-structures and their application in quantum-devices," *13th International Conference on Semiconducting and Insulating Materials, 2004. SIMC-XIII-2004.*, Beijing, China, 2004, pp. 113–118, doi: 10.1109/SIM.2005.1511398
[14] T. Wu, W. Zhang and J. Guo, "Performance Assessment of Quantum Processor Based on p-Type Semiconductor Quantum Dot Array," 2022 *IEEE International Conference on Quantum Computing and Engineering (QCE)*, Broomfield, CO, 2022, pp. 833–835, doi: 10.1109/QCE53715.2022.00135
[15] B. Xu, "Controlled growth of III-V compound semiconductor nano-structures and their application in quantum-devices," *13th International Conference on Semiconducting and Insulating Materials, 2004. SIMC-XIII-2004.*, Beijing, China, 2004, pp. 113–118, doi: 10.1109/SIM.2005.1511398
[16] W. Li, "High-Power Terahertz Quantum Cascade Lasers Based on High-Al-Composition Four Quantum Wells," *IEEE Photonics Technology Letters*, vol. 34, no. 13, pp. 671–674, 1 July, 2022, doi: 10.1109/LPT.2022.3181242
[17] J. Nichol, "Quantum information processing with semiconductor quantum dots," 2022 *6th IEEE Electron Devices Technology & Manufacturing Conference (EDTM)*, Oita, Japan, 2022, pp. 167–167, doi: 10.1109/EDTM53872.2022.9798200
[18] N. Yang and J. Guo, "A Quantum-Computing-Based Method for Solving Quantum Confinement Problem in Semiconductor," *IEEE Transactions on Electron Devices*, vol. 70, no. 3, pp. 1366–1373, March 2023, doi: 10.1109/TED.2023.3234887
[19] C. J. Chang-Hasnain and S. L. Chuang, "Slow and Fast Light in Semiconductor Quantum-Well and Quantum-Dot Devices," *Journal of Lightwave Technology*, vol. 24, no. 12, pp. 4642–4654, Dec. 2006, doi: 10.1109/JLT.2006.885767
[20] Y. Liu, "Top Grating, Surface-Emitting DFB Quantum Cascade Lasers in Continuous-Wave Operation," *IEEE Photonics Technology Letters*, vol. 27, no. 17, pp. 1829–1832, 1 Sept.1, 2015, doi: 10.1109/LPT.2015.2443780
[21] S. Bauer, "Photonic Nanostructures for Efficient Coupling and Light Enhancement in Semiconductor Quantum Technologies," 2023 *Conference on Lasers and Electro-Optics Europe & European Quantum Electronics Conference (CLEO/Europe-EQEC)*, Munich, Germany, 2023, pp. 1–1, doi: 10.1109/CLEO/Europe-EQEC57999.2023.10232738
[22] A. J. Zilkie, "Carrier Dynamics of Quantum-Dot, Quantum-Dash, and Quantum-Well Semiconductor Optical Amplifiers Operating at 1.55 $\mu {\hbox {m}} $," *IEEE Journal of Quantum Electronics*, vol. 43, no. 11, pp. 982–991, Nov. 2007, doi: 10.1109/JQE.2007.904474

[23] W. Terashima, L. Ying and H. Hirayama, "Fabrication of device structure for terahertz quantum cascade laser based on III-nitride semiconductors," *2009 34th International Conference on Infrared, Millimeter, and Terahertz Waves*, Busan, Korea (South), 2009, pp. 1–2, doi: 10.1109/ICIMW.2009.5325573

[24] Thomas Fösel, "Quantum circuit optimization with deep reinforcement learning." *arXiv preprint arXiv:2103.07585* (2021).

[25] Anjana Bhardwaj, Pradeep Kumar, B. Raj, Sunny Anand, "Design and Performance Optimization of doping-less Vertical Nanowire TFET using Gate Stack Technique", *Journal of Electronic Materials (JEMS) Springer*, Vol. 41, Issue 7, pp. 4005–4013, 2022.

[26] Jeetendra Singh, B. Raj, "Tunnel Current Model of Asymmetric MIM Structure Levying Various Image Forces to Analyze the Characteristics of Filamentary Memristor", *Applied Physics A*, Vol. 125, No. 3, pp. 203.1–203.11 Feb 2019.

[27] Candy Goyal, Jagpal Singh Ubhi and B. Raj, Low Leakage Zero Ground Noise Nanoscale Full Adder using Source Biasing Technique, *Journal of Nanoelectronics and Optoelectronics*, Vol. 14, pp. 360–370, March 2019.

[28] Gurmohan Singh, R. K. Sarin and B. Raj, "A Novel Robust Exclusive-OR Function Implementation in QCA Nanotechnology with Energy Dissipation Analysis", *Journal of Computational Electronics*, Volume 15, Issue 2, PP. 455–465, June 2016.

[29] Girish Wadhwa, Priyanka Kamboj, Balwinder Raj "Design optimisation of junctionless TFET biosensor for high sensitivity" *Advances in Natural Sciences: Nanoscience and Nanotechnology*, Vol. 10, pp. 045001 (7), (2019).

[30] Jeetendra Singh, B. Raj, "An Accurate and Generic Window function for Nonlinear Memristor Model" *Journal of Computational Electronics, Springer*, Volume 18 Issue 2, pp. 640–647 June 2019.

[31] Manjit Kaur; Neena Gupta; Sanjeev Kumar; B Raj; Arun Kumar Singh, "Comparative RF and Crosstalk Analysis of Carbon Based Nano Interconnects" *IET Circuits, Devices & Systems*, Vol.15(6), pp. 493–503, Feb 2021.

[32] Nehru Kandasamy, Firdous Ahmad, D. Ajitha, B. Raj, and Nagarjuna Telagam "Quantum Dot Cellular Automata based Scan Flip Flop and Boundary Scan Register, *IETE Journal of Research*, Volume 66, pp. 535–548, 2020.

[33] Jeetendra Singh, B. Raj, "Design and Investigation of 7T2M NVSARM with Enhanced Stability and Temperature Impact on Store/Restore Energy", *IEEE Transactions on Very Large Scale Integration Systems*, Vol. 27, Issue 6, pp. 1322–1328 June 2019.

[34] Anil Kumar Bhardwaj, Sumeet Gupta, B. Raj, Amandeep Singh, "Impact of Double Gate Geometry on the Performance of Carbon Nanotube Field Effect Transistor Structures for Low Power Digital Design", *Computational and Theoretical Nanoscience, ASP*, Vol. 16, pp. 1813–1820, 2019.

[35] Neeraj Jain, B. Raj, "Thermal Stability Analysis and Performance Exploration of Asymmetrical Dual-k underlap Spacer (ADKUS) SOI FinFET for Security and Privacy Applications", *Indian Journal of Pure & Applied Physics (IJPAP)*, Vol. 57, pp. 352–360, May 2019.

[36] Neeraj Jain and Balwinder Raj, "Dual-k Spacer Region Variation at the Drain Side of Asymmetric SOI FinFET Structure: Performance Analysis towards the Analog/RF Design Applications", communicated in "*Journal of*

Nanoelectronics and Optoelectronics", American Scientific Publishers, Vol. 14, PP. 349–359, March 2019.

[37] Jeetendra Singh, Sanjeev Sharma, B. Raj, Mamta Khosla Analysis of Barrier Layer Thickness on Performance of In1-xGaxAs Based Gate Stack Cylindrical Gate Nanowire MOSFET," *JNO, ASP*, Vol. 13, PP 1473–1477, Oct 2018.

[38] Neeraj Jain, B. Raj, "Analysis and Performance Exploration of High-k SOI FinFETs Over the Conventional Low-k SOI FinFET toward Analog/RF Design", *Journal of Semiconductors (JoS), IOP Science*, Vol. 39, Issue 12, pp. 124002-1-7, Dec 2018.

[39] Candy Goyal, Jagpal Singh Ubhi and B. Raj, A Reliable Leakage Reduction Technique for Approximate Full Adder with Reduced Ground Bounce Noise, *Journal of Mathematical Problems in Engineering, Hindawi*, Volume 2018, Article ID 3501041, 16 pages, 15 Oct 2018.

[40] Anuradha, Jeetendra Singh, B. Raj and Mamta Khosla, "Design and Performance Analysis of Nano-scale Memristor-based Nonvolatile SRAM" *Journal of Sensor Letter*, Vol. 16, pp. 798–805, Oct 2018.

[41] Girish Wadhwa, B. Raj, "Parametric Variation Analysis of Charge-Plasma-based Dielectric Modulated JLTFET for Biosensor Application" *IEEE Sensor Journal*, Vol. 18, No. 15, August 1, pp. 6070 – 6077, 2018 (SCI)

[42] Jeetendra Singh, B. Raj, "Comparative Analysis of Memristor Models for Memories Design", *JoS, IoP*, Vol. 39, No. 7, pp. 074006-1-12, July 2018.

[43] Divya Yadav, Shailesh Singh Chouhan, Santosh Kumar Vishvakarma and B. Raj, "Application Specific Microcontroller Design for IoT based WSN", *Sensor Letter, ASP*, Vol. 16, pp. 374–385, May 2018.

[44] Gurmohan Singh, R. K. Sarin and B. Raj, "Fault-Tolerant Design and Analysis of Quantum-Dot Cellular Automata Based Circuits", *IEEE/IET Circuits, Devices & Systems*, Vol. 12, pp. 638–664, 2018.

[45] Jeetendra Singh, B. Raj, "Modeling of Mean Barrier Height Levying Various Image Forces of Metal Insulator Metal Structure to Enhance the Performance of Conductive Filament Based Memristor Model", *IEEE Nanotechnology*, Vol. 17, No. 2, pp. 267–268, March 2018.

[46] Aakash Jain, Sanjeev Sharma, B. Raj, "Analysis of Triple Metal Surrounding Gate (TM-SG) III-V Nanowire MOSFET for Photosensing Application", *Opto-electronics Journal*, Volume 26, Issue 2, Pages 141–148, May 2018.

[47] Aakash Jain, Sanjeev Sharma, B. Raj, "Design and Analysis of High Sensitivity Photosensor Using Cylindrical Surrounding Gate MOSFET For Low Power Sensor Applications", *Engineering Science and Technology, an International Journal*, Volume 19, Issue 4, Pages 1864–1870, December 2016.

[48] G. Saiphani Kumar, Amandeep Singh, B. Raj, "Design and Analysis of Gate All Around CNTFET based SRAM cell Design", *Journal of Computational Electronics*, Volume 17, Issue 1, pp 138–145, March 2018.

[49] Gurinder Pal Singh, B. S. Sohi, Balwinder Raj, "Material Properties Analysis of Graphene Base Transistor (GBT) for VLSI Analog Circuits, *Indian Journal of Pure & Applied Physics (IJPAP)*, Vol. 55, pp. 896–902, Dec-2017.

[50] Sunil Kumar and B. Raj, "Estimation of Stability and Performance metric for Inward Access Transistor based 6T SRAM Cell Design using n-type/p-type DMDG-GDOV TFET" *IEEE VLSI Circuits and Systems Letter*, Volume 3, Issue 2, pp. 25–39, June 2017

[51] Shashikant Sharma, Anjan Kumar, Manisha Pattanaik and B. Raj, "Forward Body Biased Multimode Multi-Threshold CMOS Technique for Ground Bounce Noise Reduction in Static CMOS Adders" *International Journal of Information and Electronics Engineering*, pp. 567–572, Vol. 3, issue 3, 2013.
[52] Hamendra Singh, Pankaj Kumar, Balwinder Raj, "Performance Analysis of Majority Gate SET Based 1-bit Full Adder", *International Journal of Computer and Communication Engineering (IJCCE)*, ISSN: 2010-3743, Volume 2, No. 4, 2013.
[53] Anil Kumar Bhardwaj, Sumeet Gupta and B. Raj, "Investigation of Parameters for Schottky Barrier (SB) Height for Schottky Barrier Based Carbon Nanotube Field Effect Transistor Device", *Journal of Nanoelectronics and Optoelectronics*, Vol. 15, pp. 783–791, July 2020.
[54] Priya Bansal, B. Raj, "MEMRISTOR: A Versatile Nonlinear Model for Dopant Drift and Boundary Issues," *JCTN, American Scientific Publishers*, Vol. 14 (5), pp. 2319–2325, May 2017.
[55] Neeraj Jain, B. Raj, "An Analog and Digital Design Perspective Comprehensive Approach on Fin-FET (Fin-Field Effect transistor) Technology - A Review", *Reviews in Advanced Sciences and Engineering (RASE)*, Vol. 5, pp. 1–14, 2016.
[56] Balwinder Raj, A. K. Saxena and S. Dasgupta, "Analytical Modeling for the Estimation of Leakage Current and Subthreshold Swing Factor of Nanoscale Double Gate FinFET Device" *Microelectronics International*, vol. 26, pp. 53–63, 2009.
[57] Shailendra Singh Soniya, Girish Wadhwa and Balwinder Raj, "An Analytical Modeling for Dual Source Vertical Tunnel Field Effect Transistor" submitted to '*International Journal of Recent Technology and Engineering (IJRTE)*' at Volume 8 Issue 2, pp. 603–608, July 2019.
[58] Shailendra Singh, B. Raj, "Design and Analysis of Hetrojunction Vertical T-shaped Tunnel Field Effect Transistor", *Journal of Electronics Material, Springer*, Volume 48, Issue 10, pp 6253–6260, October 2019.
[59] Candy Goyal, Jagpal Singh Ubhi and B. Raj, "A Low Leakage CNTFET based Inexact Full Adder for Low Power Image Processing Applications", *International Journal of Circuit Theory and Applications, Wiley*, Volume 47, Issue 9, Pages 1446–1458, September 2019.
[60] B. Raj, A. K. Saxena and S. Dasgupta, "A Compact Drain Current and Threshold Voltage Quantum Mechanical Analytical Modeling for FinFETs" *Journal of Nanoelectronics and Optoelectronics (JNO)*, vol. 3, no. 2, pp. 163–170, 2008.
[61] Girish Wadhwa, B. Raj, "An Analytical Modeling of Charge Plasma based Tunnel Field Effect Transistor with Impacts of Gate underlap Region" *Superlattices and Microstructures*, Vol. 142, pp. 106512, June 2020.
[62] Shailendra Singh, B. Raj, "Modeling and Simulation analysis of SiGe hetrojunction Double GateVertical t-shaped Tunnel FET", *Superlattices and Microstructures* Volume 142, pp. 106496, June 2020.
[63] Neeraj Jain, Balwinder Raj, "Device and Circuit Co-Design Perspective Comprehensive Approach on FinFET Technology - A Review, *Journal of Electron Devices*, Vol. 23, 1, pp 1890–1901, 2016.
[64] Sunil Kumar and B. Raj, "Analysis of I_{ON} and Ambipolar Current for Dual-Material Gate-drain Overlapped DG-TFET," *Journal of Nanoelectronics and Optoelectronics*, Vol.11, PP. 323–333, June 2016.

[65] Naveed Anjum, Tarun Bali and B. Raj, Design And Simulation Of Handwritten Multiscript Character Recognition, *International Journal of Advanced Research in Computer and Communication Engineering*, Vol. 2 (7), pp. 2544–2549, July 2013.

[66] Karmjit Singh, B. Raj, "Performance and Analysis of Temperature Dependent Multi-Walled Carbon Nanotubes as Global Interconnects at Different Technology Nodes," *Journal of Computational Electronics, Springer*, Volume 14 Issue 2, pp. 469–476, June 2015.

[67] Sunil Kumar and B. Raj, "Compact Channel Potential Analytical Modeling of DG-TFET Based on Evanescent–Mode Approach," *Journal of Computational Electronics* Vol. 14, No. 2, pp.820–827, July 2015.

[68] Karmjit Singh, B. Raj, "Temperature Dependent Modeling and Performance Evaluation of Multi-Walled CNT and Single-Walled CNT as Global Interconnects," *Journal of Electronic Materials*, Vol. 44, Issue 12, pp 4825–4835, Dec 2015.

[69] V. K. Sharma, M. Pattanaik and B. Raj, "INDEP Approach for Leakage Reduction in Nanoscale CMOS Circuits", *International Journal of Electronics*, Vol. 102, No. 2, pp. 200–215, 2014.

[70] Karmjit Singh, B. Raj, "Influence of Temperature on MWCNT bundle, SWCNT bundle and Copper interconnects for Nanoscaled Technology nodes," *Journal of Materials Science: Materials in Electronics*, Vol. 26, No. 8, pp. 6134–6142, 2015.

[71] Naveed Anjum, Tarun Bali and B. Raj, "Design and Simulation of Handwritten Gurumukhi and Devanagri Numerical Recognition, *International Journal of Computer Applications*, Vol. 73(12), pp. 16–21, 2013.

[72] S. Khandelwal, V. Gupta, B. Raj, and R. D. Gupta, Process Variability Aware Low Leakage Reliable Nano Scale DG-FinFET SRAM Cell Design Technique, *Journal of Nanoelectronics and Optoelectronics*, Volume 10, Number 6, pp. 810–817, December 2015.

[73] V. K. Sharma, M. Pattanaik and B. Raj, "ONOFIC Approach: Low Power High Speed Nanoscale VLSI Circuits Design", *International Journal of Electronics*, Vol. 101, issue 1, pp. 61–73, 2014.

[74] S. Khandelwal, Balwinder Raj, and R. D. Gupta, FinFET based 6T SRAM Cell Design: Analysis of Performance Metric, Process Variation and Temperature Effect. *Journal of Computational and Theoretical Nanoscience, ASP, USA*, Vol. 12, pp. 2500–2506, 2015.

[75] Sumit Singh; Shekhar Yadav; Jagdeep Rahul; Anurag Srivastava; B. Raj, Impact of HfO_2 in Graded Channel Dual Insulator Double Gate MOSFET, *Journal of Computational and Theoretical Nanoscience*, Volume 12, Number 6, pp. 950–953, April 2015.

[76] Vijay Kumar Sharma, Manisha Pattanaik, B. Raj, PVT Variations Aware Low Leakage INDEP Approach for Nanoscale CMOS Circuits, *Microelectronics Reliability, Elsevier*, Vol. 54, pp 90–99, 2014.

[77] B. Raj, A. K. Saxena and S. Dasgupta, "Quantum Mechanical Analytical Modeling of Nanoscale DG FinFET: Evaluation of Potential, Threshold Voltage and Source/Drain Resistance", *Elsevier's Journal of Material Science in Semiconductor Processing*, Vol. 16, issue 4, pp. 1131–1137, 2013.

[78] Maisagalla Gopal, Siva Sankar, D. Prasad and Balwinder Raj, 8T SRAM Cell Design for Dynamic and Leakage Power Reduction, *International Journal of Computer Applications*, Vol. 71(9), pp. 43–48, June 2013.

[79] Manisha Pattanaik, B. Raj, Shashikant Sharma and Anjan Kumar, "Diode Based Trimode Multi-Threshold CMOS Technique for Ground Bounce Noise Reduction in Static CMOS Adders", *Advanced Materials Research*, vol. 548, pp 885–889, 2012.
[80] Balwinder Raj, A. K. Saxena and S. Dasgupta, "Nanoscale FinFET Based SRAM Cell Design: Analysis of Performance metric, Process variation, Underlapped FinFET and Temperature effect" *IEEE Circuits and System Magazine*, vol. 11, issue 2, pp. 38–50, 2011.
[81] V. K. Sharma, M. Pattanaik & Balwinder Raj, "Leakage Current ONOFIC Approach for Deep Submicron VLSI Circuit Design", *International Journal of Electrical, Computer, Electronics and Communication Engineering*, Vol. 7, issue 4, pp. 239–244, 2013.
[82] Parminder Kaur, Sandeep Singh Gill and B. Raj, "Comparative Analysis of OFETs Materials and Devices for Sensor Applications", *Journal of Silicon*, Vol 14, pp. 4463–4471, 2022.
[83] Sanjeev Kumar Sharma, Parveen Kumar, Balwant Raj and B. Raj, "$In_{1-x}Ga_xAs$ Double Metal Gate-Stacking Cylindrical Nanowire MOSFET for highly sensitive Photo detector", *Journal of Silicon* Vol.14, PP. 3535–3541, 2022.
[84] B. Raj, A. K. Saxena and S. Dasgupta, "Analytical Modeling of Quasi Planar Nanoscale Double Gate FinFET with Source/Drain Resistance and field dependent carrier mobility: A quantum mechanical study" *Journal of Computer (JCP)*, vol. 4, issue 9, PP.1–8, 2009.
[85] S. Bhushan, S. Khandelwal, and B. Raj, (2012). Analyzing Different Mode FinFET based Memory Cell at Different Power Supply for Leakage Reduction, *Seventh International Conference on Bio-Inspired Computing: Theories and Application, (BIC-TA 2012) Advances in Intelligent Systems and Computing*, Volume 202, 89–100.
[86] Jeetendra Singh, B. Raj, "Temperature Dependent Analytical Modeling and Simulations of Nanoscale Memristor, *Journal: Engineering Science and Technology, an International Journal*, Vol. 21, pp. 862–868, Oct 2018.
[87] Shradhya Singh, Shashi Bala, Balwant Raj, B. Raj, "Improved Sensitivity of Dielectric Modulated Junctionless Transistor for Nanoscale Biosensor Design", *Sensor Letter, ASP*, Vol. 18, pp. 328–333, Apl 2020.
[88] Vivek Kumar, Santosh Kumar Vishvakarma, and B. Raj, "Design and Performance Analysis of ASIC for IoT Applications" *Sensor Letter ASP*, Vol. 18, PP. 31–38, Jan 2020.
[89] Akanksha Jaiswal, R. K. Sarin, B. Raj and Shikha Sukhija, "A Novel Circular Slotted Microstrip-fed Patch Antenna with Three Triangle Shape Defected Ground Structure for Multiband Applications", *Advanced Electromagnetic (AEM)*, Vol. 7, Issue 3, PP. 56–63, Aug 2018.
[90] Girish Wadhwa, B. Raj, "Label Free Detection of Biomolecules using Charge-Plasma-Based Gate Underlap Dielectric Modulated Junctionless TFET" *Journal of Electronic Materials (JEMS)*, Volume 47, Issue 8, pp. 4683–4693, August 2018.
[91] Gurmohan Singh, R. K. Sarin and B. Raj, "Design and Performance Analysis of a New Efficient Coplanar Quantum-Dot Cellular Automata Adder", *Indian Journal of Pure & Applied Physics (IJPAP)*, Vol. 55, pp. 97–103, February 2017.

[92] Parminder Kaur, Vikas Pandey and B. Raj, “Comparative Study of Efficient Design, Control and Monitoring of Solar Power using IoT”, *Sensor Letter, ASP* Vol. 18, PP. 419–426, May 2020.
[93] Anil Kumar Bhardwaj, Sumeet Gupta and B. Raj, Development & Analysis of compact model for Double Gate Schottky Barrier CNTFET, *Journal of Nanoelectronics and Optoelectronics, ASP*, Vol. 15, pp. 1199–1208, Aug 2020.
[94] Girish Wadhwa, Priyanka Kamboj; Jeetendra Singh, B. Raj, “Design and Investigation of Junctionless DGTFET for Biological Molecule Recognition” *Transactions on Electrical and Electronic Materials, Springer*, Vol.22, pp. 282–289, 2021.
[95] Sachin Kumar Verma, Shailendra Singh, Girish Wadhwa and B. Raj, “Detection of Biomolecules using Charge-Plasma based Gate Underlap Dielectric Modulated Dopingless TFET” *Transactions on Electrical and Electronic Materials (TEEM)*, Vol. 21, pp. 528–535, June 2020.
[96] Neeraj Jain, B. Raj, “Impact of Underlap Spacer Region Variation on Electrostatic and Analog/RF Performance of Symmetrical High-K SOI FinFET at 20 nm Channel Length”, *Journal of Semiconductors (JoS)*, Vol. 38, Issue 12, pp. 122002, Dec 2017.
[97] Shailendra Singh, B. Raj,” Analytical Modeling and Simulation Analysis of T-Shaped III-V Heterojunction Vertical T-FET”, *Superlattices and Microstructures*, Vol. 147, pp. 106717, Nov 2020.
[98] Gurmohan Singh, R. K. Sarin and B. Raj, “Design and Analysis of Area Efficient QCA Based Reversible Logic Gates”, *Journal of Microprocessors and Microsystems*, Vol. 52, pp. 59–68, May 2017.
[99] Shailendra Singh, B. Raj, Parametric Variation Analysis on Hetero-Junction Vertical t-Shape TFET for Supressing Ambipolar Conduction, *Indian Journal of Pure and Applied Physics*, Vol. 58,pp. 478–485 June 2020.
[100] Shailendra Singh Soniya, Girish Wadhwa and Balwinder Raj, Design and Analysis of Dual Source Vertical Tunnel Field Effect Transistor for high performance, *Transactions on Electrical and Electronics Materials*, Vol. 21, pp. 74–82, Oct 2019.
[101] Manjit Kaur; Neena Gupta; Sanjeev Kumar; B. Raj; Arun Kumar Singh, “RF Performance Analysis of Intercalated Graphene Nanoribbon Based Global Level Interconnects” *Journal of Computational Electronics*, Vol. 19, pp. 1002–1013, June 2020
[102] Girish Wadhwa, B. Raj, “Design and Performance Analysis of Junctionless TFET Biosensor for high sensitivity” *IEEE Nanotechnology*, Vol.18, pp. 567 - 574, 2019.
[103] Jeetendra Singh, B. Raj, Enhanced Nonlinear Memristor Model Encapsulating Stochastic Dopant Drift, *JNO, ASP* 14, 958–963 (2019)
[104] Mohit Kumar, Tarun Chaudhary; Balwinder Raj, “Design and Comparative Analysis of Ferroelectric Nanowire with Dielectric HfO_2 and $Al_2O_3$for Low Power Applications”, *Journal of Electronic Materials*, Vol. 53, pp. 6785–6791, 2024
[105] Yogesh Thakur, Balwant Raj and Balwinder Raj, “Design of Pentacene Thin-Film Transistor Based Hydrogen Gas Sensor for High Sensitivity”, *ECS Journal of Solid State Science and Technology*, Vol. 13, pp. 047005–047012, 24 Apl 2024

[106] Manoharan Arun Kumar, Kamal Batcha Mohamed Ismail, Shanmugam Mahalingam, Balwinder Raj, Junghwan Kim, Raji Atchudan, "Recent Advances in Two-Dimensional Nanomaterials and Nanostructures for Healthcare Monitoring", *ACS Sensors*, Vol.9, issue 4, PP-1–8, 2 April 2024
[107] Mohamed Ismail Kamal Batcha; Arun Kumar Manoharan; Shanmugam Mahalingam; Raj Balwinder, "Carbon Fiber-Reinforced Polymers for Energy Storage Applications", *Journal of Energy Storage*, Vol. 84, Part B, pp. 110931–110939, 20 April 2024.
[108] Shahraan Hussain, Balwinder Raj, Divya Sharma "Design and Parametric Optimization of Lead Free Perovskite based Tandem Solar Cell for High Efficient", *Engineering Research Express*, Vol. 6, pp. 015315–0153122, 13 March 2024.
[109] Parveen Kumar, Balwinder Raj, Girish Wadhwa, Balwinder Singh and Raj Kumar "Design and Analysis of Junctionless based Gate All Around N+ Doped Layer Nanowire TFET Biosensor", *ECS Journal of Solid State Science and Technology*, Vol 13, PP 017002–017010, 15 Jan 2024
[110] Divya Sharma, Rajesh Mehra and Balwinder Raj, "Enhancement in Efficiency of Methyl Ammonium Tin Iodide Based Perovskite Solar Cell Using SCAPS-1 D", *Nano*,, Vol. 18, No. 14, PP. 2350109, 14 Dec 2023.
[111] Parveen Kumar, Sanjeev Kumar Sharma, Balwant Raj and Balwinder Raj, "Analysis of Device Parameter Variations in In1-xGaxAs based Gate Stacked Double Metal Surrounding Gate Nanowire MOSFET", *TEEM*, Vol. 24, pp. 570–578, Oct 2023
[112] Anjana Bhardwaj, Pradeep Kumar, Balwinder Raj, Naveen Kumar, Sunny Anand, Design and Optimization of Vertical Nanowire Tunnel FET with Electrostatic Doping, *Engineering Research Express*, Vol.5, PP. 045025–045033, 2023
[113] Yogesh Thakur, Balwinder Raj, Sandeep Singh Gill, "Design and Performance Analysis of Pentacene Organic Field Effect Transistor with High-K dielectric Materials, *Journal of Optoelectronics and Advanced Materials – Rapid Communications*, Vol. 17, No. 7–8, July-August 2023.
[114] Parveen Kumar and Balwinder Raj, "Design and Simulation of Junctionless Nanowire Tunnel Field Effect Transistor Biosensor for High Sensitivity", *Microelectronic Journal*, Vol. 137, PP. 105826–105834, July 2023.
[115] Priya Bansal, B. Raj, "Memristor Modeling and Analysis for Linear Dopant Drift Kinetics," *Journal of Nanoengineering and Nanomanufacturing*, Vol. 6, pp. 1–7, 2016.
[116] Ashima, D. Vaithiyanathan, B. Raj, "Performance Analysis of Charge Plasma induced Graded Channel Si nanotube", *Journal of Engineering Research (JER)*, EMSME Special Issue pp. 146–154 Aug 2021.
[117] Abhishek Singh Tomar, Vijay Kumar Magraiya, and B. Raj, Scaling of Access and Data Transistor for High Performance DRAM Cell Design, *Quantum Matter*, Vol. 2, pp. 412–416 Oct. 2013.
[118] Neeraj Jain, B. Raj, Parasitic Capacitance and Resistance Model Development and Optimization of Raised Source/Drain SOI FinFET Structure for Analog Circuit Applications, *Journal of Nanoelectronics and Optoelectronins*, Vol. 13, pp. 531–539, 2018.

[119] Shradhya Singh, S. K. Vishvakarma and B. Raj, "Analytical Modeling of Split-Gate Junction-Less Transistor for a Biosensor Application", *Sensing and Bio-sensing*, Vol. 18, pp. 31–36, April 2018.

[120] Maisagalla Gopal, and Balwinder Raj, Low Power 8T SRAM Cell Design for High Stability Video Applications, *ITSI Transaction on Electrical and Electronics Engineering*, Vol. 1(5), pp. 91–97, 2013.

[121] Balwinder Raj, Jatin Mitra, Deepak Kumar Bihani, V. Rangharajan, A. K. Saxena and S. Dasgupta, "Analysis of Noise Margin, Power and Process Variation for 32 nm FinFET Based 6T SRAM Cell" *Journal of Computer (JCP)*, vol. 5, issue 6, pp. 1–8, 2010.

[122] Divya Sharma, Rajesh Mehra and B. Raj, "Comparative Analysis of Photovoltaic Technologies for High Efficiency Solar Cell Design", *Superlattices and Microstructures*, Volume 153, pp. 106861, May 2021.

[123] Pawandeep Kaur, Avtar Singh Buttar, B. Raj, "A comprehensive Analysis of Nanoscale Transistor based Biosensor: A Review, *Indian Journal of Pure and Applied Physics*, Vol. 59, pp. 304–318, April 2021.

[124] Divya Yadav, Balwant Raj and B. Raj, "Design and Simulation of Low Power Microcontroller for IoT Applications", *Journal of Sensor Letters*, Vol. 18, pp. 401–409, May 2020.

[125] Shailendra Singh, B. Raj, "A 2-D Analytical Surface Potential and Drain current Modeling of Double-Gate Vertical t-shaped Tunnel FET", *Journal of Computational Electronics*, Vol. 19, pp. 1154–1163, 2020.

Chapter 5

Nanowire-based biosensors

Unveiling growth to applications

Mohit Kumar Srivastava, Preeti Agarwal Mittal, Shailendra Singh, Vivek Kumar, Arti Saxena, and Man Mohan Shukla

5.1 INTRODUCTION

The further developments of nanotechnology catalyze critical innovations in other fields, especially in biosensing. Further investigation into the anomalous properties of nanomaterials, more specifically nanowires, hastened the advancement of highly sensitive and efficient biosensors. Such nanoscale materials, on account of a structure with an enormously high ratio of surface-to-volume, find high utility to interact with targeted analytes as applied in monitoring real-time events in different settings. This nanowire-based biosensor novel represents a critical step forward toward multifunctional devices for rapid detection. Such applications address critical healthcare, environmental, and food security needs effectively. Wearable technology will be a promising area of future opportunities in personalized medicine and proactive health management. This introduction will set the context for analyzing growth trajectories and practical implications of nanowire-based biosensors, which can help to establish their transformative potential in current research and industry [1–9].

5.1.1 Definition of nanowire-based biosensors

The further developments of nanotechnology catalyze critical innovations in other fields, especially in biosensing. Further investigation into the anomalous properties of nanomaterials, more specifically nanowires, hastened the advancement of highly sensitive and efficient biosensors. Such nanoscale materials, on account of a structure with an enormously high ratio of surface-to-volume, find high utility to interact with targeted analytes as applied in monitoring real-time events in different settings [10–18]. The novelty of this nanowire-based biosensor marks a crucial step forward in rapid detection toward multifunctional devices. These applications address some critical healthcare, environmental, and food security needs efficiently. Wearable technology is an area where promising future opportunities can be realized with personalized medicine and proactive health management. This introduction will serve as a premise for analyzing growth trajectories and practical

DOI: 10.1201/9781003634645-5

implications of nanowire-based biosensors, which would help to define their transformative potential in current research and industry [19–28].

5.1.2 Importance of biosensors in modern science

Improvements in biosensor technology have transformed the mode of diagnostics and environmental monitoring in modern science, which marks an important change. Biosensors that use nanowires demonstrate exceptional sensitivity and specificity that enables the identification of disease-related biomarkers such as cancer. Silicon nanochannel field-effect transistor (FET) biosensors exhibit single molecule detection capabilities, which enable them to provide essential diagnostic data for breast cancer patient monitoring [3]. Scientists pursue carbon-nanotube-based biosensors because they want to enhance melanoma detection accuracy while reducing both costs and patient invasiveness [4]. Biosensors demonstrate essential value in both medical research and clinical operations through their ability to solve problems within dynamic scientific environments [29–37].

5.1.3 Overview of nanowire technology

Research continues to yield that nanowires are to be applied to biosensor technology. This application of such miniaturized structures offers very apt detection of biological signals with extreme sensitivity and precision, considering their extraordinary electrical, optical, and mechanical properties. For example, by using nanowires in electrochemical sensors, the sensors will have to work together with the desired biomolecules; this enhances the detection process with faster response times [38–46]. They also provide flexibility that would allow new designs to be incorporated into other fields of biosensing, including checks on the environment and health diagnostics. New developments in the creation of nanowire sensors seem promising for real-time monitoring, which is crucial for immediate medical assistance. As mentioned above, "nanotools, such as nanobiosensors, facilitate the development of high-tech agricultural farms," which indicates their potential use in various fields [5]. Moreover, recent research focuses on the development of flexible sensors for various environmental conditions [6].

5.1.4 Historical context of biosensor development

Decades of evolution of biosensors have seen a radical shift in biomedical diagnostics and environmental monitoring, brought about by technological progress and the increased demand for more efficient detection methods [47–56]. The designs of early biosensors were basically based on the traditional electrode system, but that changed with the introduction of nanomaterials, which further enhanced their sensitivity and specificity. With the emergence of materials such as graphene quantum dots, rapid, economical, and highly accurate immunosensing techniques became possible, which opened the way for their application in various diseases, including cancer and cardiovascular disorders [7]. Moreover, with the advancements in nanotechnology, these systems were integrated into portable devices to meet the demands of point-of-care testing [8]. Such developments are not only telling the history of biosensors but also underlining the potential of such devices in changing the face of healthcare diagnostics and monitoring in the face of increasingly digital health.

5.1.5 Purpose and scope of the chapter

In the age of rapid technological breakthroughs, research into nanowire-based biosensors has taken center stage. This paper explores the synthesis, functionality, and practical applications of these biosensors, underlining their transformative impact in various areas from healthcare to environmental monitoring [57–65]. The objective is to describe how nanowires, as critical components, improve the sensitivity and specificity of biosensing technologies to overcome the existing limitations in traditional sensor methodologies. Discussion will be extended to commercialization challenges of such advanced technologies and suggest solutions to integrate them into market applications. As noted in latest analyses, for the guidance of research efforts for sustainable growth of the sector, concise roadmaps should be created to guide ([9]). This paper aims to join the larger-scale discourse regarding what the implications or future directions in nanotechnology toward biosensing will be [10].

5.2 FUNDAMENTALS OF NANOWIRE TECHNOLOGY

Advances in nanowire technology have catalyzed significant growth in the field of biosensing, allowing innovative approaches to the detection of biomolecules with unmatched sensitivity [66–75]. Through the unique properties of nanowires, researchers have been able to fabricate sensors that exhibit improved performance metrics such as rapid response times and low detection limits. These properties are essential in several applications, especially in diagnostics, where early disease detection may significantly improve patient outcome. Additionally, the combination of different materials, such as metals and conducting polymers, has allowed for the creation of nanowire sensors that can be designed to respond to specific targets, as seen in the electrochemical growth of metal nanowires in laboratory experiments [13]. The combination of nanotechnology with biosensing not only enhances detection capabilities but also supports the shift toward portable, point-of-care devices, addressing the exigent need for efficient health monitoring solutions [12].

5.2.1 Definition and characteristics of nanowires

Nanotechnology development in nanowires is highly regarded, yielding substantial advancement, which results from miniaturization of electronic components. The diameter is on a nanoscale while having a highly extended length; this has exhibited impressive electrical, thermal, and optical properties distinct from the corresponding bulk material. Most notably, high surface-to-volume ratio means higher reactivity and hence highly desirable for use in biosensing technologies, where sensitivity and specificity are important factors [76–88]. This feature allows for the incorporation of nanowires into field-effect transistor (FET) biosensors to detect biomolecules at ultra-low concentrations, an essential requirement for cancer diagnostics [14]. Moreover, the flexibility in their composition—ranging from metals to semiconductors—enables tailoring of their functionality, making them suitable for continuous health monitoring systems, which require long-term biocompatibility and reliability [26]. Nanowires, therefore, are a critical part of the development of biosensor technology [89–97].

5.2.2 Methods of nanowire synthesis

The versatility of these methods has led to significant advances in development principles, making it possible to customize synthesis methods for specific applications. The predominant methods to synthesize nanowires include chemical vapor deposition (CVD), electrospinning, and template-generation methods that each circumvent some capabilities in controlling aspects of morphology and size. CVD enabled the precise growth of silver nanowires that exhibited superior physical and optical properties compared with bulk materials, thereby increasing the utility of nanowires in biosensing applications [15]. The electrochemical production methods rely on the fundamental tenets of electrochemistry to establish functional sensors that detect biological interactions predicated on advances in electrochemical impedance spectroscopy (EIS) [16]. Syntheses methods create advances to the performance of nanowire-based devices enabling progressively broader multi-field application [98–107].

5.2.3 Material types used in nanowire fabrication

Diverse materials are of prime importance in the fabrication of nanowires, as their physical and chemical properties directly influence the efficacy of nanowire-based biosensors. Among the materials used, silicon stands out because of its excellent electrical properties and compatibility with existing semiconductor technologies, which enable high sensitivity in biosensing applications. Other than that, metal nanowires, particularly silver and gold, have drawn interest due to their unique conducting properties and their ability to induce localized surface plasmon resonance, which enables detection

[108–115]. Several nanostructured materials, including carbon nanotubes and graphene, have been discovered to possess mechanical strength and electrical conductivity, hence suitable for use in flexible high-performance biosensors. Table 5.1 shows different materials used in nanowire fabrication. The synergy of these materials not only broadens the scope of applications but also allows for the development of multifunctional sensors capable of detecting a wide array of clinically relevant analytes, underscoring their importance in the realm of biomedical diagnostics [17–19].

5.2.4 Properties that enhance biosensor performance

Nanowire-based biosensors exhibit greatly enhanced utility due to the inherent properties of nanomaterials that contribute to their heightened sensitivity and specificity in detecting biological analytes. The high surface-to-volume ratios characteristic of nanowires enable efficient interactions with target molecules [116–125]. This further promotes single molecule detection, which plays a critical role in clinical diagnostics. Furthermore, the versatility of introducing various nanomaterials into biosensing platforms allows for a synergistic effect that can be used in conjunction with the unique properties of each material to expand the range of detectable analytes, including nucleic acids and proteins. Table 5.2 summarizes different

Table 5.1 Materials used in nanowire fabrication

Material	*Properties*	*Applications*	*Year introduced*
Silicon	Excellent electronic properties, abundant and cost-effective.	Used in semiconductor devices and sensors.	1970
Gold	High conductivity and biocompatibility.	Utilized in biomedical sensors and diagnostics.	2000
Zinc Oxide (ZnO)	Wide bandgap semiconductor with piezoelectric properties.	Applied in UV light sensors and biosensors.	2005
Silver	Excellent antimicrobial properties and electrical conductivity.	Used in biosensing applications and coatings.	2010
Graphene	High electrical conductivity and mechanical strength.	Utilized in advanced biosensing and electronic applications.	2013

Table 5.2 Properties enhancing biosensor performance

Property	*Description*	*Impact*	*Source*
Sensitivity	Ability to detect low concentrations of biological analytes	Higher sensitivity allows for earlier detection of diseases.	Journal of Nanobiotechnology, 2022
Selectivity	The ability to distinguish target analytes from nontargets.	Increased selectivity reduces false positives and negatives.	Sensors and Actuators B: Chemical, 2023
Stability	Long-term performance consistency of the biosensor.	Enhanced stability leads to reliable performance in real-world applications.	Analytical Chemistry, 2023
Response Time	The time taken to generate a measurable response after exposure to the analyte.	Shorter response times improve the usability of biosensors in critical scenarios.	Biosensors and Bioelectronics, 2022
Biocompatibility	Compatibility of the biosensor materials with biological systems.	Higher biocompatibility is crucial for in vivo applications.	Advanced Healthcare Materials, 2023

properties of enhancing biosensor performance. Recent advances in silicon nanochannel field-effect transistor biosensors are of great promise for outstanding performance, owing to their small dimensions and fast fabrication processes, which enhance throughput and reduce costs, thus paving the way for broad clinical utility [20]. Collectively, these characteristics mark a giant leap in biosensor design that could lead to the development of new applications in health care [21].

5.2.5 Challenges in nanowire production

The complicated environment of nanowire production has several major issues that impede the progress of nanowire-based biosensors, particularly maintaining control over the synthesis process to manufacture nanowires of consistent diameter, length, and composition. Consistent characteristics will benefit sensor performance, as variation in these standards will impact the transduction capabilities of biosensors, ultimately limiting sensitivity and accuracy [22]. In addition to precise control over the production of uniform nanowires, the logistical problem of scaling up production techniques when

Table 5.3 Challenges in nanowire production

Challenge	*Impact*	*Statistics*
Material Purity	High defect rates leading to inconsistent performance	Over 30% of synthesized nanowires show defects due to impurities
Scalability	Difficulty in transitioning from lab scale to industrial production	Only 15% of nanowire production methods are scalable for commercial use
Cost-Effectiveness	Expensive raw materials and processes limit widespread use	Production costs can exceed $500 per gram for high-quality nanowires
Uniformity	Variability in size and aspect ratio affects sensor performance	Up to 25% variation in diameter among produced nanowires

moving from the lab to introducing nanowires into an industrial capacity is another issue, as these adjustments may require additional expense, not to mention possible limitations on mass production [23]. Table 5.3 highlights the primary challenges in nnanowire production. Moreover, the integration of nanowires into current biosensing architecture requires consideration and optimization as interactions with the surrounding materials may affect sensor capacity. With consideration to the nature of these challenges, these are key issues to consider for the successful use of nanowires in real-world biosensing applications.

5.3 MECHANISMS OF BIOSENSING WITH NANOWIRES

The complicated environment of nanowire production has several major issues that impede the progress of nanowire-based biosensors, particularly maintaining control over the synthesis process to manufacture nanowires of consistent diameter, length, and composition. Consistent characteristics will benefit sensor performance, as variation in these standards will impact the transduction capabilities of biosensors, ultimately limiting sensitivity and accuracy [22]. In addition to precise control over the production of uniform nanowires, the logistical problem of scaling up production techniques when moving from the lab to introducing nanowires into an industrial capacity is another issue, as these adjustments may require additional expense, not to mention possible limitations on mass production [23]. Table 5.3 highlights the primary challenges in nanowire production. Moreover, the integration of nanowires into current biosensing architecture requires consideration and optimization as interactions with the surrounding materials may affect sensor capacity. With consideration to the nature of these challenges, these are key issues to consider for the successful use of nanowires in real-world biosensing applications [126–132].

5.3.1 Principles of biosensing technology

The use of biosensing technology, in diverse disciplines, has increasingly become pivotal primarily because it offers rapid and accurate diagnostics essential for health monitoring and environmental safety. At the core of this technology is the principles of selectivity, sensitivity, and reliability that are enhanced through the use of nanowire-based structures. This assists the transduction process by enabling the interaction between larger surface areas for the analyses to differentiate effectively between target and nontargeted molecules, such as using impedance spectroscopy in nanowire scaffolds to monitor their activities has just been proven where nanowire-based systems allow highly selective sensing with high sensitivities toward molecules like breast cancer cells [26]. The progress in biosensors has also been highlighted by the need for cheaper, handheld sensing methodologies that can evaluate water quality immediately, particularly toward the identification of pathogens and other contaminants [27]. This being the case, the introduction of nanowire technology is not only desirable but necessary for biosensing applications [133–142].

5.3.2 Interaction of nanowires with biological molecules

The development of nanowire-based biosensors is increasingly recognized as a pivotal advancement in the detection of biological molecules due to their unique structural properties and enhanced surface area. Nanostructures facilitate effective interactions with biomolecules, thereby allowing specific recognition events that are essential for the accurate sensing mechanism. Since nanowires are excellent conductors and can be functionalized very well, receptor molecules can be immobilized, thereby enhancing the overall sensitivity of biosensors. This integration of nanowires into biosensing technology has led to major advancements in analytical performance, which makes them very suitable for medical diagnostics and environmental monitoring applications. Methods like electrochemical impedance spectroscopy take advantage of these benefits by allowing the sensitive detection of biomolecular interactions at the nanoscale, showcasing enormous possibilities for use of nanowire-based systems in research and applications [28, 29]. Ultimately, the interaction of nanowires with biological molecules represents a paradigmatic change in the concept of the biosensor to meet the present limitations of sensitivity and specificity in detection.

5.3.3 Signal transduction mechanisms in biosensors

Crafting good biosensing technology involves understanding the key interaction between bioreception and signal transduction, so that the desired measurements are both accurate and reliable. Biosensors convert biorecognition responses into electrical signal responses, and bioreception is not a signal until the transduction mechanism absorbs it and changes it into a signal [143–150]. The transduction mechanisms can absorb biorecognition signals (the conversion of a biological response into measurable reponse) and translate them into measurable responses, without sacrificing sensitivity or reproducibility, especially relevant for clinical or environmental biosensing. The advances in nanowire-based biosensors have significantly improved the transduction mechanisms by miniaturizing these processes and assimilating a range of nanomaterials with a range of nanosensing technologies into functional sensing systems, leading to improvement in performance metrics [30]. The high surface area and electrical conductivity of nanowires allow for fast responses to target analytes, which are necessary for real-time analyses. Thus, understanding the signal transduction processes in biosensors widens the avenues for biosensing applications and suggests solutions for challenges with developing cost-effective and efficient sensor technologies for commercial use, as both flexible and reliable biosensing technologies remain elusive. [31].

5.3.4 Role of surface modifications in biosensing

Strategic surface modification significantly enhances the effectiveness of nanowire-based biosensors for tailored interactions with biomolecules. Improved biocompatibility and specificity improve the detection sensitivity toward the targeted analytes. For instance, functional groups or bioreceptors can be deposited on the surface of nanowires, causing an increase in the sensitivity and selectivity toward certain biomolecules, as proven by recent findings using various nanomaterials for healthcare [32]. Furthermore, the possibility of self-sustained gas sensors, as described in the new methodologies, shows how organic surface functionalization can maximize the electronic interfaces to achieve unprecedented selectivity toward individual gas species, thus enhancing the overall sensor performance [33]. In this regard, the role of surface modifications in biosensing is pivotal because it bridges the gap between nanostructure design and practical applications in biomedical diagnostics, which ultimately contributes to the reliability of biosensing technologies.

5.3.5 Comparison of nanowire biosensors with traditional biosensors

The evolution of biosensors has seen great improvements as researchers explore alternate technologies that yield enhanced sensitivity and specificity. Among such developments, nanowire biosensors are becoming an

Table 5.4 Comparison of nanowire biosensors and traditional biosensors

Features	*Nanowire biosensors*	*Traditional biosesnors*
Sensitivity	High	Moderate
Size	Nanoscale (typically 1–100 nm)	Microscale (typically >1 μm)
Response time	Fast (milliseconds)	Slow (seconds to minutes)
Cost	Higher due to fabrication complexity	Generally lower
Specificity	Higher specificity due to surface modifications	Lower specificity
Applications	Diagnostics, environmental monitoring, medical therapies	Clinical diagnostics, food safety, environmental testing

interesting alternative to traditional biosensors because of their structural uniqueness and superior performance features in comparison with conventional counterparts. Generally, traditional biosensors have larger sensing elements that may be less sensitive and have slower response times. In contrast, nanowire biosensors leverage nanoscale dimensions to offer a high surface area-to-volume ratio that provides for the sensitivity of detection on single molecules while allowing for real-time responses. Moreover, through the use of nanomaterials such as silicon nanochannels, innovations have been introduced with diagnostic capabilities through recent literature on applications in the field of cancer diagnostics and food safety [34]. Table 5.4 lists the compares the various factor of nanowire biosensors and traditional biosensors. This extraordinary sensitivity together with the feasibility of miniaturization makes nanowire biosensors a game changer within the realm of biosensing; therefore, reshaping the limits for future deployments.

5.4 APPLICATIONS OF NANOWIRE-BASED BIOSENSORS

With properties such as a high surface area and conductivity, nanowire-based biosensors are revolutionizing the landscape of biomedical diagnostics and environmental monitoring. Advanced sensing devices that can detect a multitude of biologically relevant molecules, such as pathogens, nucleic acids, and toxins, in real time with remarkable sensitivity, are becoming increasingly important in various fields. Table 5.5 lists various applications of nanowire-based biosensors. The application of nanowires in biosensors also enables label-free detection, where processing times significantly decrease while offering enhanced multiplexing capabilities, allowing for simultaneous analysis of numerous targets. Other recent studies identify Fin-field-effect transistor biosensors as alternatives to the issue of reproducibility and the noise levels, which are typical challenges in traditional sensors, opening gates to precision in diagnostics. In addition, the flexibility of

Table 5.5 Applications of nanowire-based biosensors

Application	*Description*	*Example*	*Source*
Medical Diagnostics	Used for detecting biomarkers associated with diseases.	Early detection of cancer through specific biomarker identification [35].	Journal of Science, 2022
Environmental Monitoring	Utilized to sense pollutants and toxins in air and water.	Real-time detection of heavy metals in water supplies [36].	Environmental Science & Technology, 2023
Food Safety	Employed to detect pathogens and contaminants in food products.	Rapid identification of *E. coli* in food samples.	Food Control, 2023
Wearable Health Devices	Integrated into wearables for continuous health monitoring.	Monitoring glucose levels in diabetic patients.	IEEE Transactions on Biomedical Engineering, 2022
Drug Delivery Systems	Facilitates targeted delivery of therapeutics.	Nanowires used in localized cancer treatment.	Journal of Nanobiotechnology, 2023

nanomaterials has provided a platform where various detection mechanisms can coexist, thus underlining the huge potential for early disease screening and point-of-care testing.

5.4.1 Medical diagnostics and health monitoring

Some important developments in this regard have recently been made related to biosensor technology, nanowire devices in particular, which have created a new breakthrough in medical diagnosis and health monitoring systems. These sensors are highly sensitive, transforming all the detection mechanisms, allowing continuous monitoring of various critical health indicator parameters and diagnosis of diseases very early. Such is the promise of silicon nanochannel FET biosensors, which achieve single-molecule detection, giving a new footing to personalized medicines and targeted treatments. Such technologies not only enhance diagnostic accuracy but also reduce costs and improve manufacturing scalability, as shown by the ability to fabricate multiplexed sensor arrays with high yield and speed [37]. Additionally, the diversity of nanomaterials provides a route toward developing sensing platforms that can identify a wide variety of clinically relevant molecules, ranging from cancer antigens to pathogens, significantly advancing point-of-care testing and health monitoring systems [38]. This integration brings the future of medical care to us, promising new inventions for effective management of patients.

5.4.2 Environmental monitoring and detection

Advances in environmental monitoring rely more and more on nanowire-based biosensor technologies. These sensors are highly sensitive to a wide range of environmental pollutants, and real-time assessments are essential for public health and ecological preservation. As noted in new studies recently, the silicon nanochannel field-effect transistor (FET) biosensors show excellent examples of the use of nanowires to achieve an adequately high sensitivity to the analytes; they detect pollutants at single-molecule levels. The effectiveness of such sensors is also augmented by the diversity of different nanomaterials, making it easier to monitor a large number of environmental factors, such as toxins and pathogens, in various media, such as air and water [39]. This integration of nanowire-based biosensors not only enhances detection capabilities but also supports ongoing efforts in sustainable environmental management.

5.4.3 Food safety and quality control

Advanced biosensing technologies are changing the food safety and quality control landscape. Consumer awareness of foodborne illnesses and food contamination are on the rise, as is the rapid development of biosensors based on nanowires. Biosensing nanowires, for example, offer the tangible possibility for rapid and portable detection of pathogens, toxins, and chemical residues, and therefore improve safety and quality at any point of the supply chain. Improved sensitivity and specificity mean that contaminants would be detected at lower concentrations limiting the health risk to the consumer. Additionally, the integration of nanomaterials will now assist with sensor performance and rapidly monitor food quality in a live manner. The changes brought about by nanowire-based biosensors will not only meet the required regulations of our time but also realizes a sustainable food production system. This fundamentally opens the door to new innovations that can bring tremendous benefits to public health [40]. Overall, there is considerable potential to benefit food safety protocols in general development through enhancing the integration of nanotechnology and biosensing.

5.4.4 Drug discovery and pharmaceutical applications

Utilizing nanowire-based biosensors in drug discovery illustrates a disruptive change for pharmaceutical applications that improve the accuracy and speed of diagnostics. Numerous studies highlight the ways to exploit biosensors that detect clinically relevant molecules, such as nucleic acids and drugs, that allows researchers to identify therapeutic targets and biomarkers. For instance, there has been continual progress toward developing nanomaterials that have the unique ability to detect different analytes encapsulated within complex biological microenvironments, like blood and urine, thus strengthening the progress toward early disease screening [41]. Another recent example is the development of impedance sensing technology using nanowire platforms, which are helpful for evaluating cancer cell responses to drugs, as well as assessing the metabolic state and breast cancer cell responses to drugs in real time [42]. These types of inventions are advantageous for our drug mechanisms and ushering in a path toward personalized medicine, eventually yielding improved outcomes for patients through tailored treatment.

5.4.5 Future potential applications in various fields

As nanotechnology keeps unfolding itself, the scale of nanowire-based biosensors is very soon going to span across significant segments. Since nanowire biosensors detect pollution at extraordinary sensitivities that even surpass other available methods and hence enable responses that are ecological, they present promise in various significant sectors; more importantly, integration of nanowires within wearable health technologies promises to monitor real-time physiology necessary for treating chronic conditions and tailored treatment. Research shows that from molecular-level applications to higher-level ecosystem-based frameworks, accurate monitoring of soil health and crop conditions will make transformative impacts in agricultural productivity and food security [43]. On the other hand, as techniques for fabricating nanowires improve, these nanowires will be cemented in applications ranging from smart textiles to advanced diagnostic capabilities in medicine. Future nanowire-based biosensors open up vast areas of implications for comprehensive advancements in many fields.

5.5 RECENT ADVANCES AND INNOVATIONS

The ongoing evolution of sensor technology is marked by significant strides in nanowire-based biosensors, which enhance our capabilities in health monitoring and diagnostics. Recent advances have focused on improving the efficacy and practicality of these devices, particularly through the integration of nanowires in flexible electronics. This innovation enables the creation of wearable sweat sensors, which offer noninvasive methods for real-time health monitoring by analyzing sweat composition for biomarkers indicative of various health conditions [44]. While the potential applications are promising, challenges remain in sensor performance, such as ensuring reliable measurements and improving user comfort. Consequently, the development of nanowire-based biosensors requires a multidisciplinary approach that encompasses materials science, device engineering, and clinical validation. The ongoing research in this field not only strives to address current limitations but also opens pathways for tailored healthcare solutions, ultimately advancing precision medicine.

Figure. 5.1 shows barchart, which presents an overview of recent advances, challenges, and future directions in technology. It highlights the values associated with various subcategories within these three main categories, showcasing the importance of tailored healthcare solutions and advancements in precision medicine as well as the challenges related to user comfort and reliable measurements.

5.5.1 Novel nanowire materials and structures

As nanotechnology continues to inform new discoveries in biosensing applications, the value of new nanowire materials and structures is becoming increasingly clear. These materials have unique properties, such as high aspect ratios and electrical conductivity, which increase detection in sensor technology. The use of nanowires in biosensors will not only increase the sensitivity and response times but also provide platforms to detect numerous biomolecules and environmental toxins. In addition, their tunable surfaces allow for functionalization with specific biomolecules and thus targeted sensing. The use of such nanowire architectures will be the basis for developing electrochemical sensors that will be highly efficient. There has been a lot of recent literature that describes the role of nanomaterials applied to electrochemical impedance spectroscopy, especially for pathogen detection and monitoring environmental contaminants [45]. New attempts to utilize nanowires applied to biosensing formats have the potential of overcoming existing constraints and having real impact on biomedical diagnostics and environmental applications.

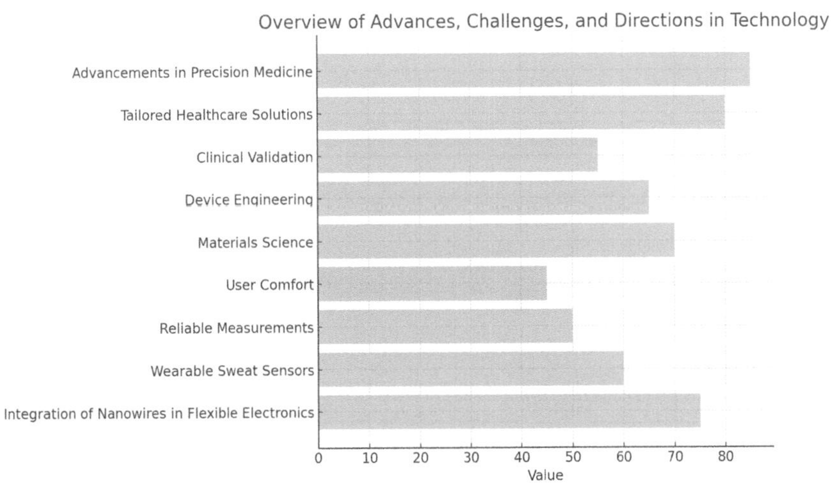

Figure 5.1 Overview of aadvances, challenges, and directions in technology.

5.5.2 Integration of nanowires with microfluidics

The possibility of combining nanowires with microfluidic systems is an exciting development in the world of biosensing applications. As researchers develop new ways of detecting diseases, the integration of these two technologies is having incredible potential to improve the sensitivity and speed of biosensors. Nanowires provide high surface area and high conductivity to use for rapid and specific binding events to help detect target target biomolecules at low concentrations. Recent studies indicate that nanowire-based biosensors can effectively detect Escherichia colis 16S rRNA at concentrations as low as 10 fM (femtomolar), demonstrating their effectiveness in real-world applications like pathogen detection in food and water [46]. Moreover, the fusion of microfluidics allows for the miniaturization and portability of diagnostic tools, aligning with the goals of tiny medicine to facilitate early diagnosis and treatment. Thus, this integration paves the way for the development of advanced biosensing technologies.

Figure 5.2 chart presents an analysis of key factors in detection technology, highlighting various subcategories and their corresponding values on a scale of 0 to 100. Each bar represents a different aspect, such as sensitivity, application areas, and future potential, providing a clear comparison of their effectiveness or importance.

5.5.3 Advances in signal amplification techniques

The development of nanowire-based biosensors has repeatedly emphasized the role of signal amplification methods for improving detection sensitivity,

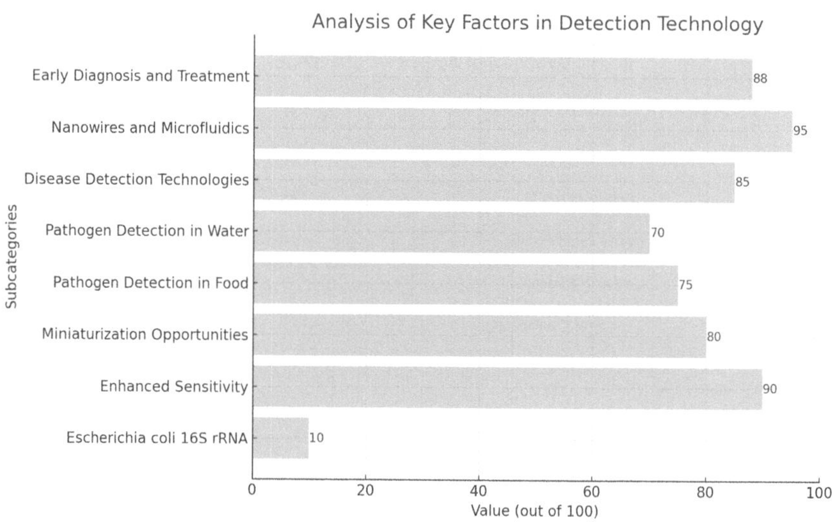

Figure 5.2 Analysis of key factors in detection technology.

thereby driving advances in clinical diagnostics. Nanowire biosensors capitalize on their unique characteristics in the transformation and measurable output of biological signals. The use of signal amplification methods, whether electrical or optical, facilitates the detection of low-cost target analytes, which is essential for the diagnosis of early stages of disease. As an example, silicon nanochannel field-effect transistor (FET) biosensors use high surface-to-volume ratios to allow the detection of single molecules and represents a powerful analytical tool for cancer diagnostics [47]. Other similar devices add nanomaterial diversity that contributes to build sensor systems with multipurpose functions in many different applications, from pathogen detection to biomarker identification. Simply put, progress in signal amplification methods continues to refine the arena of biosensing technology in creative ways to solve challenges in healthcare and beyond.

5.5.4 Development of portable biosensing devices

Recent developments in biosensing technology highlight an important move towards portable devices designed to support real-time diagnostics and monitoring of environments. Employing novel materials, such as nanowires, these devices have shown excellent sensitivity and specificity to detect a range of biological and chemical agents. For example, one detection platform was identified as portable and able to detect pathogenic microbes at a limit of detection as low as 10 fM. Additionally, silicon nanochannel field-effect transistor biosensors are developing into a promising direction as they can provide high speeds, lower manufacturing costs, and sensitivity like those already established [48]. Biosensing products such as these enhance access to important diagnostics for use in multiple settings, including food safety and health care, and furthermore, represent a step toward "biosensing products" that will one day be ubiquitous for daily use, while support a personalized, proactive approach to health and environmental management.

5.5.5 Case studies of successful nanowire biosensor applications

Conducting case studies enables researchers to effectively demonstrate how nanowire biosensors are making a change in the world with its specific use for tailored applications. In healthcare, nanowire biosensors are enabling real-time monitoring of disease markers, allowing for rapid diagnostics and individualized treatment plans, thus indicating the potential of biosensors to advance health care delivery. In parallel, with the growing pressure to monitor the environment for toxic substances, nanowire biosensors have worked in real-time detection of the most dangerous of substances, which can allow for swift action in addressing adverse environmental exposure and public safety. Additionally, significant new applications of nanotechnology in agriculture, including targeted nutrients for crops and pathogen identification, demonstrate the widespread usefulness of nanowire biosensors in improving crop yield and sustainability [49]. All these examples show how the integration of nanowires is playing a role in operating biosensor systems to support advancements in a variety of sectors ultimately resulting in improved health outcomes and environmental sustainability.

5.6 CONCLUSIONS

This study encompasses the innovative development and implications of nanowire-based biosensors across multiple domains in investigation. As highlighted across the telemetry and biosensor development work, recent advances in nanowires have produced ultrasensitive and selective biosensing platforms with potential applications in cancer diagnostics and health biomarker monitoring in real time. The example of bioceramic titanate and its use with nanowires to detect cancer cells captures the unique functionality of tissue differentiation of cell types through impedance signals like never before. Future opportunities arise with single polymer nanowires with customized functionalization for specificity detection and cardiac biomarkers demonstrates continued opportunities for novel biosensors. As research unfolds, these biosensor formats can mark a bold new direction and catalyst for new standards in diagnostics and utilitarian applications in personalized medicine with relevance to research and clinical practice.

5.6.1 Summary of key findings

The research into biosensors based on nanowires has made significant progress toward applications for cellular studies and real-time monitoring. One of the advancements shows that the devices can differentiate not only between cell lineages but cancer cell lineages based on its impedance signals that statistically were different from the control in cancer research, which will have health care staff be able to see selectivity and confidence in diagnostics. Interestingly, the biosensors will work reliably in multiple contexts, such as fermentation because they can measure redox potential to monitor utilizing real-time metabolic activity, avoiding interference, and enhancing effectiveness and control. All of which demonstrate why nanowire biosensors can be utilized in many different biological contexts. Nanowire biosensors continue to be an area of immense potential in research studies related to cancer research and industrial biotechnology. The unmitigated integrated sensor technologies are promising a new way of doing diagnostics and production, which opens the door for countless health and environmental innovative gadgets, and other novel applications all while demonstrating new versatility. Thus, continuous coverage of the research and development of nanowire biosensors merits consideration. Figure 5.3 indicates the different

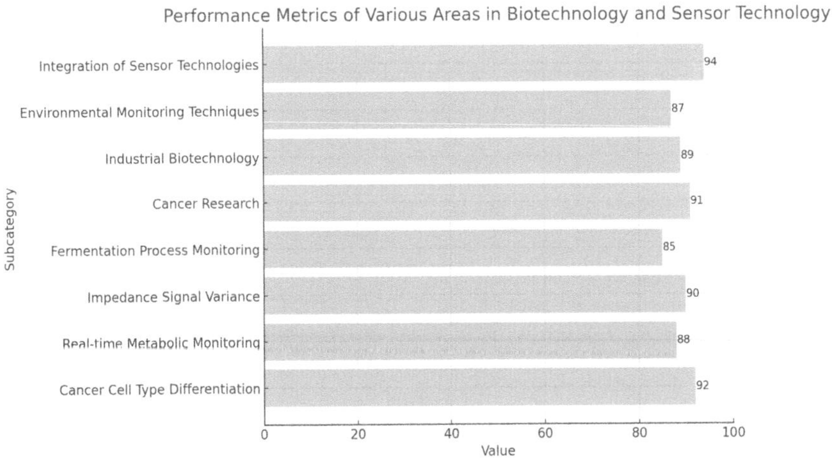

Figure 5.3 Performance metrics in various areas of biotechnology and sensor technology.

performance angles for different sectors in biotechnology and sensor technology. Each subcategory shown in bars has a value in relation to effectiveness or efficiency, where value is represented in the bars, with greater values having better performance.

5.6.2 Implications for future research

The investigation of biosensors based on nanowires has made great strides toward developing applications for cellular research and real-time monitoring. One of the breakthroughs shows that the devices can distinguish not just between cell types, but between cancer cell types based on its impedance signals that statistically differed from the control in cancer research, which will allow health care staff to see selectivity and reliability in diagnostics. Interestingly, the biosensors will work reliably in multiple contexts, such as fermentation because they can measure redox potential to monitor utilizing real-time metabolic activity, avoiding interference, and enhancing effectiveness and control. All of which demonstrate why nanowire biosensors can be utilized in many different biological contexts. Nanowire biosensors continue to be an area of immense potential in research studies related to cancer research and industrial biotechnology. The unmitigated integrated sensor technologies are promising a new way of doing diagnostics and production, which opens the door for countless health and environmental innovative gadgets, and other novel applications all while demonstrating new versatility. Thus, continuous coverage of the research and development of nanowire biosensors merits consideration. Figure 5.4 indicates the

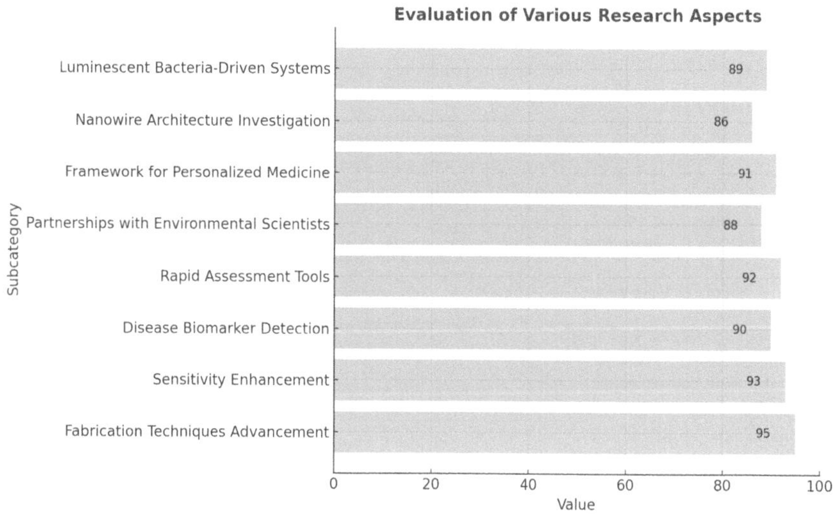

Figure 5.4 Evaluation of various research aspects.

evaluation of various research aspects. Each subcategory shown in bars has a value in relation to effectiveness or efficiency, where value is represented in the bars, with greater values having better utilization.

5.6.3 Potential impact on various industries

The emergence of nanowire-based biosensors may be life altering for many organizations and sectors of society, particularly in the contexts of human health, and food safety. These sensors deliver rapid and sensitive detection of biomolecules, such as toxins or pathogens, can lead to improved diagnostics, making intervention possible in a manner that promotes a substantial improvement in patient care outcomes. To give a better sense of what is occurring in the food industry, with the developments of intelligent logistics systems for food preservation, the introduction of biosensors enables real-time monitoring of environmental conditions to counteract spoilage or waste, and may allow for real-time solutions to offer fresh food that has been preserved for longer time periods. This effort provides a greater amount of shelf life for the food producer to release their product to market and the consumer to manage their perishables. It is important to note that biosensors based on nanowire technologies are also versatile technologies that can be produced cost effectively and scaled to coverage to access different application sectors, even the potential accomodations of previously hard to access biosensors in working environments.to deliver useful information while safety and efficiencies metrics in human health and economic terms are improved on a global scale. The aforementioned advances of biosensors will individually, and collectively, have major consequences.

5.6.4 Limitations and challenges ahead

While there are advancements in nanowire-based biosensors, there are still serious limitations and challenges that need to be addressed if the technology is to be optimized. One of the challenges is the development of biosensors with improved sensitivity and specificity to allow for a wider variety of applications, especially for biosensors used in complex biological samples due to multiple nonspecific interactions that can lead to false-positive results. In addition, the technical issues related to the uniform fabrication of nanostructures can vastly limit the scalability of these production methods and reliability of results. There are also an increasing philosophies for the development of signal amplification methods or creative detection solutions, which creates additional avenues for research, yet adds to the complexity of the developed biosensors. It is important to understand these challenges, as alluded to in the discussion on silicon nanochannel field-effect transistor biosensors, that show promise but face barriers to sensitivity and application, and similarly, the need for new signal transduction methods within surface-based detection platforms give context to the research developments happening in this space.

5.6.5 Final thoughts on the future of nanowire-based biosensors

The future of nanowire-based biosensors is pretty exciting as these high-tech devices continue to be developed with advancements made in nanotechnology. They have the potential to disrupt a variety of fields, including medical diagnostics, environmental monitoring, and food safety, as a result of their unique properties and capabilities. One example of this potentially disruptive technology is biosensors that rely on nanowires, which can provide unprecedented sensitivity and ability to detect biomolecules at low concentrations and high specificity. While work is still being done to improve the scalability and cost-effectiveness of the production of nanowires, there is hope that this technology will be readily adopted for clinical and commercial use in the near future. In addition, there are a variety of wearable technologies becoming available that can provide continuous health data, so the field of biosensing for health information and healthcare in general has great potential to diffuse in such applications and deliver health data in real time to facilitate proactive healthcare. While researchers continue to create more advanced materials and different configurations for nanowire-based biosensors to solve real-world problems, the use of this technology for nanowire-based biosensing will continue to grow in scope in order to address global challenges while advancing progress and technological advancement.

REFERENCES

1. Seyed Saman Nemati, Gholamreza Dehghan, Nader Sheibani, Y. Abdi, "Ion-Sensitive Field-Effect Transistor-Based Biosensors: The Sources of Gates and Their Sensitive Layers—A Review", *IEEE Sensors Journal*, 2024.
2. Junyang Huo, "Evaluation of the Impact of Intelligent Logistics Systems Based on Biosensors on Food Preservation EffEct", *Molecular & Cellular Biomechanics*, 2024.
3. Yu Chen, Shyamsunder Erramilli, Mi K. Hong, P. Mohanty, et al., "Field Effect Transistor Nanosensor for Breast Cancer Diagnostics", 2012.
4. Zeynep Altintas, Muqsit Pirzada, "Nanomaterials for Healthcare Biosensing Applications", 2019.
5. K. Goldaeva, T. Pleshakova, Y. D. Ivanov, "Nanowire-based biosensors for Solving BioMedical Problems."*Biomeditsinskaia khimiia*, 2024.
6. Yingying Li, Yuankun Zhao, Yiyang Du, Xuechun Ren, He Ding, Zhimin Wang, "Recent Advances in the Development and Applications of Luminescent Bacteria-Based Biosensors", *Luminescence: the journal of Biological and Chemical Luminescence*, 2024.
7. Hanan Alismail, "Impedance Sensing of Cancer Cells Directly on Sensory Bioscaffolds of Bioceramics Nanofibers", *ScholarWorks@UARK*, 2018.
8. Innam Lee, "Biosensing of Cardiac Biomarkers Using Single Polyaniline Nanowires", 2012.
9. Anisha Banda, Leyla Esfandiari, Gayane Kocharyan, Lorenzini, et al., "*PCR-Independent Detection of Bacterial Species-Specific 16S rRNA at 10 fM by a Pore-Blockage Sensor.*", eScholarship, University of California, 2016.
10. César Pascual García, Renaud Leturcq, Wouter Olthuis, Rani, et al., "*A high aspect ratio Fin-Ion Sensitive Field Effect Transistor: compromises towards better electrochemical bio-sensing*", American Chemical Society (ACS), 2019.
11. L.M.T. Dicks, M.B. Maas, W.J. Perold, "Review: Biosensors for the Detection of Escherichia coli", *African Journals Online (AJOL)*, 2017.
12. Balasubramanian Arya, Bayoudh Banada, Boehm Berti, Chai, et al., "New Trends in Impedimetric Biosensors for the Detection of Foodborne Pathogenic Bacteria", *Molecular Diversity Preservation International (MDPI)*, 2012.
13. Harold Monbouquette, Nosang V. Myung, Richard P. Vasquez, Wang, et al., "Nanowire Growth for Sensor Arrays", *SPIE-Intl Soc Optical Eng*, 2003.
14. Stephanie Klinghammer, "Nanostructured Biosensors with DNA-Based Receptors for Real-Time Detection of Small Analytes", 2023.
15. S. N. Faisal, S. Kim, H. J. Lee, Lee, et al., "*Bioaffinity Detection of Pathogens on Surfaces*", Elsevier BV, 2010.
16. Marc Bockrath, Isabella Kierk, Maria Landi, "Novel Nano-Biosensors for Life Science Systems and thEir Applications in Early, Accurate, and Non-Invasive Melanoma and Other Types of Cancer Detection", *Institute of Electrical and Electronics Engineers (IEEE)*, 2007.
17. Hend S. Magar, Rabeay Y. A. Hassan, Ashok Mulchandani, "Electrochemical Impedance Spectroscopy (EIS): Principles, Construction, and Biosensing Applications", *Sensors*, 2021.
18. V. Naresh, Jeong Yong Lee, "A Review on Biosensors and Recent Development of Nanostructured Materials-Enabled Biosensors", *Sensors*, 2021.

19. Chou-Yi Hsu, Ahmed Mahdi Rheima, Zainab sabri Abbas, Muhammad Usman Faryad, Mustafa M. Kadhim, Usama S. Altimari, Ashour H. Dawood, Alaa Dhari Jawad Al-Bayati, Zainab Talib Abed, Rusul Saeed Radhi, Asala Salam Jaber, Safa K. Hachim, Farah K. Ali, Zaid H. Mahmoud, Ghobad Behzadi Pour, Ehsan Kianfar, Nanowires Properties and Applications: A Review Study, *South African Journal of Chemical Engineering*, Volume 46, 2023, pp. 286–311.
20. Heebo Ha, Chiara Amicucci, Paolo Matteini, Byungil Hwang, "Mini Review of SynthesiS Strategies of Silver Nanowires and Their Applications", *Colloids and Interface Science Communications*, 2022.
21. Aizawa Ahn, Amit Bhattacharya, Berry Azzazy, Borisov Biju, et al., "Tiny Medicine: Nanomaterial-Based Biosensors", *Molecular Diversity Preservation International (MDPI)*, 2009.
22. Md Yifen Shang, Kamrul Hasan, Golam Jalal Ahammed, Mengqi Li, Hanqin Yin, Jie Zhou, "Applications of Nanotechnology in Plant Growth and Crop Protection: A Review", *Molecules*, 2019.
23. Yu-Cheng Syu, Wei-En Hsu, Chih-Ting Lin, "Review—Field-Effect Transistor Biosensing: Devices and Clinical Applications", *ECS Journal of Solid State Science and Technology*, 2018, https://doi.org/10.1149/2.0291807jss
24. Su-Ting Han, Haiyan Peng, Qi-Jun Sun, S. Venkatesh, K.S. Chung, Siu Chuen Lau, Ye Zhou, et al., "An Overview of the Development of Flexible Sensors", *Advanced Materials*, 2017.
25. Jihong Min, Jiaobing Tu, Changhao Xu, Heather Lukas, Soyoung Shin, Yiran Yang, Samuel A. Solomon, et al., "Skin-Interfaced Wearable Sweat Sensors for Precision Medicine", *Chemical Reviews*, 2023.
26. Yifei Luo, Mohammad Reza Abidian, Jong-Hyun Ahn, Deji Akinwande, Anne M. Andrews, Markus Antonietti, Zhenan Bao, et al., "Technology Roadmap for Flexible Sensors", *ACS Nano*, 2023.
27. Camilla Baj-Rossi, Wayne Burleson, Sandro Carrara, Cavallini, et al., "Fully Integrated Biochip Platforms for Advanced Healthcare", *ScholarWorks@ UMass Amherst*, 2012.
28. L. Caccamo, O. Casals, W. Daum, Fàbrega, et al., "*Novel Approaches towards Highly Selective Self-Powered Gas Sensors*", Elsevier BV, 2015.
29. Saehyun Kim, Pengju Li, Bozhi Tian, "Beyond 25 Years of Biomedical Innovation in Nano-Bioelectronics", 2024.
30. Ahmed Touhami, "Biosensors and Nanobiosensors: Design and Applications", *ScholarWorks @ UTRGV*, 2014.
31. Malik Shiza, Khalid Muhammad, Yasir Waheed, "Nanotechnology: A Revolution in Modern Industry", *Molecules*, 2023.
32. Zeynep Altintas, Bhargav D. Mansuriya, "Graphene Quantum Dot-Based Electrochemical Immunosensors for Biomedical Applications", 2019.
33. Vincent Callegari, Sophie Demoustier-Champagne, A. Dinescu, Duvail, et al., *Conjugated Polymer and Hybrid Polymer-Metal Single Nanowires: Correlated Characterization and Device Integration*, IntechOpen, 2010.
34. Siyang Shen, "*Microbial Fuel Cell-Based Biosensor for Redox Potential Detection in Fermentation Process*", University of Saskatchewan Library, 2022.
35. Ali Bodaghi, Nadia Fattahi, Ali Ramazani, "Biomarkers: Promising and Valuable Tools Towards Diagnosis, Prognosis and Treatment of COVID-19 and Other Diseases", *Heliyon*, Volume 9, Issue 2, 2023.

36. Shraddha Chauhan, Digvijay Dahiya, Vikas Sharma, Nawaz Khan, Deepshi Chaurasia, Ashok Kumar Nadda, Sunita Varjani, Ashok Pandey, Preeti Chaturvedi Bhargava, “Advances from Conventional to Real Time Detection of Heavy Metal(loid)s for Water Monitoring: An Overview of Biosensing Applications”, *Chemosphere*, Volume 307, Part 4, 2022,
37. Matthew C. Cole, Paul J.A. Kenis, “Multiplexed Electrical Sensor Arrays in Microfluidic Networks”, *Sensors and Actuators B: Chemical*, Volume 136, Issue 2, 2009.
38. Ali Aghababai Beni, Hadi Jabbari, “Nanomaterials for Environmental Applications, *Results in Engineering*”, Volume 15, 2022.
39. Kerry Ann Brown, Nikhil Srinivasapura Venkateshmurthy, Cherry Law, Francesca Harris, Suneetha Kadiyala, Bhavani Shankar, Sailesh Mohan, Dorairaj Prabhakaran, Cécile Knai, “Moving Towards Sustainable Food Systems: A Review of Indian Food Policy Budgets, *Global Food Security*”, Volume 28, 2021.
40. Mansi Sharma, Priyanka Mahajan, Abdullah Saad Alsubaie, Virat Khanna, Surjeet Chahal, Abhinay Thakur, Ankush Yadav, Atul Arya, Amanpreet Singh, Gulab Singh, “Next-Generation Nanomaterials-Based Biosensors: Real-Time Biosensing Devices For Detecting Emerging Environmental Pollutants”, *Materials Today Sustainability*, Volume 29, 2025.
41. Sumit Malik, Joginder Singh, Rohit Goyat, Yajvinder Saharan, Vivek Chaudhry, Ahmad Umar, Ahmed A. Ibrahim, Sheikh Akbar, Sadia Ameen, Sotirios Baskoutas, Nanomaterials-Based Biosensor and Their Applications: A Review, *Heliyon*, Volume 9, Issue 9, 2023.
42. Malgorzata Roszkowska, “Multilevel Mechanisms of Cancer Drug Resistance”, *International Journal of Molecular Sciences* 25, no. 22, 2025.
43. Yahya Faqir, Abdul Qayoom, Elizabeth Erasmus, Marietjie Schutte-Smith, Hendrik G. Visser, “A Review on the Application of Advanced Soil and Plant Sensors in the Agriculture Sector”, *Computers and Electronics in Agriculture*, Volume 226, 2024.
44. G. Backiyalakshmi, U. Snekhalatha, Anela L. Salvador, “Recent Advancements in Non-Invasive Wearable Electrochemical Biosensors for Biomarker Analysis – A Review”, *Analytical Biochemistry*, Volume 692, 2024.
45. Raghvendra Kumar Mishra, Kartikey Verma, Deepa Sethi Singh, “Defect Engineering in Nanomaterials: Impact, Challenges, and Applications”, *Smart Materials in Manufacturing*, Volume 2, 2024.
46. Jialin Zhang, Jiajia Wang, Xiaoqing Zhang, Fengjiao He, “Rapid Detection of Escherichia Coli Based on 16S rDNA Nanogap Network Electrochemical Biosensor”, *Biosensors and Bioelectronics*, Volume 118, 2018, Pages 9–15.
47. Sumit Malik, Joginder Singh, Rohit Goyat, Yajvinder Saharan, Vivek Chaudhry, Ahmad Umar, Ahmed A. Ibrahim, Sheikh Akbar, Sadia Ameen, Sotirios Baskoutas, “Nanomaterials-Based Biosensor and Their Applications: A Review”, *Heliyon*, Volume 9, Issue 9, 2023.
48. M. Poorna Sundari, G. Lakshmi Priya, “Nano-Biosensors with Subthreshold Swing Tunnel Field Effect Transistor: A Cutting-Edge Review”, *Sensing and Bio-Sensing Research*, Volume 44, 2024.
49. Rittick Mondal, Paulami Dam, Joydeep Chakraborty, Mathew L. Paret, Ahmet Katı, Sevde Altuntas, Ranit Sarkar, Suvankar Ghorai, Debnirmalya Gangopadhyay, Amit Kumar Mandal, Azamal Husen, “Potential of

Nanobiosensor in Sustainable Agriculture: The State-of-Art", *Heliyon*, Volume 8, Issue 12, 2022.
50. Anjana Bhardwaj, Pradeep Kumar, B. Raj, Sunny Anand, "Design and Performance Optimization of doping-less Vertical Nanowire TFET using Gate Stack Technique", *Journal of Electronic Materials (JEMS) Springer*, Vol. 41, Issuse 7, PP. 4005–4013, 2022.
51. Jeetendra Singh, B. Raj, "Tunnel Current Model of Asymmetric MIM Structure Levying Various Image Forces to Analyze the Characteristics of Filamentary Memristor", *Applied Physics A, Springer*, Vol. 125, No.3, pp. 203.1 to 203.11 Feb 2019.
52. Candy Goyal, Jagpal Singh Ubhi and B. Raj, Low Leakage Zero Ground Noise Nanoscale Full Adder using Source Biasing Technique, "*Journal of Nanoelectronics and Optoelectronics*", Vol. 14, pp. 360–370, March 2019.
53. Gurmohan Singh, R. K. Sarin and B. Raj, "A Novel Robust Exclusive-OR Function Implementation in QCA Nanotechnology with Energy Dissipation Analysis", *Journal of Computational Electronics*, Volume 15, Issue 2, pp. 455–465, June 2016.
54. Girish Wadhwa, Priyanka Kamboj, Balwinder Raj "Design Optimisation of Junctionless TFET Biosensor for High Sensitivity" *Advances in Natural Sciences: Nanoscience and Nanotechnology*, Vol. 10, pp. 045001 (7), (2019).
55. Priya Bansal, B. Raj, "Memristor Modeling and Analysis for Linear Dopant Drift Kinetics," *Journal of Nanoengineering and Nanomanufacturing*, Vol. 6, pp. 1–7, 2016.
56. Ashima, D. Vaithiyanathan, B. Raj, Performance Analysis of Charge Plasma induced Graded Channel Si nanotube, *Journal of Engineering Research (JER), EMSME* Special Issue pp. 146–154 Aug 2021.
57. Abhishek Singh Tomar, Vijay Kumar Magraiya, and B. Raj, Scaling of Access and Data Transistor for High Performance DRAM Cell Design, *Quantum Matter*, Vol. 2, pp. 412–416 Oct. 2013.
58. Neeraj Jain, B. Raj, "Parasitic Capacitance and Resistance Model Development and Optimization of Raised Source/Drain SOI FinFET Structure for Analog Circuit Applications, *Journal of Nanoelectronics and Optoelectronins*, Vol. 13, pp. 531–539, 2018.
59. Shradhya Singh, S. K. Vishvakarma and B. Raj, "Analytical Modeling of Split-Gate Junction-Less Transistor for a Biosensor Application", *Sensing and Bio-sensing, Elsevier*, Vol. 18, pp. 31–36, April 2018.
60. Maisagalla Gopal, and Balwinder Raj, Low Power 8T SRAM Cell Design for High Stability Video Applications, *ITSI Transaction on Electrical and Electronics Engineering*, Vol. 1(5), pp. 91–97, 2013.
61. Balwinder Raj, Jatin Mitra, Deepak Kumar Bihani, Rangharajan V, A.K. Saxena and S. Dasgupta, "Analysis of Noise Margin, Power and Process Variation for 32 nm FinFET Based 6T SRAM Cell" *Journal of Computer (JCP)*, vol. 5, issue 6, PP.1–8, 2010.
62. Divya Sharma, Rajesh Mehra and B. Raj, "Comparative Analysis of Photovoltaic Technologies for High Efficiency Solar Cell Design", *Superlattices and Microstructures, Elsevier*, Volume 153, PP. 106861, May 2021.
63. Pawandeep Kaur, Avtar Singh Buttar, B. Raj, "A comprehensive Analysis of Nanoscale Transistor based Biosensor: A Review, *Indian Journal of Pure and Applied Physics*, Vol. 59, pp. 304–318, April 2021.

64. Divya Yadav, Balwant Raj and B. Raj, "Design and Simulation of Low Power Microcontroller for IoT Applications", *Journal of Sensor Letters*, Vol. 18, pp. 401–409, May 2020.
65. Shailendra Singh, B. Raj," A 2-D Analytical Surface Potential and Drain current Modeling of Double-Gate Vertical t-shaped Tunnel FET", *Journal of Computational Electronics, Springer*, Vol. 19, pp. 1154–1163, 2020.
66. Jeetendra Singh, B. Raj, "An Accurate and Generic Window function for Non-linear Memristor Model" *Journal of Computational Electronics*, Volume 18 Issue 2, pp. 640–647 June 2019.
67. Manjit Kaur; Neena Gupta; Sanjeev Kumar; B. Raj; Arun Kumar Singh, "Comparative RF and Crosstalk Analysis of Carbon Based Nano Interconnects" *IET Circuits, Devices & Systems*, Vol.15(6), pp. 493–503, Feb 2021.
68. Nehru Kandasamy, Firdous Ahmad, D. Ajitha, B. Raj, and Nagarjuna Telagam "Quantum Dot Cellular Automata based Scan Flip Flop and Boundary Scan Register, *IETE Journal of Research*, Volume 66, pp. 535–548, 2020.
69. Jeetendra Singh, B. Raj, "Design and Investigation of 7T2M NVSARM with Enhanced Stability and Temperature Impact on Store/Restore Energy", *IEEE Transactions on Very Large Scale Integration Systems*, Vol. 27, Issue 6, PP.1322–1328 June 2019.
70. Anil Kumar Bhardwaj, Sumeet Gupta, B Raj, Amandeep Singh, "Impact of Double Gate Geometry on the Performance of Carbon Nanotube Field Effect Transistor Structures for Low Power Digital Design", *Computational and Theoretical Nanoscience, ASP*, Vol. 16, pp. 1813–1820, 2019.
71. Neeraj Jain, B. Raj, "Thermal Stability Analysis and Performance Exploration of Asymmetrical Dual-k underlap Spacer (ADKUS) SOI FinFET for Security and Privacy Applications", *Indian Journal of Pure & Applied Physics (IJPAP)*, Vol. 57, pp. 352–360, May 2019.
72. Neeraj Jain and Balwinder Raj, "Dual-k Spacer Region Variation at the Drain Side of Asymmetric SOI FinFET Structure: Performance Analysis towards the Analog/RF Design Applications", communicated in "*Journal of Nanoelectronics and Optoelectronics*", Vol. 14, pp. 349–359, March 2019.
73. Jeetendra Singh, Sanjeev Sharma, B. Raj, Mamta Khosla, Analysis of Barrier Layer Thickness on Performance of In1-xGaxAs Based Gate Stack Cylindrical Gate Nanowire MOSFET," *JNO, ASP*, Vol. 13, pp. 1473–1477, Oct 2018.
74. Neeraj Jain, B. Raj, "Analysis and Performance Exploration of High-k SOI FinFETs Over the Conventional Low-k SOI FinFET toward Analog/RF Design", *Journal of Semiconductors (JoS)*, Vol.39, Issue 12, pp. 124002-1-7, Dec 2018.
75. Candy Goyal, Jagpal Singh Ubhi and B. Raj, A Reliable Leakage Reduction Technique for Approximate Full Adder with Reduced Ground Bounce Noise, *Journal of Mathematical Problems in Engineering*, Volume 2018, Article ID 3501041, 16 pages, 15 Oct 2018.
76. Anuradha, Jeetendra Singh, B. Raj and Mamta Khosla, "Design and Performance Analysis of Nano-scale Memristor-based Nonvolatile SRAM" *Journal of Sensor Letter*, Vol. 16, PP 798–805, Oct 2018.
77. Girish Wadhwa, B. Raj, "Parametric Variation Analysis of Charge-Plasma-based Dielectric Modulated JLTFET for Biosensor Application" *IEEE Sensor Journal*, Vol. 18, No. 15, August 1, pp. 6070 – 6077, 2018 (SCI)

78. Jeetendra Singh, B. Raj, "Comparative Analysis of Memristor Models for Memories Design", *JoS, IoP*, Vol. 39, No. 7, PP. 074006-1-12, July 2018.
79. Divya Yadav, Shailesh Singh Chouhan, Santosh Kumar Vishvakarma and B. Raj, "Application Specific Microcontroller Design for IoT based WSN", *Sensor Letter, ASP*, Vol. 16, pp. 374–385, May 2018.
80. Gurmohan Singh, R. K. Sarin and B. Raj, "Fault-Tolerant Design and Analysis of Quantum-Dot Cellular Automata Based Circuits", *IEEE/IET Circuits, Devices & Systems*, Vol. 12, pp. 638–664, 2018.
81. Jeetendra Singh, B. Raj, "Modeling of Mean Barrier Height Levying Various Image Forces of Metal Insulator Metal Structure to Enhance the Performance of Conductive Filament Based Memristor Model", *IEEE Nanotechnology*, Vol. 17, No. 2, pp. 267–268, March 2018.
82. Aakash Jain, Sanjeev Sharma, B. Raj, "Analysis of Triple Metal Surrounding Gate (TM-SG) III-V Nanowire MOSFET for Photosensing Application", *Opto-electronics Journal*, Volume 26, Issue 2, Pages 141–148, May 2018.
83. Aakash Jain, Sanjeev Sharma, B. Raj, "Design and Analysis of High Sensitivity Photosensor Using Cylindrical Surrounding Gate MOSFET For Low Power Sensor Applications", *Engineering Science and Technology, an International Journal, Elsevier's*, Volume 19, Issue 4, pp. 1864–1870, December 2016.
84. G. Saiphani Kumar, Amandeep Singh, B. Raj, "Design and Analysis of Gate All Around CNTFET based SRAM cell Design", *Journal of Computational Electronics, Springer*, Volume 17, Issue 1, pp 138–145, March 2018.
85. Gurinder Pal Singh, B. S. Sohi, Balwinder Raj, "Material Properties Analysis of Graphene Base Transistor (GBT) for VLSI Analog Circuits, *Indian Journal of Pure & Applied Physics (IJPAP)*, Vol. 55, pp. 896–902, Dec-2017.
86. Sunil Kumar and B. Raj, "Estimation of Stability and Performance metric for Inward Access Transistor based 6T SRAM Cell Design using n-type/p-type DMDG-GDOV TFET" *IEEE VLSI Circuits and Systems Letter*, Volume 3, Issue 2, pp. 25–39, June 2017
87. Shashikant Sharma, Anjan Kumar, Manisha Pattanaik and B. Raj, "Forward Body Biased Multimode Multi-Threshold CMOS Technique for Ground Bounce Noise Reduction in Static CMOS Adders" *International Journal of Information and Electronics Engineering*, pp. 567–572, Vol. 3, issue 3, 2013.
88. Hamendra Singh, Pankaj Kumar, Balwinder Raj, "Performance Analysis of Majority Gate SET Based 1-bit Full Adder", *International Journal of Computer and Communication Engineering (IJCCE)*, IACSIT Press Singapore, ISSN: 2010-3743, Volume 2, No. 4, 2013.
89. Anil Kumar Bhardwaj, Sumeet Gupta and B. Raj, "Investigation of Parameters for Schottky Barrier (SB) Height for Schottky Barrier Based Carbon Nanotube Field Effect Transistor Device", *Journal of Nanoelectronics and Optoelectronics, ASP*, Vol. 15, PP. 783–791, July 2020.
90. Priya Bansal, B. Raj, "Memristor: A Versatile Nonlinear Model for Dopant Drift and Boundary Issues," *JCTN, American Scientific Publishers*, Vol. 14 (5), PP. 2319–2325, May 2017.
91. Neeraj Jain, B. Raj, "An Analog and Digital Design Perspective Comprehensive Approach on Fin-FET (Fin-Field Effect transistor) Technology - A Review", *Reviews in Advanced Sciences and Engineering (RASE), ASP*, Vol. 5, PP. 1–14, 2016.

92. Balwinder Raj, A. K. Saxena and S. Dasgupta, "Analytical Modeling for the Estimation of Leakage Current and Subthreshold Swing Factor of Nanoscale Double Gate FinFET Device" *Microelectronics International, UK*, vol. 26, pp. 53–63, 2009.
93. Shailendra Singh Soniya, Girish Wadhwa and Balwinder Raj, "An Analytical Modeling for Dual Source Vertical Tunnel Field Effect Transistor" *submitted to 'International Journal of Recent Technology and Engineering (IJRTE)'* at Volume 8 Issue 2, pp. 603–608, July 2019.
94. Shailendra Singh, B. Raj, "Design and Analysis of Hetrojunction Vertical T-Shaped Tunnel Field Effect Transistor", *Journal of Electronics Material, Springer*, Volume 48, Issue 10, pp. 6253–6260, October 2019.
95. Candy Goyal, Jagpal Singh Ubhi and B. Raj, "A Low Leakage CNTFET based Inexact Full Adder for Low Power Image Processing Applications", *International Journal of Circuit Theory and Applications, Wiley*, Volume 47, Issue 9, pp. 1446–1458, September 2019.
96. B. Raj, A. K. Saxena and S. Dasgupta, "A Compact Drain Current and Threshold Voltage Quantum Mechanical Analytical Modeling for FinFETs" *Journal of Nanoelectronics and Optoelectronics (JNO), USA*, vol. 3, no. 2, pp. 163–170, 2008.
97. Girish Wadhwa, B. Raj, "An Analytical Modeling of Charge Plasma based Tunnel Field Effect Transistor with Impacts of Gate underlap Region" *Superlattices and Microstructures, Elsevier*, Vol. 142, pp.106512, June 2020.
98. Shailendra Singh, B. Raj,"Modeling and Simulation analysis of SiGe heterojunction Double GateVertical t-shaped Tunnel FET", *Superlattices and Microstructures*, Volume 142, pp. 106496, June 2020.
99. Neeraj Jain, Balwinder Raj, "Device and Circuit Co-Design Perspective Comprehensive Approach on FinFET Technology - A Review, *Journal of Electron Devices*, Vol. 23, No. 1, pp. 1890–1901, 2016.
100. Sunil Kumar and B. Raj, "Analysis of I_{ON} and Ambipolar Current for Dual-Material Gate-drain Overlapped DG-TFET," *Journal of Nanoelectronics and Optoelectronics*, Vol. 11, pp. 323–333, June 2016.
101. Naveed Anjum, Tarun Bali and B. Raj, Design and Simulation of Handwritten Multiscript Character Recognition, *International Journal of Advanced Research in Computer and Communication Engineering*, Vol. 2 (7), pp. 2544–2549, July 2013.
102. Karmjit Singh, B. Raj, "Performance and Analysis of Temperature Dependent Multi-Walled Carbon Nanotubes as Global Interconnects at Different Technology Nodes," *Journal of Computational Electronics*, Volume 14 Issue 2, pp. 469–476, June 2015.
103. Sunil Kumar and B. Raj, "Compact Channel Potential Analytical Modeling of DG-TFET Based on Evanescent–Mode Approach," *Journal of Computational Electronics* Vol. 14, No. 2, pp.820–827, July 2015.
104. Karmjit Singh, B. Raj, "Temperature Dependent Modeling and Performance Evaluation of Multi-Walled CNT and Single-Walled CNT as Global Interconnects," *Journal of Electronic Materials*, Vol. 44, Issue 12, pp 4825–4835, Dec 2015.
105. V.K. Sharma, M. Pattanaik and B. Raj, "INDEP Approach for Leakage Reduction in Nanoscale CMOS Circuits", *International Journal of Electronics*, Vol. 102, No. 2, pp. 200–215, 2014.

106. Karmjit Singh, B. Raj, "Influence of Temperature on MWCNT bundle, SWCNT bundle and Copper interconnects for Nanoscaled Technology nodes," *Journal of Materials Science: Materials in Electronics*, Vol. 26, No. 8, pp. 6134–6142, 2015.
107. Naveed Anjum, Tarun Bali and B. Raj, "Design and Simulation of Handwritten Gurumukhi and Devanagri Numerical Recognition, *International Journal of Computer Applications*, Vol. 73(12), pp. 16–21, 2013.
108. S. Khandelwal, V. Gupta, B. Raj, and R. D. Gupta, Process Variability Aware Low Leakage Reliable Nano Scale DG-FinFET SRAM Cell Design Technique, *Journal of Nanoelectronics and Optoelectronics*, Volume 10, Number 6, pp. 810–817, December 2015.
109. V.K. Sharma, M. Pattanaik and B. Raj, "ONOFIC Approach: Low Power High Speed Nanoscale VLSI Circuits Design", *International Journal of Electronics*, Vol. 101, issue 1, pp. 61–73, 2014.
110. S. Khandelwal, Balwinder Raj, and R. D. Gupta, FinFET based 6T SRAM Cell Design: Analysis of Performance Metric, Process Variation and Temperature Effect. *Journal of Computational and Theoretical Nanoscience*, Vol. 12, pp. 2500–2506, 2015.
111. Sumit Singh; Shekhar Yadav; Jagdeep Rahul; Anurag Srivastava; B. Raj, Impact of HfO_2 in Graded Channel Dual Insulator Double Gate MOSFET, *Journal of Computational and Theoretical Nanoscience, American Scientific Publishers*, Volume 12, Number 6, pp. 950–953, April 2015.
112. Vijay Kumar Sharma, Manisha Pattanaik, B. Raj, PVT variations aware low leakage INDEP Approach for Nanoscale CMOS Circuits, *Microelectronics Reliability, Elsevier*, Vol. 54, pp 90–99, 2014.
113. B. Raj, A. K. Saxena and S. Dasgupta, "Quantum Mechanical Analytical Modeling of Nanoscale DG FinFET: Evaluation of Potential, Threshold Voltage and Source/Drain Resistance", *Elsevier's Journal of Material Science in Semiconductor Processing, Elsevier*, Vol. 16, issue 4, pp. 1131–1137, 2013.
114. Maisagalla Gopal, Siva Sankar, D. Prasad and Balwinder Raj, 8T SRAM Cell Design for Dynamic and Leakage Power Reduction, *International Journal of Computer Applications*, Vol. 71(9), pp. 43–48, June 2013.
115. Manisha Pattanaik, B. Raj, Shashikant Sharma and Anjan Kumar, "Diode Based Trimode Multi-Threshold CMOS Technique for Ground Bounce Noise Reduction in Static CMOS Adders", *Advanced Materials Research*, vol. 548, pp. 885–889, 2012.
116. Balwinder Raj, A. K. Saxena and S. Dasgupta, "Nanoscale FinFET Based SRAM Cell Design: Analysis of Performance metric, Process variation, Underlapped FinFET and Temperature effect" *IEEE Circuits and System Magazine*, vol. 11, issue 2, pp. 38–50, 2011.
117. V.K. Sharma, M. Pattanaik & Balwinder Raj, "Leakage Current ONOFIC Approach for Deep Submicron VLSI Circuit Design", *International Journal of Electrical, Computer, Electronics and Communication Engineering, World Academy of Sciences, Engineering and Technology*, Vol. 7, issue 4, pages 239–244, 2013.
118. Parminder Kaur, Sandeep Singh Gill and B. Raj, "Comparative Analysis of OFETs Materials and Devices for Sensor Applications", *Journal of Silicon, Springer*, Vol 14, PP. 4463–4471, 2022.

119. Sanjeev Kumar Sharma, Parveen Kumar, Balwant Raj and B. Raj, "$In_{1-x}Ga_xAs$ Double Metal Gate-Stacking Cylindrical Nanowire MOSFET for highly sensitive Photo detector", *Journal of Silicon, Springer* Vol.14, PP. 3535–3541, 2022.
120. B. Raj, A. K. Saxena and S. Dasgupta, "Analytical Modeling of Quasi Planar Nanoscale Double Gate FinFET with Source/Drain Resistance and field dependent carrier mobility: A quantum mechanical study" *Journal of Computer (JCP),Academy Publisher, FINLAND*, vol. 4, issue 9, pp. 1–8, 2009.
121. S. Bhushan, S. Khandelwal and B. Raj, (2012). Analyzing Different Mode FinFET based Memory Cell at Different Power Supply for Leakage Reduction, Seventh International Conference on Bio-Inspired Computing: Theories and Application, *(BIC-TA 2012) Advances in Intelligent Systems and Computing*, Volume 202, 89–100.
122. Jeetendra Singh, B. Raj, "Temperature Dependent Analytical Modeling and Simulations of Nanoscale Memristor, *Journal: Engineering Science and Technology, an International Journal, Elsevier's*, Vol. 21, PP- 862–868, Oct 2018.
123. Shradhya Singh, Shashi Bala, Balwant Raj, B. Raj, "Improved Sensitivity of Dielectric Modulated Junctionless Transistor for Nanoscale Biosensor Design", *Sensor Letter, ASP*, Vol. 18, pp. 328–333, 2020.
124. Vivek Kumar, Santosh Kumar Vishvakarma, and B. Raj, "Design and Performance Analysis of ASIC for IoT Applications" *Sensor Letter ASP*, Vol. 18, PP. 31–38, Jan 2020.
125. Akanksha Jaiswal, R. K. Sarin, B. Raj and Shikha Sukhija, "A Novel Circular Slotted Microstrip-fed Patch Antenna with Three Triangle Shape Defected Ground Structure for Multiband Applications", *Advanced Electromagnetic (AEM)*, Vol. 7, Issue 3, PP. 56–63, 2018.
126. Girish Wadhwa, B. Raj, "Label Free Detection of Biomolecules using Charge-Plasma-Based Gate Underlap Dielectric Modulated Junctionless TFET", *Journal of Electronic Materials (JEMS), Springer*, Volume 47, Issue 8, pp 4683–4693, August 2018.
127. Gurmohan Singh, R. K. Sarin and B. Raj, "Design and Performance Analysis of a New Efficient Coplanar Quantum-Dot Cellular Automata Adder", *Indian Journal of Pure & Applied Physics (IJPAP)*, Vol. 55, pp. 97–103, February 2017.
128. Parminder Kaur, Vikas Pandey and B. Raj, "Comparative Study of Efficient Design, Control and Monitoring of Solar Power using IoT", *Sensor Letter, ASP* Vol. 18, pp. 419–426, May 2020.
129. Mohit Kumar, Tarun Chaudhary; Balwinder Raj, "Design and Comparative Analysis of Ferroelectric Nanowire with Dielectric HfO_2 and $Al_2O_3$for Low Power Applications", *Journal of Electronic Materials*, Vol. 53, pp. 6785–6791, 2024
130. Yogesh Thakur, Balwant Raj and Balwinder Raj, "Design of Pentacene Thin-Film Transistor Based Hydrogen Gas Sensor for High Sensitivity", *ECS Journal of Solid State Science and Technology*, Vol. 13, pp. 047005–047012, 24 April 2024.
131. Manoharan Arun Kumar, Kamal Batcha Mohamed Ismail, Shanmugam Mahalingam, Balwinder Raj, Junghwan Kim, Raji Atchudan, "Recent Advances in Two-Dimensional Nanomaterials and Nanostructures for Healthcare Monitoring", *ACS Sensors*, Vol. 9, issue 4, pp. 1–8, 2 April 2024

132. Mohamed Ismail Kamal Batcha; Arun Kumar Manoharan; Shanmugam Mahalingam; Raj Balwinder, "Carbon Fiber-Reinforced Polymers for Energy Storage Applications", *Journal of Energy Storage*, Vol. 84, Part B, pp. 110931–110939, 20 April 2024.
133. Shahraan Hussain, Balwinder Raj, Divya Sharma "Design and Parametric Optimization of Lead Free Perovskite based Tandem Solar Cell for High Efficient", *Engineering Research Express*, Vol.6, PP. 015315–0153122, 13 March 2024
134. Parveen Kumar, Balwinder Raj, Girish Wadhwa, Balwinder Singh and Raj Kumar "Design and Analysis of Junctionless based Gate All Around N+ Doped Layer Nanowire TFET Biosensor", *ECS Journal of Solid State Science and Technology*, Vol 13, PP 017002–017010, 15 Jan 2024
135. Divya Sharma, Rajesh Mehra and Balwinder Raj, "Enhancement in Efficiency of Methyl Ammonium Tin Iodide based perovskite solar cell using SCAPS-1 D", *Nano, World Scientific Publisher*, Vol. 18, No. 14, PP. 2350109, 14 Dec 2023.
136. Parveen Kumar, Sanjeev Kumar Sharma, Balwant Raj and Balwinder Raj, "Analysis of Device Parameter Variations in In1-xGaxAs based Gate Stacked Double Metal Surrounding Gate Nanowire MOSFET", *TEEM Springer*, Vol.24, PP. 570–578, Oct 2023
137. Anjana Bhardwaj, Pradeep Kumar, Balwinder Raj, Naveen Kumar, Sunny Anand, "Design and Optimization of Vertical Nanowire Tunnel FET with Electrostatic Doping, *Engineering Research Express, IoP*, Vol.5, PP. 045025–045033, 2023
138. Yogesh Thakur, Balwinder Raj, Sandeep Singh Gill, "Design and Performance Analysis of Pentacene Organic Field Effect Transistor with High-K dielectric Materials, *Journal of Optoelectronics and Advanced Materials – Rapid Communications*, Vol. 17, No. 7–8, July-August 2023 SCI
139. Parveen Kumar and Balwinder Raj, "Design and Simulation of Junctionless Nanowire Tunnel Field Effect Transistor Biosensor for High Sensitivity", *Microelectronic Journal Elsevier*, Vol. 137, PP. 105826–105834, July 2023, SCIE
140. Anil Kumar Bhardwaj, Sumeet Gupta and B. Raj, Development & Analysis of compact model for Double Gate Schottky Barrier CNTFET, *Journal of Nanoelectronics and Optoelectronics, ASP*, Vol. 15, pp. 1199–1208, Aug 2020.
141. Girish Wadhwa, Priyanka Kamboj; Jeetendra Singh, B. Raj, "Design and Investigation of Junctionless DGTFET for Biological Molecule Recognition" *Transactions on Electrical and Electronic Materials, Springer*, Vol.22, PP. 282–289, 2021.
142. Sachin Kumar Verma, Shailendra Singh, Girish Wadhwa and B. Raj, "Detection of Biomolecules using Charge-Plasma based Gate Underlap Dielectric Modulated Dopingless TFET" *Transactions on Electrical and Electronic Materials (TEEM), Springer*, Vol. 21, PP. 528–535, June 2020.
143. Neeraj Jain, B. Raj, "Impact of Underlap Spacer Region Variation on Electrostatic and Analog/RF Performance of Symmetrical High-k SOI FinFET at 20 nm Channel Length", *Journal of Semiconductors (JoS), IOP Science*, Vol.38, Issue 12, PP- 122002, Dec 2017.

144. Shailendra Singh, B. Raj," Analytical Modeling and Simulation Analysis of T-Shaped III-V Heterojunction Vertical T-FET", *Superlattices and Microstructures, Elsevier*, Vol. 147, PP. 106717, Nov 2020.
145. Gurmohan Singh, R. K. Sarin and B. Raj, "Design and Analysis of Area Efficient QCA Based Reversible Logic Gates", *Journal of Microprocessors and Microsystems, Elsevier*, Vol. 52, PP.59–68, May 2017.
146. Shailendra Singh, B. Raj,"Parametric Variation Analysis on Hetero-Junction Vertical t-Shape TFET for Supressing Ambipolar conduction", *Indian Journal of Pure and Applied Physics*, Vol. 58,pp. 478–485 June 2020.
147. Shailendra Singh Soniya, Girish Wadhwa and Balwinder Raj, Design and Analysis of Dual Source Vertical Tunnel Field Effect Transistor for High Performance, *Transactions on Electrical and Electronics Materials*, Vol. 21, Pages 74–82, Oct 2019.
148. Manjit Kaur; Neena Gupta; Sanjeev Kumar; B. Raj; Arun Kumar Singh, "RF Performance Analysis of Intercalated Graphene Nanoribbon Based Global Level Interconnects" *Journal of Computational Electronics*, Vol. 19, pp. 1002–1013, June 2020
149. Girish Wadhwa, B. Raj, "Design and Performance Analysis of Junctionless TFET Biosensor for high sensitivity" *IEEE Nanotechnology*, Vol. 18, pp. 567–574, 2019.
150. Jeetendra Singh, B. Raj, "Enhanced Nonlinear Memristor Model Encapsulating Stochastic Dopant Drift, *JNO, ASP* 14, 958–963 (2019)

Chapter 6

Nanowire-FET

Role in healthcare applications

Anita Shukla, Ankit Jain, Shailendra Singh, Puspraj Singh Chauhan, and Imran Ullah Khan

6.1 INTRODUCTION

Nanostructures create 2D layers, 1D nanowires (NWs), or 0D nanoclusters by shrinking bulk materials in one, two, or three dimensions, respectively [1]. Over time, considerable advancements have been made in nanowires, which are incredibly thin structures on the nanoscale. Nanowires, nanofibers, nanotubes, nanorods, and whiskers generally represent single-dimensional nanostructures. It was first reported in 1998 that silicon nanowires may be successfully synthesized by a growing process. Eventually, a group of researchers released a number of articles detailing their advancements in the preparation of Si and other semiconductor nanowires and the development of lab-scale devices using them, like silicon nanowire transistors. The majority of methods for fabricating nanowires are based on the vapor-liquid-solid (VLS) crystal growth mechanism [1–8].

Synthesized nanomaterials, such as carbon nanotubes, nanoparticles, and quantum dots, including nanowires, have contributed significantly in different sectors, including biological sensors, in the past ten years. On comparing with bulk materials, nanoparticles' high surface area-to-volume ratio significantly boosts their sensitivity. The usage in drug discovery and the detection of viruses, DNA, and biomarkers are the main topics of this review. Nanomaterials are also employed in photonics, data storage, chemical and biological sensing, drug delivery, and energy storage [2, 3]. Further advancements in the creation, description, and incorporation of nanowires may lead to innovations in a number of domains, including flexible electronics and biomedicine [4]. Due to their small size and adaptable characteristics, nanowires have great promise for an extensive range of technological applications [9–16].

Nanowires are highly effective structures for modern nanoelectronic devices, offering the ability to precisely control their diameter during fabrication [17]. In recent years, scaling of MOSFET technology has enabled integration of a large number of field-effect transistors (FETs) onto a single chip. Gordon Moore's suggestion that the number of transistors on a chip will double every two years has proven accurate. The International

DOI: 10.1201/9781003634645-6

Technology Roadmap for Semiconductors (ITRS) forecasts that complementary metal-oxide-semiconductor (CMOS) technology scaling will continue for the foreseeable future. Over the past four decades, advances in integrated circuit (IC) technology have brought transistor sizes down to the nanometer scale with the introduction of 22 nm process technologies. Today, transistors are typically fabricated using gate lengths ranging from 22 nm to 45 nm. Leading companies like Intel, AMD, and IBM have showcased p-channel FETs with gate lengths as small as 6 nm. Nanowires, with their one-dimensional structure, are essential for the development of various nanoscale semiconductor devices, including FETs, photodetectors, LEDs, and biosensors. Nanowires, created through techniques such as electrospinning and vapor deposition, possess remarkable mechanical, optical, and electrical properties that make them highly suited for an extensive range of applications, particularly in the fields of nanoscale electronics, drug delivery, and chemical sensing. The precision provided by advanced synthesis methods, including vapor-liquid-solid (VLS) and template-assisted techniques, allows for custom fabrication of nanowires with specific characteristics. This capability provides opportunities to apply nanowires for intracellular signaling, providing an efficient pathway for specific drug delivery, both of which are necessary to improve precision medicine [18–26].

Biomedical areas where nanowires have biopharmaceutical applications are particularly exciting; biomedical imaging, neural interfacing, and tNo. engineering are just a few examples. As far as potential multipurpose applications are concerned, nanowires have tremendous potential; notwithstanding the issues of biocompatibility and the feasibility of manufacturing, these challenges could be addressed because nanowires could provide multifaceted characteristics and functionality, leading to innovative diagnostic, biosensing, and regenerative medicine methodologies. In addition, nanowires are expected to play an important role in the next generation of point-of-care medical technologies [126–134].

In drug delivery, nanowires display great promise for precision therapy. For example, iron oxide nanowires can hold chemotherapeutic agents, and then heat the nanowires with an external magnetic field for targeted chemotherapy for cancer [5]. In addition to their potential to provide targeted treatment, nanowires, such as nanowire arrays, are able to provide continuous and controlled drug release in a manner that elicits consistent therapeutic effects [6]. Other uses for nanowires that hold great promise include nanowires for neurotrophic drug delivery targeting brain inflammation and hollow nanowires for delivering insulin for diabetes treatments, with possibilities for self-regulated treatments [7, 8]. Furthermore, nanowires are being explored for their ability to accelerate tNo. regeneration, with growth factors released from the nanowires promoting wound healing and tNo. repair [9]. Nanowire stents have further applications in the vascular field, as they stimulate endothelial cell growth, which facilitates the formation of new blood vessels [10]. Even though nanowire drug delivery systems are still in the beginning stages, they have the potential to reduce systemic side effects and allow for personalized therapies for diseases such as cancer, cardiovascular diseases, and diabetes, which will lead to better patient outcomes [11].

Because of their remarkable properties, specifically their high sensitivity and ability for label-free detection, nanowires have excellent opportunities to develop biomedical imaging modalities. Their attributes enable accurate quantification of biomarkers, allowing the diagnosis of disease at earlier stages and with more reliability [12]. An example is iron oxide nanowires, which can serve as MRI contrast agents to aid the rate of non-invasive imaging, thus providing an improvement in imaging versus surrounding tissues [13]. Additionally, nanowires can help enhance guided photosensitive tumor resections by eliminating larger portions of tumor during surgical procedures when they are combined with proteins, quantum dots, and/or fluorophores [14]. Fluorescent nanowires allow for continuous in vitro monitoring of gene regulation, which also relates to cellular metabolic processes or efficiency, via oligonucleotide delivery tracking [15]. Nanowire neural interfaces represent another domain of this integration for training purposes, where higher resolution or a more accurate design could enable one to study single-neuron calorimetry, improving our ability to measure metabolic processes, while also improving methodology in characterizing neuronal networks refractory to potential neurochemical systems modeling of neurological disease [16]. Albeit these methods are yet to be fully investigated, they represent potential milestones in an otherwise biomedical imaging field that remains deficient in possibilities for standardized diagnostic utility for neuronal mapping, improved imaging of targeted cancers, or diagnosing disease molecular signatures at all [6].

Nanowires have the potential to model the behavior of cells in artificial tissues, therefore being important for tNo. engineering. Nanowire scaffolds can also be arranged by electrospinning and 3D bioprinting to control the linear growth of cells and orientation of the extracellular matrix deposition by specific axes [27]. These wire structures have the potential to be designed and applied individually or stacked to create complex biostructures, with all possible design decisions affecting the tNo. adaptation process by controlling its mechanical properties, degradation rates, and bioactivity variables [28]. Furthermore, nanowire structures could provide clues to stem cells, guiding the path to a specific cell lineage [29]. When engineered surface nanostructures are manipulated in their application, they may represent the in vivo cellular microenvironment, influencing cell adhesion, proliferation, differentiation, and other critical life cycle behaviors, further refining the engineering process. It has been demonstrated in a number of different areas in tNo. engineering, including neural, cardiac, cartilage, and bone, that nanowire structures can generate biomimetic signals that encourage tNo. regeneration instead of fibrosis or scarring [30, 31]. Although these advances are promising, it is important to note that we do not know all of the potential risks, such as cytotoxicity, and additional studies need to be done to ensure the safety of nanowire biomimetic scaffolds in the clinical setting [32–41].

Future revolutions in nanotechnology and healthcare might be influenced by the multifunctional characteristics of nanowires [42]. This smooth integration into biological systems has the ability to impact the technologies of biosensing and diagnostics altogether [43]. For example, the ability to increase the efficacy of regenerative medicine, when coupled with controlled stem cell proliferation, by using nanowire scaffolds could have vast implications in tNo. repair and repair solutions. Additionally, the development of a nanowire thermoresponsive system is providing possibilities for cancer therapies, such as therapies that can be targeted by temperature change [44]. In addition, their precise and adaptable drug delivery mechanisms could pave the way for next-generation point-of-care treatments, offering personalized and timely interventions. Nanowires can be a possible tool to accelerate clinical translation of nanomedicine as developments progress, but achieving their full potential in healthcare applications would require resolving issues with biocompatibility and scalable manufacturing [32–41, 45–57].

6.2 BASICS OF FIELD-EFFECT NANOWIRE TRANSISTORS

Traditional planar field-effect transistors (FETs), made up of gate, source, drain, and body, gave rise to nanowire FETs, a kind of sensor (Figure 6.1). On the body, metallic materials are used to build the source and drain at micro- or nanoscale. Gate is a thin layer of isolation, such as SiO2, positioned between the source and drain [58–67]. The conductivity between source and drain is regulated by the gate's creation of changes in electric potential. Usually, an external voltage triggers these potential changes. Furthermore, by adhering to the gate, species that are chemically or biologically charged can affect the gate's potential and conductivity. Decades ago, the buildup of charged species was hypothesized as the basis for this detection technique [68]. Though the low sensitivity of planar gate FET sensors necessitates large sample volumes for detection, limiting their use in many applications.

Nanowires and receptors, respectively, take the place of conventional doped channels and gates in nanowire field-effect transistors (FETs) (Figure 6.1a). Notwithstanding structural variations, the fundamental idea behind nanowire FETs is the same as that of conventional FETs: external changes brought on by charged species regulate the conductivity of the nanowires, which serve as the doped channels [69–83]. Silicon and silica nanowires are the most often utilized varieties of semiconductor nanowires, although they can be made from a variety of materials. Their strong compatibility with conventional complementary metal-oxide-semiconductor (CMOS)

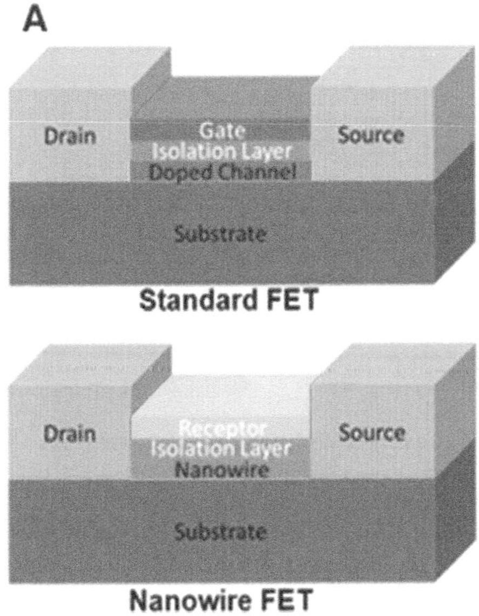

Figure 6.1 Diagrams showing the distinctions between conventional and nanowire field-effect transistors provide an understanding of field-effect transistors (FETs).

technology, which has long been the mainstay of contemporary semiconductor manufacture, is the main reason for this. Both silicon and silica nanowires' usefulness in device applications is further increased by their capacity to naturally form an oxide layer that acts as an efficient isolation layer on their surface and by how simple it is to alter their surfaces [84–92].

While biomolecules themselves are charged, simply accumulating on the isolation layer of a nanowire does not alter the conductivity of the nanowire, since the silicon and silica surfaces are not inherently sensitive to biomolecules [93]. To overcome this limitation, it is essential to functionalize the surface of the nanowires with specific receptors that can bind the target charged species, such as DNA, RNA, viruses, and other biomolecules. The functionalization process involves modifying the nanowire surfaces so that they can chemically interact with and capture these biomolecules. The process of functionalization creates a bridge between the nanowire surface, which is coated with an isolation layer, and the specific biomolecules being detected (Figure 6.2).

Once the receptors on the nanowire bind to the biomolecules, they induce an electric field in the nanowires. This, in turn, alters the conductivity of the nanowires, which is the core principle behind the detection mechanism. In this way, the receptors on the nanowire FETs effectively play a role similar to the gate in traditional FETs by converting the input signal in this case, the binding of charged species) into a change in the device's conductivity. As a

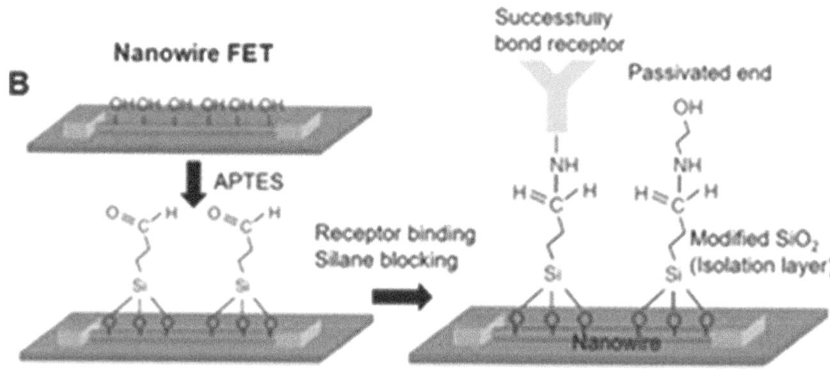

Figure 6.2 Conceptual overview of field-effect transistors (FET).

result, the presence of specific biomolecules can be detected through the electrical response of the nanowires [94–102].

A typical process for functionalizing the nanowire FETs is illustrated in Figure 6.1b, where 3-aminopropyltriethoxysilane (APTES) is used as a receptor for DNA, peptide nucleic acid (PNA), and antibodies. APTES reacts with the silicon surface to convert the silicon-oxygen bonds to a silane chemistry layer (Si–O–Si–X). This silane layer can also be modified or linked with a specific receptor meant for specific biomolecules [103–114]. The functionalization step is fundamental, as it allows for the nanowire FETs to recognize and sense selective target molecules with sensitivity and specificity for biosensing purposes. With these modifications, the nanowire FETs can be functionalized for a variety of diagnostics and analyses and will have an important role to play in healthcare, environmental monitoring, and chemical analysis.

6.3 MOS TRANSISTOR

The field-effect transistor's (FET) working principle was first proposed in 1930. Since then, MOSFETs have been used extensively in integrated circuits (ICs). As circuit complexity increases and device dimensions decrease, IC technology has developed [103–111]. Discrete atomistic effects (DTE) in doping, bandgap problems, quantum confinement at the interface, and tunneling through the gate terminal are examples of small-scale effects that have surfaced. These difficulties result in issues like thermal effects and dispersion, especially when high-power densities (HPD) are present.

The MOSFET continues to be the core component of microprocessors and VLSI circuits. In CMOS technology, the gate terminal is composed of n-type and p-type polysilicon with silicide, which reduces the series resistance of the gate terminal. The high circuit density in CMOS technology is managed by shallow trench isolation (STI) technology. Deep and shallow implants are combined to generate the source and drain. Because doping concentration affects a MOSFET's threshold voltage, which decreases as doping increases, MOSFETs with shorter gate lengths can be manufactured [112–120].

6.3.1 Scaling of MOSFET

Ever since the MOSFET was invented, integrated circuit technology has seen rapid and exponential growth, largely driven by the downscaling of MOSFET dimensions, a concept first proposed in 1974. Two primary scaling methods are used: constant voltage scaling and constant field scaling. The packing density of MOSFET devices has steadily increased over the last forty-five years. According to the International Technology Roadmap for Semiconductors (ITRS), the gate length would be lowered to 7 nm by 2018. As device dimensions have reached the nanoscale, new challenges in modeling and simulating carrier transport have emerged, particularly due to the influence of quantum mechanical effects. The most commonly used model for simulating carrier transport is derived from the Boltzmann transport equation. In recent years, transistor performance has improved with reductions in size, supply voltage, and gate oxide thickness. As MOSFET dimensions continue to approach the nanometer scale, quantum physics phenomena must be considered to account for new effects. To maintain continuous improvement, scaling and modifying the devices is essential. At these reduced scales, various phenomena, known as short-channel effects, become significant and must be addressed for the continued advancement of transistor technology [18–25, 121, 122].

6.3.2 Advance MOSFET structure

According to the ITRS, semiconductor devices will need channel lengths of about 10 nm by 2015 due to the ongoing downscaling of field-effect transistors. As new semiconductor materials emerge, novel nanoscale devices have been developed, necessitating further scaling down to achieve 10 nm gate lengths. The advancement of multi-gate field-effect transistors, such as double-gate MOSFETs, Pi-gate FETs, omega-gate MOSFETs, tri-gate MOSFETs, and surrounding-gate MOSFETs, calls for further progress in fabrication techniques to support these innovations [26, 123–127].

6.3.3 Double-gate (DG) MOSFET

An ultra-thin silicon layer is positioned between the front gate and the rear gate, the two gate terminals of a double-gate MOSFET (DG MOSFET). Both symmetric and asymmetric DG MOSFET architectures have been created. These devices can operate in two modes: three-terminal and four-terminal.

Many research groups have developed analytical and compact models to describe the behavior of DG MOSFETs, advancing the understanding of their performance and potential applications in modern semiconductor technology.

6.3.4 Surrounding-gate MOSFET

Gate terminal of a surrounding-gate MOSFET completely encircles the channel, offering enhanced channel control and a notable decrease in leakage current. Numerous surrounding-gate MOSFET models have been created to facilitate circuit modeling. In order to improve device performance, a charge plasma process is employed for constructing a graded channel, while a doping-less technique has been introduced to design graded-channel nanowire MOSFETs. According to studies, nanowire FETs perform better than DG MOSFETs in terms of sensitivity and response, which has led to their proposal for usage in photosensors. Furthermore, a three-gate metal nanowire MOSFET has been created especially for photo-sensing applications.

Comparative studies have indicated that carbon nanotube (CNT) FETs offer superior performance compared to nanowire FETs, demonstrating enhanced efficiency and faster response times in various applications.

6.3.5 FinFET

FinFET is a variant of the double-gate MOSFET and serves as an effective alternative to conventional MOSFETs. Its fabrication is relatively straightforward because of its simple structure. Various experiments have shown that FinFETs experience fewer short-channel effects compared to traditional MOSFETs. Researchers have developed different FinFET structures, including symmetric and asymmetric designs, to optimize performance and address specific challenges in semiconductor technology. These advancements demonstrate the potential of FinFETs to improve device efficiency and scalability in modern electronics.

6.4 ISSUES AT NANOSCALE LEVEL

With transistor channel lengths below 50 nm, new ways of fabricating devices are required to address problems encountered by varying physical phenomena. At the nanoscale, classical physics fails to adequately characterize electron behaviors, leading to issues such as short-channel effects. The number of quantum tunneling events where charge carriers can penetrate thin oxide layers from source to drain is a real problem, but in addition to quantum tunneling, potential changes in threshold voltage due to quantum confinement can impose further limitations on electronic device performance.

Material dimensions clearly affect their electronic behaviors. However, a suitable method for the fabrication of nanowires < 10 nm is not yet commonplace. As technology nodes continue to advance toward sub-10 nm dimensions, new strategies to control and manage charge carrier transport will be needed, specifically for nanostructures such as nanotubes and nanowires. Building seriously contributes toward furthering these devices and seeing respect for better understanding the basic physics that govern the disposition and characteristics of one-dimensional nanostructures.

To push the boundaries of what is possible at the nanoscale, new material classes, fabrication approaches, and a better understanding of quantum mechanical behaviors will be required to continue scaling and improve the performance of our future electronics.

6.5 NANOWIRE FOR BIOMEDICAL APPLICATIONS

Sensors that are extremely sensitive allow for disease detection when diseases are present in their earliest stages and timely diagnosis. At the nanoscale, when the size of the material drops down to the nanometer order ($\sim 10^{-9}$ m), the substance's physical and chemical characteristics are going to be largely influenced by factors such as the high surface area-to-volume ratio and quantum effects, which are totally different from macroscopic materials. As the size of a material decreases, the ratio of surface area to volume increases, meaning that a much larger proportion of that material's structure is in contact with the target. Because there is a bigger surface area, it will allow more interactions with the surrounding environment as well. Also, in nanostructures, electrons flow under the influence of quantum effects and these quantum effects confine their behavior and interactions. There is high significance for nanowires for biomedical applications when thinking about everything from disease diagnosis to drug discovery, because they are very useful for identifying and characterizing chemical and biological species. Surprisingly, semiconductor nanowires directly measure electrical signals coming from bio/chemical interactions without labeling. This means that nanowires therefore represent an extremely potent tool for the continuation of the evolution of diagnostic processes and research into therapeutics.

6.6 IDENTIFYING VIRUSES

Infections are the main cause of disease in humans, and there is an increasing fear of the use of infections as biological weapons. Improvement of the detection of viruses can enhance the health and safety of the public. Silicon nanowire sensors can detect many harmful viruses, such as dengue, influenza A, H3N2, H1N1, and HIV. Silicon nanowire sensors can detect viruses at the single-virus level. Specific antibodies that target a virus are selected and then attached to the surface of a nanowire FET sensor, which is fabricated, in order to detect a target virus. When a specific virus comes into contact with the sensor, the antibodies attach to the nanowires and this changes the conductivity of the nanowires.

Detection of smallpox is one of the most important uses of nanowire sensors, as it is a highly fatal disease and poses a significant biological weapon threat. Nanowires have been used for more than detecting smallpox; they have detected many other harmful viruses, including dengue, H3N2, H1N1, and HIV. The antibodies over the surface of the nanowire used for detection attach to targeted viruses in a precise, detecting way that changes the electrical conductivity of the nanowire.

Additionally, nanowires have been used in Ebola detection, with a recent study achieving a limit of detection (LOD) of 6.25 nm after only 30 minutes of incubation—six orders of magnitude better than conventional ELISA methods. As the need for real-time, point-of-care diagnostics grows, nanoelectronics will increasingly play a critical role in the detection and monitoring of emerging and epidemic diseases [128].

6.7 DETECTION OF BIOMARKERS

Recent developments in genomics research have led to the discovery of several biomarkers that show great promise for detecting infections. Nonetheless, considering the variability of infections such as cancer, relying on a single biomarker is often not sufficient to yield suitable diagnostic results. This No. has created an increasing need to turn to multiple biomarkers for improved detection. One major advancement is using p-type silicon nanowire sensors as one of the first to enable electrical detection for proteins in solution. This innovation is extremely valuable for real-time diagnostics and monitoring disease biomarkers, as it presents high sensitivity and specificity aspects relative to the detection process summary map in diagnostic systems (as depicted in Figure 6.3). In recent years, numerous research groups have focused on developing integrated systems capable of continuous or direct detection of infections from biological samples, such as blood.

These platforms have not only been able to detect a particular class of biomarker but have also expanded to different types of biomarkers and biomarkers associated with cardiovascular health and other diseases. For instance, Shahad et al. [129] created a nanowire-based diagnostic system that uses the breath volatome as input for the detection of disease. Breath analysis provides non-invasive detection of disease biomarkers, which offers a valuable mechanism for early diagnosis or monitoring of infections. This work identified the tremendous impact of nanowire-based biosensors that may be able to change the landscape of medical diagnostics using quicker, more precise, and less invasive detection techniques.

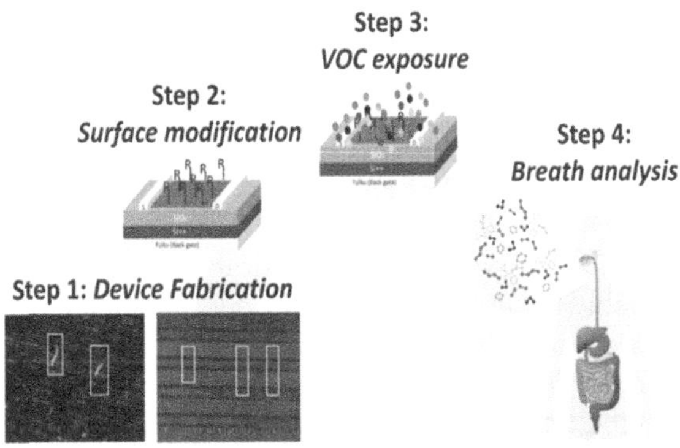

Figure 6.3 Biomarker detection using a nanowire FET sensor.

6.8 DETECTION OF DNA AND RNA

Nanowire field-effect transistor (FET) sensors are an efficient and sensitive platform to quantify specific DNA sequences. The sensors use reciprocal single-stranded peptide nucleic acid (PNA) sequences as receptors (probes); complementary DNA strands on the surface of silicon nanowire FETs bind with PNA receptors and provide an extremely specific identification of DNA and RNA sequences whose detection limits extend to femtoliter concentrations. As mentioned before, nanowire sensors provide more efficient identification and detection of DNA compared to traditional approaches such as quartz crystal microbalance, surface plasmon resonance, and nanoparticle-enhanced resonance.

More than simply using nanowires to detect DNA, other applications include the detection of nucleic acid-associated proteins and protein-protein interactions. And ultimately develop sensors to detect RNA to enhance genetic applications.

Nanowire biosensors have been applied to virus biosensing in the biotechnology industry, improving diagnostic detection speed and sensitivity. Nanowire-based DNA sensors for cancer biomarkers, such as telomerase and carcinoembryonic antigen, are one area of application. The current development stage of multiplexing, measuring multiple markers at the same time, appears promising because of its ability to alter the standard or norm of clinical diagnostic procedures. Being able to measure a variety of analytes from a single sample simultaneously would change the pressures on clinical diagnostics due to the speed and accuracy shown. Nanowire sensors combined with machine learning (a branch of artificial intelligence) and natural language processing could implement several steps in the diagnostic protocol, creating a more automated and timely diagnostic procedure with real-time data analysis. Combining the nanowire sensors, machine learning, and natural language processing could provide new levels of disease detection and treatment in providing a path for precision medicine.

6.9 MEDICATION DISCOVERY

Nanowire FET biosensors offer many desirable qualities that make them particularly suitable for drug discovery and medical diagnostics. It is possible to detect low concentrations of biological markers through these devices, a critical feature in developing new drugs. Nanowires can also be used as drug carriers as their unique properties allow targeted therapy to local parts of the body. An ability to accurately control drug release makes them ideal therapeutic agents for the treatment of certain disease types, fictionalization, and personalized medicine.

Nanowire FET biosensors can not only be useful in drug delivery but also in making and measuring electrophysiological signals, thereby allowing real-time monitoring of biological activity. For example, monitoring cellular response to a drug or monitoring changes in signaling pathways that occur due to a drug, if these were used alongside an advanced monitoring system, could provide researchers with detailed information about how a drug works or its toxicity. Nanowire FET biosensors, therefore, are not only a powerful tool in drug discovery but also a means to enable the development of better and more effective therapeutic approaches.

6.10 CONCLUSIONS

The intersection of nanotechnology and healthcare is accelerating faster than anticipated, generating tremendous opportunities for novel advances in nanomedicine. Continued experimentation with and development of nanowires and aspects unique to them prepare us to tackle relevant issues while enhancing human health. Utilizing nanowire field-effect transistor (FET) biosensors in particular is proving to be a tremendous resource for both biomedical research and clinical applications. They offer a remarkable sensitivity and specificity for an extensive variety of diagnostic and therapeutic applications.

Nanowire FET biosensors will have a huge advantage and be prominently useful in drug discovery. These sensors can be used to detect an extensive number of biologically relevant markers and could help find a drug target that can be quantified at the molecular level. Researchers are also exploring the use of nanowires as drug delivery vehicles by creating targeted drug delivery systems. These systems could potentially improve the efficacy of a therapeutic treatment while minimizing side effects in the patient. These sensors can create and detect electrophysiological signals, and they can also measure cellular responses in real time. The ability to monitor cellular responses makes these sensors an immediate necessity for measuring drug effects in relevant cellular and molecular systems.

In this chapter, we discuss a more complex structure of a MOSFET and its role in healthcare, focusing specifically on the role of nanowire FETs in biosensing. We discussed the applications of these technologies in detecting viruses, identifying biomarkers, analyzing DNA, and drug discovery. Nanowires, with an enormous aspect ratio and controlled synthesis methods, are changing the face of biomedicine by making possible nanoscale electronics, localized drug delivery, and tNo. regeneration. These advances will fundamentally increase the potential impact of healthcare technologies, offering a potential pathway toward personalized medicine and next-generation diagnostics.

REFERENCES

[1] Zhang, A., Zheng, G., Lieber, C.M. Emergence of nanowires. *NanoScience and Technology*, pp. 1–13, 2016. doi: 10.1007/978-3-319-41981-7_1

[2] Raman, S., Sankar, A.R., Sindhuja, M. Advances in Silicon Nanowire Applications in Energy Generation, Storage, Sensing, and Electronics: A Review. *Nanotechnology*, Vol. 34, No. 18, 2023. doi: 10.1088/1361-6528/acb320

[3] Azam, Z., Singh, A. *Various Applications of Nanowires*. Advances in Computer and Electrical Engineering Book Series (pp. 17–53). 2021.

[4] Mukhtar, A., Wu, K., Cao, X., Gu, L. Magnetic Nanowires in Biomedical Applications. *Nanotechnology*, Vol. 31, No. 43, 2020. doi: 10.1088/1361-6528/aba1ba

[5] Karaca, G.Y., Kuralay, F., Uygun, E., Özaltin, K., Demirbuken, S.E., Garipcan, B., Öksuz, L., Öksüz, A.U. Gold–Nickel Nanowires as Nanomotors for Cancer Marker Biodetection and Chemotherapeutic Drug Delivery. *ACS Appl. Nano Mater.*, Vol. 4, No. 4, pp. 3377–3388, 2021. doi: 10.1021/acsanm.0c03145

[6] Li, X., Mo, J., Fang, J., Xu, D., Yang, C., Zhang, M., Li, H., Xie, X., Hu, N., Liu, F. Vertical Nanowire Array-Based Biosensors: Device Design Strategies and Biomedical Applications. *J. Mater. Chem. B.*, Vol. 8, No. 34, pp. 7609–7632. doi: 10.1039/d0tb00990c

[7] Zhang, A., Lee, J., Lieber, C.M. Nanowire-Enabled Bioelectronics. *Nano Today*, p. 38, 2021. doi: 10.1016/j.nantod.2021.101135

[8] Shen, D., Yu, H., Khan, A., Haq, F., Chen, X., Huang, Q., Teng, L. Recent Progress in Design and Preparation of Glucose-Responsive Insulin Delivery Systems. *J. Contr. Release.*, 321, pp. 236–258, 2020. doi: 10.1016/j.jconrel.2020.02.014

[9] Zeng, J., Geng, X., Tang, Y., Xiong, Z., Zhu, Y., Chen, X. Flexible Photothermal Biopaper Comprising Cu2+-Doped Ultralong Hydroxyapatite Nanowires and Black Phosphorus Nanosheets for Accelerated Healing of Infected Wound. *Chem. Eng. J.*, pp. 437, 2022. doi: 10.1016/j.cej.2022.135347

[10] Feng, Y., Wu, M., Zhang, H., He, X.T., Li, H., Chen, D., Jiang, H., Chang, J., Dong, Z., Yang, C. Mild-Temperature Photothermal Assisted CuSi Nanowires for Promoting Infected Wound Healing. *Front. Bioeng. Biotechnol.*, p. 11 2023. doi: 10.3389/fbioe.2023.1158007

[11] Kumar, R., Aadil, K.R., Mondal, K., Mishra, Y.K., Oupický, D., Ramakrishna, S., Kaushik, A. Neurodegenerative Disorders Management: State-of-Art and Prospects of Nano-Biotechnology. *Crit. Rev. Biotechnol.*, Vol. 42, No. 8, pp. 1180–1212, 2021. doi: 10.1080/07388551.2021.1993126

[12] Gaidi, M., Daoudi, K., Tlili, A., Columbus, S., Leblanc-Lavoie, J., Ramachandran, K., Suleiman, B.M., Alhazaa, A., Khakani M.A.E. Fast, Highly Sensitive and Label Free Detection of Small Genetic Sequence Difference of DNA Using Novel Surface-Enhanced Raman Spectroscopy Nanostructured Sensor. *Sensing and Bio-Sensing Research*, p. 32, 2021. doi: 10.1016/j.sbsr.2021.100406

[13] Cao, X., Gu, L., Hu, S., Mukhtar, A., Wu, K. FE/MN Multilayer Nanowires as High-Performance T1-T2 Dual Modal MRI Contrast Agents. *Materials*, Vol. 14, No. 9, pp. 2238, 2021. doi: 10.3390/ma14092238

[14] Li, Z., Zhang, C., Zhang, X., Sui, J., Jin, L., Lin, L., Fu, Q., Lin, H., Song, J. NIR-II Functional Materials for Photoacoustic Theranostics. *Bioconjugate Chem.*, Vol. 33, No. 1, pp. 67–86, 2022. doi: 10.1021/acs.bioconjchem.1c00520

[15] Li, C., Luo, S., Wang, J., Shen, Z., Wu, Z. Nuclease-Resistant Signaling Nanostructures Made Entirely of DNA Oligonucleotides. *Nanoscale*, Vol. 13, No. 15, pp. 7034–7051, 2021. doi: 10.1039/d1nr00197c

[16] Liu, R., Lee, J., Tchoe, Y., Prè, D., Bourhis, A.M., D'Antonio-Chronowska, A., Robin, G., Lee, S., Ro, Y.G., Vatsyayan, R., Tonsfeldt, K.J., Hossain, L., Phipps, M.L., Yoo, J., Nogan, J., Martinez, J.S., Frazer, K.A., Bang, A.G., Dayeh, S.A. Ultra-Sharp Nanowire Arrays Natively Permeate, Record, and Stimulate Intracellular Activity in Neuronal and Cardiac Networks. *Adv. Funct. Mater.*, Vol. 32, No. 8, 2021. doi: 10.1002/adfm.202108378

[17] Tian, J., Paterson, T., Zhang, J., Li, Y., Ouyang, H., Asencio, Í.O., Hatton, P.V., Zhao, Y., Li, Z. Enhanced Antibacterial Ability of Electrospun PCL Scaffolds Incorporating ZNO Nanowires. *Int. J. Mol. Sci.*, Vol. 24, No. 19, 2023. doi: 10.3390/ijms241914420

[18] Kumar, Parveen, Sharma, Sanjeev Kumar, Raj, Balwant, Raj, B. Analysis of Device Parameter Variations in In1-xGaxAs Based Gate Stacked Double Metal Surrounding Gate Nanowire MOSFET. *TEEM Springer*, Vol. 24, pp. 570–578, 2023.

[19] Bhardwaj, A., Kumar, P., Raj, B., Kumar, N., Anand, S. Design and Optimization of Vertical Nanowire Tunnel FET With Electrostatic Doping. *Engineering Research Express*, Vol. 5, pp. 045025–045033, 2023.

[20] Thakur, Y., Raj, B., Gill, S.S. Design and Performance Analysis of Pentacene Organic Field Effect Transistor with High-K dielectric Materials. *Journal of Optoelectronics and Advanced Materials – Rapid Communications*, Vol. 17, No. 7–8, 2023.

[21] Kumar, Parveen, Raj, B. Design and Simulation of Junctionless Nanowire Tunnel Field Effect Transistor Biosensor for High Sensitivity. *Microelectronic Journal Elsevier*, Vol. 137, pp. 105826–105834, 2023.

[22] Jain, N., Raj, B. An Analog and Digital Design Perspective Comprehensive Approach on Fin-FET (Fin-Field Effect transistor) Technology–A Review. *Reviews in Advanced Sciences and Engineering (RASE)*, Vol. 5, pp. 1–14, 2016.

[23] Raj, B., Saxena, A.K., Dasgupta, S. Analytical Modeling for the Estimation of Leakage Current and Subthreshold Swing Factor of Nanoscale Double Gate FinFET Device *Microelectronics International*, Vol. 26, pp. 53–63, 2009.

[24] Soniya, Shailendra Singh, Wadhwa, Girish, Raj, B. An Analytical Modeling for Dual Source Vertical Tunnel Field Effect Transistor. *International Journal of Recent Technology and Engineering (IJRTE)*, Vol. 8, No. 2, pp. 603–608, 2019.

[25] Singh, Shailendra, Raj, B. Design and Analysis of Hetrojunction Vertical T-Shaped Tunnel Field Effect Transistor. *Journal of Electronics Material*, Vol. 48, No. 10, pp 6253–6260, 2019.

[26] Goyal, C., Ubhi, J.S., Raj, B. A Low Leakage CNTFET Based Inexact Full Adder for Low Power Image Processing Applications. *International Journal of Circuit Theory and Applications*, Vol. 47, No. 9, pp. 1446–1458, 2019.

[27] Lee, J.W., Chae, S., Oh, S., Kim, S.H., Meeseepong, M., Choi, K.H., Jeon, J., Lee, N., Song, S.Y., Lee, J.H., Choi, J. Bio-Essential Inorganic Molecular Nanowires as a Bioactive Muscle Extracellular-Matrix-Mimicking Material. *ACS Appl. Mater. Interfaces*, Vol. 13, No. 33, pp. 39135–39141, 2021. doi: 10.1021/acsami.1c12440

[28] Kharaghani, D., Tajbakhsh, Z., Nam, P.D., Kim, I.S. Application of Nanowires for Retinal Regeneration. *Regen. Med.*, Vol. 119, 2020.

[29] Lăbușcă, L., Danceanu, C., Minuti, A.E., Herea, D., Ghemes, A., Rotărescu, C., Dragoș-Pînzaru, O., Ţibu, M., Grigoraş, M., Chiriac, H., Lupu, N. Magnetic Nanowires Substrate Increases Adipose-Derived Mesenchymal Cells Osteogenesis. *Sci. Rep.*, Vol. 12, No. 1, 2022. doi: 10.1038/s41598-022-21145-z

[30] Damiati, L.A., Tsimbouri, M.P., Hernández, V.L., Jayawarna, V., Ginty, M., Childs, P., Xiao, Y., Burgess, K., Wells, J., Sprott, M.R., Meek, R.D., Li, P., Oreffo, R.O., Nobbs, A.H., Ramage, G., Su, B., Salmerón-Sánchez, M., Dalby, M.J. Materials-Driven Fibronectin Assembly on Nanoscale Topography Enhances Mesenchymal Stem Cell Adhesion, Protecting Cells From Bacterial Virulence Factors and Preventing Biofilm Formation. *Biomaterials*, Vol. 280, No. 121263, 2022. doi: 10.1016/j.biomaterials.2021.121263

[31] Dvir, T., Timko, B.P., Kohane, D.S., Langer, R. Nanotechnological Strategies for Engineering Complex Tissues. *Nano-Enabled Medical Applications*, pp. 351–382, 2020. doi: 10.1038/nnaNo.2010.246

[32] Bansal, Priya, Raj, B. Memristor Modeling and Analysis for Linear Dopant Drift Kinetics. *Journal of Nanoengineering and Nanomanufacturing*, Vol. 6, pp. 1–7, 2016.

[33] Ashima, Vaithiyanathan, D., Raj, B. Performance Analysis of Charge Plasma induced Graded Channel Si Nanotube. *Journal of Engineering Research (JER)*, EMSME Special, Issue. pp. 146–154, 2021.

[34] Tomar, Abhishek Singh, Magraiya, Vijay Kumar, Raj, B., Scaling of Access and Data Transistor for High Performance DRAM Cell Design. *Quantum Matter*, Vol. 2, pp. 412–416, 2013.

[35] Jain, N., Raj, B. Parasitic Capacitance and Resistance Model Development and Optimization of Raised Source/Drain SOI FinFET Structure for Analog Circuit Applications. *Journal of Nanoelectronics and Optoelectronins*, Vol. 13, pp. 531–539, 2018.

[36] Singh, Shradhya, Vishvakarma, S.K., Raj, B. Analytical Modeling of Split-Gate Junction-Less Transistor for a Biosensor Application. *Sensing and Bio-Sensing*, Vol. 18, pp. 31–36, 2018.

[37] Gopal, Maisagalla, Raj, B., Low Power 8T SRAM Cell Design for High Stability Video Applications. *ITSI Transaction on Electrical and Electronics Engineering*, Vol. 1, No. 5, pp. 91–97, 2013.
[38] Raj, B., Mitra, Jatin, Bihani, Deepak Kumar, Rangharajan, V., Saxena, A.K., Dasgupta, S. Analysis of, Noise Margin, Power and Process Variation for 32 nm FinFET Based 6T SRAM Cell. *Journal of Computer (JCP)*, Vol. 5, No. 6, pp. 1–8, 2010.
[39] Sharma, Divya, Mehra, Rajesh, Raj, B. Comparative Analysis of Photovoltaic Technologies for High Efficiency Solar Cell Design. *Superlattices and Microstructures*, Vol. 153, pp. 106861, 2021.
[40] Kaur, Pawandeep, Buttar, Avtar Singh, Raj, B. A comprehensive Analysis of Nanoscale Transistor Based Biosensor: A Review, *Indian Journal of Pure and Applied Physics*, Vol. 59, pp. 304–318, 2021.
[41] Yadav, Divya, Raj, Balwant, Raj, B. Design and Simulation of Low Power Microcontroller for IoT Applications. *Journal of Sensor Letters*, Vol. 18, pp. 401–409, 2020.
[42] Mishra, A.C., Chaubey, R., Srivastava, S.P. Implementation of Nanotechnology in Healthcare: Immense Challenges and Opportunities. *Biol. Sci.*, Vol. 2, No. 3, 2022. doi: 10.55006/biolsciences.2022.2303
[43] Bhardwaj, A., Kumar, P., Raj, B., Anand, S. Design and Performance Optimization of Doping-Less Vertical Nanowire TFET using Gate Stack Technique. *Journal of Electronic Materials (JEMS)*, Vol. 41, No. 7, pp. 4005–4013, 2022.
[44] Bayat, H., Alhmoud, H., Raoufi, M., Voelcker, N.H., Schönherr, H. Geometrical Constraints of Poly (Diethylene Glycol Methyl Ether Methacrylate) Brushes on Spherical Nanoparticles and Cylindrical Nanowires: Implications for Thermos-Responsive Brushes on Nanoobjects. *ACS Appl. Nano Mater.*, Vol. 3, No. 4, pp. 3693–3705, 2020. doi: 10.1021/acsanm.0c00401
[45] Singh, Jeetendra, Raj, B. Tunnel Current Model of Asymmetric MIM Structure Levying Various Image Forces to Analyze the Characteristics of Filamentary Memristor. *Applied Physics A*, Vol. 125, No.3, pp. 203.1–203.11, 2019.
[46] Goyal, Candy, Ubhi, Jagpal Singh, Raj, B. Low Leakage Zero Ground, Noise Nanoscale Full Adder Using Source Biasing Technique. *Journal of Nanoelectronics and Optoelectronics*, Vol. 14, pp. 360–370, 2019.
[47] Singh, Gurmohan, Sarin, R.K., Raj, B. A Novel Robust Exclusive-OR Function Implementation in QCA Nanotechnology With Energy Dissipation Analysis. *Journal of Computational Electronics*, Vol. 15, No. 2, pp. 455–465, 2016.
[48] Wadhwa, Girish, Kamboj, P., Raj, B. Design Optimisation of Junctionless TFET Biosensor for High Sensitivity. *Advances in Natural Sciences: Nanoscience and Nanotechnology*, Vol. 10, No. 7, pp. 045001, 2019.
[49] Singh, Shailendra, Raj, B. A 2-D Analytical Surface Potential and Drain Current Modeling of Double-Gate Vertical t-Shaped Tunnel FET. *Journal of Computational Electronics*, Vol. 19, pp. 1154–1163, 2020.
[50] Singh, Jeetendra, Raj, B. An Accurate and Generic Window Function for, Non-linear Memristor Model. *Journal of Computational Electronics*, Vol. 18, No. 2, pp. 640–647, 2019.
[51] Kaur, M., Gupta, N., Kumar, S., Raj, B., Singh, A.K. Comparative RF and Crosstalk Analysis of Carbon Based Nano Interconnects. *IET Circuits, Devices & Systems*, Vol. 15, No. 6, pp. 493–503, 2021.

[52] Kandasamy, N., Ahmad, F., Ajitha, D., Raj, B., Telagam, Nagarjuna Quantum Dot Cellular Automata Based Scan Flip Flop and Boundary Scan Register, *IETE Journal of Research*, Vol. 66, pp. 535–548, 2020.

[53] Singh, Jeetendra, Raj, B. Design and Investigation of 7T2M NVSARM With Enhanced Stability and Temperature Impact on Store/Restore Energy. *IEEE Transactions on Very Large Scale Integration Systems*, Vol. 27, No. 6, pp. 1322–1328, 2019.

[54] Bhardwaj, Anil Kumar, Gupta, Sumeet, Raj, B., Singh, Amandeep Impact of Double Gate Geometry on the Performance of Carbon Nanotube Field Effect Transistor Structures for Low Power Digital Design. *Computational and Theoretical Nanoscience*, Vol. 16, pp. 1813–1820, 2019.

[55] Jain, N., Raj, B. Thermal Stability Analysis and Performance Exploration of Asymmetrical Dual-k Underlap Spacer (ADKUS) SOI FinFET for Security and Privacy Applications. *Indian Journal of Pure & Applied Physics (IJPAP)*, Vol. 57, pp. 352–360, 2019.

[56] Jain, N., Raj, B. Dual-k Spacer Region Variation at the Drain Side of Asymmetric SOI FinFET Structure: Performance Analysis Towards the Analog/RF Design Applications. *Journal of Nanoelectronics and Optoelectronics*, Vol. 14, pp. 349–359, 2019.

[57] Singh, Jeetendra, Sharma, Sanjeev, Raj, B., Khosla, M. Analysis of Barrier Layer Thickness on Performance of In1-xGaxAs Based Gate Stack Cylindrical Gate Nanowire MOSFET. *JNO*, Vol. 13, pp. 1473–1477, 2018.

[58] Jain, N., Raj, B. Analysis and Performance Exploration of High-k SOI FinFETs Over the Conventional Low-k SOI FinFET Toward Analog/RF Design. *Journal of Semiconductors (JoS)*, Vol. 39, No. 12, p. 124002-1-7, 2018.

[59] Goyal, Candy, Ubhi, Jagpal Singh, Raj, B., A Reliable Leakage Reduction Technique for Approximate Full Adder with Reduced Ground Bounce Noise. *Journal of Mathematical Problems in Engineering*, Vol. 2018, 3501041, 2018.

[60] Anuradha, J.S., Raj, B., Khosla, M. Design and Performance Analysis of Nano-scale Memristor-Based, Nonvolatile SRAM. *Journal of Sensor Letter*, Vol. 16, pp. 798–805, 2018.

[61] Wadhwa, Girish, Raj, B. Parametric Variation Analysis of Charge-Plasma-Based Dielectric Modulated JLTFET for Biosensor Application. *IEEE Sensor Journal*, Vol. 18, No. 15, pp. 6070–6077, 2018.

[62] Singh, Jeetendra, Raj, B. Comparative Analysis of Memristor Models for Memories Design. *JoS*, Vol. 39, No. 7, pp. 074006-1-12, 2018.

[63] Yadav, Divya, Chouhan, Shailesh Singh, Vishvakarma, Santosh Kumar, Raj, B. Application Specific Microcontroller Design for IoT Based WSN. *Sensor Letter*, Vol. 16, pp. 374–385, 2018.

[64] Singh, Gurmohan, Sarin, R.K., Raj, B. Fault-Tolerant Design and Analysis of Quantum-Dot Cellular Automata Based Circuits. *IEEE/IET Circuits, Devices & Systems*, Vol. 12, pp. 638–664, 2018.

[65] Singh, Jeetendra, Raj, B. Modeling of Mean Barrier Height Levying Various Image Forces of Metal Insulator Metal Structure to Enhance the Performance of Conductive Filament Based Memristor Model. *IEEE Nanotechnology*, Vol. 17, No. 2, pp. 267–268, 2018.

[66] Jain, Aakash, Sharma, Sanjeev, Raj, B. Analysis of Triple Metal Surrounding Gate (TM-SG) III-V Nanowire MOSFET for Photosensing Application. *Opto-electronics Journal*, Vol. 26, No. 2, pp. 141–148, 2018.

[67] Jain, Aakash, Sharma, Sanjeev, Raj, B. Design and Analysis of High Sensitivity Photosensor Using Cylindrical Surrounding Gate MOSFET For Low Power Sensor Applications. *Engineering Science and Technology, an International Journal*, Vol. 19, No. 4, pp. 1864–1870, 2016.
[68] Bergveld, P. Development, Operation, and Application of the Ion-Sensitive Field-Effect Transistor as a Tool for Electrophysiology. *IEEE Trans. Biomed. Eng.*, Vol. 19, pp. 342–351, 1972.
[69] Saiphani Kumar, G., Singh, Amandeep, Raj, B. Design and Analysis of Gate All Around CNTFET Based SRAM Cell Design. *Journal of Computational Electronics*, Vol. 17, No. 1, pp 138–145, 2018.
[70] Singh, Gurinder Pal, Sohi, B.S., Raj, B. Material Properties Analysis of Graphene Base Transistor (GBT) for VLSI Analog Circuits. *Indian Journal of Pure & Applied Physics (IJPAP)*, Vol. 55, pp. 896–902, 2017.
[71] Sunil, K., Raj, B. Estimation of Stability and Performance Metric for Inward Access Transistor based 6T SRAM Cell Design Using n-Type/p-Type DMDG-GDOV TFET *IEEE VLSI Circuits and Systems Letter*, Vol.3, No. 2, pp. 25–39, 2017.
[72] Sharma, S., Kumar, A., Pattanaik, Manisha, Raj, B. Forward Body Biased Multimode Multi-Threshold CMOS Technique for Ground Bounce, Noise Reduction in Static CMOS Adders. *International Journal of Information and Electronics Engineering*, pp. 567–572, Vol. 3, No. 3, 2013.
[73] Singh, H., Kumar, P., Raj, B. Performance Analysis of Majority Gate SET Based 1-bit Full Adder. *International Journal of Computer and Communication Engineering (IJCCE)*, Vol. 2, No. 4, 2013.
[74] Bhardwaj, Anil Kumar, Gupta, Sumeet, Raj, B. Investigation of Parameters for Schottky Barrier (SB) Height for Schottky Barrier Based Carbon Nanotube Field Effect Transistor Device. *Journal of Nanoelectronics and Optoelectronics*, Vol. 15, pp. 783–791, 2020.
[75] Bansal, P., Raj, B. Memristor: A Versatile Nonlinear Model for Dopant Drift and Boundary Issues. *JCTN*, Vol. 14, No. 5, pp. 2319–2325, 2017.
[76] Anjum, N., Bali, T., Raj, B., Design and Simulation of Handwritten Multiscript Character Recognition. *International Journal of Advanced Research in Computer and Communication Engineering*, Vol. 2, No. 7, pp. 2544–2549, 2013.
[77] Singh, K. Performance and Analysis of Temperature Dependent Multi-Walled Carbon Nanotubes as Global Interconnects at Different Technology Nodes, *Journal of Computational Electronics*, Vol. 14, No. 2, pp. 469–476, 2015.
[78] Kumar, Sunil, Raj, B. Compact Channel Potential Analytical Modeling of DG-TFET Based on Evanescent–Mode Approach. *Journal of Computational Electronics*, Vol. 14, No. 2, pp.820–827, 2015.
[79] Singh, K., Raj, B. Temperature Dependent Modeling and Performance Evaluation of Multi-Walled CNT and Single-Walled CNT as Global Interconnects. *Journal of Electronic Materials*, Vol. 44, No. 12, pp 4825–4835, 2015.
[80] Sharma, V.K., Pattanaik, M., Raj, B. INDEP Approach for Leakage Reduction in Nanoscale CMOS Circuits. *International Journal of Electronics*, Vol. 102, No.2, pp. 200–215, 2014.
[81] Singh, K., Raj, B. Influence of Temperature on MWCNT Bundle, SWCNT Bundle and Copper Interconnects for Nanoscaled Technology Nodes. *Journal*

of Materials Science: Materials in Electronic, Vol. 26, No. 8, pp. 6134–6142, 2015.

[82] Anjum, N., Bali, T., Raj, B. Design and Simulation of Handwritten Gurumukhi and Devanagri, Numerical Recognition. *International Journal of Computer Applications*, Vol. 73, No. 12, pp. 16–21, 2013.

[83] Khandelwal, S., Gupta, V., Raj, B., Gupta, R. D. Process Variability Aware Low Leakage Reliable Nano Scale DG-FinFET SRAM Cell Design Technique. *Journal of Nanoelectronics and Optoelectronics*, Vol. 10, No. 6, pp. 810–817, 2015.

[84] Sharma, V.K., Pattanaik, M., Raj, B. ONOFIC Approach: Low Power High Speed Nanoscale VLSI Circuits Design. *International Journal of Electronics*, Vol. 101, No. 1, pp. 61–73, 2014.

[85] Khandelwal, S., Raj, B., Gupta, R. D., FinFET based 6T SRAM Cell Design: Analysis of Performance Metric, Process Variation and Temperature Effect. *Journal of Computational and Theoretical Nanoscience*, Vol. 12, pp. 2500–2506, 2015.

[86] Singh, S., Yadav, S., Rahul, J., Srivastava, A., Raj, B. Impact of HfO_2 in Graded Channel Dual Insulator Double Gate MOSFET, *Journal of Computational and Theoretical Nanoscience*, Vol. 12, No. 6, pp. 950–953, 2015.

[87] Sharma, Vijay Kumar, Pattanaik, Manisha, Raj, B., PVT Variations Aware Low Leakage INDEP Approach for Nanoscale CMOS Circuits. *Microelectronics Reliability*, Vol. 54, pp 90–99, 2014.

[88] Raj, B., Saxena, A.K., Dasgupta, S. Quantum Mechanical Analytical Modeling of Nanoscale DG FinFET: Evaluation of Potential, Threshold Voltage and Source/Drain Resistance. *Elsevier's Journal of Material Science in Semiconductor Processing*, Vol. 16, No. 4, pp. 1131–1137, 2013.

[89] Gopal, Maisagalla, Prasad, Siva Sankar D., Raj, B. 8T SRAM Cell Design for Dynamic and Leakage Power Reduction. *International Journal of Computer Applications*, Vol. 71, No. 9, pp. 43–48, 2013.

[90] Pattanaik, M., Raj, B., Sharma, S., Kumar, Anjan. Diode Based Trimode Multi-Threshold CMOS Technique for Ground Bounce, Noise Reduction in Static CMOS Adders. *Advanced Materials Research*, Vol. 548, pp 885–889, 2012.

[91] Raj, B., Saxena, A.K., Dasgupta, S. Nanoscale FinFET Based SRAM Cell Design: Analysis of Performance Metric, Process Variation, Underlapped FinFET and Temperature Effect *IEEE Circuits and System Magazine*, Vol. 11, No. 2, pp. 38–50, 2011.

[92] Sharma, V.K., Pattanaik, M., Raj, B. Leakage Current ONOFIC Approach for Deep Submicron VLSI Circuit Design. *International Journal of Electrical, Computer, Electronics and Communication Engineering*, Vol. 7, No. 4, pp. 239–244, 2013.

[93] Perez, J.M., Josephson, L., Weissleder, R. Use of Magnetic Nanoparticles as Nanosensors to Probe for Molecular Interactions. *ChemBioChem*, Vol. 5, pp. 261–264, 2004.

[94] Kaur, Parminder, Gill, Sandeep Singh, Raj, B. Comparative Analysis of OFETs Materials and Devices for Sensor Applications. *Journal of Silicon*, Vol 14, pp. 4463–4471, 2022.

[95] Sharma, Sanjeev Kumar, Kumar, Parveen, Raj, Balwant, Raj, B. $In_{1-x}Ga_xAs$ Double Metal Gate-Stacking Cylindrical Nanowire MOSFET for Highly Sensitive Photo Detector. *Journal of Silicon*, Vol. 14, pp. 3535–3541, 2022.

[96] Raj, B., Saxena, A.K., Dasgupta, S. Analytical Modeling of Quasi Planar Nanoscale Double Gate FinFET With Source/Drain Resistance and Field Dependent Carrier Mobility: A Quantum Mechanical Study. *Journal of Computer (JCP)*, Vol. 4, No. 9, pp. 1–8, 2009.

[97] Bhushan, S., Khandelwal, S., Raj, B. Analyzing Different Mode FinFET based Memory Cell at Different Power Supply for Leakage Reduction, Seventh International Conference on Bio-Inspired Computing: Theories and Application, (BIC-TA 2012). *Advances in Intelligent Systems and Computing*, Vol. 202, pp. 89–100, 2013.

[98] Singh, J., Raj, B. Temperature Dependent Analytical Modeling and Simulations of Nanoscale Memristor. *Journal: Engineering Science and Technology, an International Journal*, Vol. 21, pp. 862–868, 2018.

[99] Singh, Shradhya, Bala, Shashi, Raj, Balwant, Raj, B. Improved Sensitivity of Dielectric Modulated Junctionless Transistor for Nanoscale Biosensor Design. *Sensor Letter*, Vol. 18, pp. 328–333, 2020.

[100] Kumar, Vivek, Vishvakarma, Santosh Kumar, Raj, B. Design and Performance Analysis of ASIC for IoT Applications. *Sensor Letter*, Vol. 18, pp. 31–38, 2020.

[101] Jaiswal, A., Sarin, R.K., Raj, B., Sukhija, S. A Novel Circular Slotted Microstrip-Fed Patch Antenna With Three Triangle Shape Defected Ground Structure for Multiband Applications. *Advanced Electromagnetic (AEM)*, Vol. 7, No. 3, pp. 56–63, 2018.

[102] Wadhwa, G., Raj, B. Label Free Detection of Biomolecules using Charge-Plasma-Based Gate Underlap Dielectric Modulated Junctionless TFET. *Journal of Electronic Materials (JEMS)*, Vol. 47, No. 8, pp 4683–4693, 2018.

[103] Singh, G., Sarin, R.K., Raj, B. Design and Performance Analysis of a New Efficient Coplanar Quantum-Dot Cellular Automata Adder. *Indian Journal of Pure & Applied Physics (IJPAP)*, Vol. 55, pp. 97–103, 2017.

[104] Kaur, P., Pandey, V., Raj, B. Comparative Study of Efficient Design, Control and Monitoring of Solar Power Using IoT. *Sensor Letter*, Vol. 18, pp. 419–426, 2020.

[105] Bhardwaj, Anil Kumar, Gupta, Sumeet, Raj, B., Development & Analysis of Compact Model for Double Gate Schottky Barrier CNTFET. *Journal of Nanoelectronics and Optoelectronics*, Vol. 15, pp. 1199–1208, 2020.

[106] Wadhwa, G., Kamboj, P., Singh, J., Raj, B. Design and Investigation of Junctionless DGTFET for Biological Molecule Recognition. *Transactions on Electrical and Electronic Materials*, Vol. 22, pp. 282–289, 2021.

[107] Verma, S.K., Singh, S., Wadhwa, G., Raj, B. Detection of Biomolecules using Charge-Plasma Based Gate Underlap Dielectric Modulated Dopingless TFET *Transactions on Electrical and Electronic Materials (TEEM)*, Vol. 21, pp. 528–535, 2020.

[108] Jain, N., Raj, B. Impact of Underlap Spacer Region Variation on Electrostatic and Analog/RF Performance of Symmetrical High-k SOI FinFET at 20 nm Channel Length. *Journal of Semiconductors (JoS)*, Vol. 38, No. 12, pp. 122002, 2017.

[109] Singh, S., Raj, B. Analytical Modeling and Simulation analysis of T-shaped III-V heterojunction Vertical T-FET. *Superlattices and Microstructures*, Vol. 147, pp. 106717, 2020.

[110] Singh, G., Sarin, R.K., Raj, B. Design and Analysis of Area Efficient QCA Based Reversible Logic Gates. *Journal of Microprocessors and Microsystems*, Vol. 52, pp. 59–68, 2017.

[111] Singh, S., Raj, B. Parametric Variation Analysis on Hetero-Junction Vertical t-Shape TFET for Supressing Ambipolar Conduction. *Indian Journal of Pure and Applied Physics*, Vol. 58, pp. 478–485, 2020.

[112] Soniya, Shailendra Singh, Wadhwa, Girish, Raj, B. Design and Analysis of Dual Source Vertical Tunnel Field Effect Transistor for High Performance. *Transactions on Electrical and Electronics Materials*, Vol. 21, pp. 74–82, 2019.

[113] Kaur, M., Gupta, N., Kumar, S., Raj, B., Singh, A.K. RF Performance Analysis of Intercalated Graphene Nanoribbon Based Global Level Interconnects. *Journal of Computational Electronics*, Vol. 19, pp. 1002–1013, 2020.

[114] Wadhwa, G., Raj, B. Design and Performance Analysis of Junctionless TFET Biosensor for High Sensitivity *IEEE Nanotechnology*, Vol. 18, pp. 567–574, 2019.

[115] Singh, J., Raj, B. Enhanced, Nonlinear Memristor Model Encapsulating Stochastic Dopant Drift. *JNO* 14, 958–963, 2019.

[116] Kumar, M., Chaudhary, T., Raj, B. Design and Comparative Analysis of Ferroelectric Nanowire With Dielectric HfO_2 and $Al_2O_3$for Low Power Applications. *Journal of Electronic Materials*, Vol. 53, pp. 6785–6791, 2024.

[117] Thakur, Yogesh, Raj, Balwant, Raj, B. Design of Pentacene Thin-Film Transistor Based Hydrogen Gas Sensor for High Sensitivity. *ECS Journal of Solid State Science and Technology*, Vol. 13, pp. 047005–047012, 2024.

[118] Kumar, M.A., Ismail, K.B.M., Mahalingam, S., Raj, B., Kim, J., Atchudan, R. Recent Advances in Two-Dimensional Nanomaterials and Nanostructures for Healthcare Monitoring. *ACS Sensors*, Vol. 9, No. 4, pp. 1–8, 2024.

[119] Batcha, M.I.K., Manoharan, A.K., Mahalingam, S., Raj, B. Carbon Fiber-Reinforced Polymers for Energy Storage Applications. *Journal of Energy Storage*, Vol. 84, No. Part B, pp. 110931–110939, 2024.

[120] Hussain, S., Raj, B., Sharma, D. Design and Parametric Optimization of Lead Free Perovskite Based Tandem Solar Cell for High Efficient. *Engineering Research Express*, Vol. 6, pp. 015315–0153122, 2024.

[121] Kumar, P., Raj, B., Wadhwa, G., Singh, B., Kumar, Raj. Design and Analysis of Junctionless based Gate All Around N+ Doped Layer Nanowire TFET Biosensor. *ECS Journal of Solid State Science and Technology*, Vol 13, pp. 017002–017010, 2024.

[122] Sharma, D., Mehra, R., Raj, B. Enhancement in Efficiency of Methyl Ammonium Tin Iodide Based Perovskite Solar Cell Using SCAPS-1 D. *Nano*, Vol. 18, No. 14, pp. 2350109, 2023.

[123] Raj, B., Saxena, A.K., Dasgupta, S. A Compact Drain Current and Threshold Voltage Quantum Mechanical Analytical Modeling for FinFETs. *Journal of Nanoelectronics and Optoelectronics (JNO)*, Vol. 3, No. 2, pp. 163–170, 2008.

[124] Wadhwa, G., Raj, B. An Analytical Modeling of Charge Plasma Based Tunnel Field Effect Transistor With Impacts of Gate Underlap Region. *Superlattices and Microstructures*, Vol. 142, pp. 106512, 2020.

[125] Singh, Shailendra, Raj, B. Modeling and Simulation Analysis of SiGe Hetrojunction Double Gate Vertical t-Shaped Tunnel FET. *Superlattices and Microstructures*, Vol. 142, pp. 106496, 2020.

[126] Jain, N., Raj, B. Device and Circuit Co-Design Perspective Comprehensive Approach on FinFET Technology–A Review, *Journal of Electron Devices*, Vol. 23, No. 1, pp 1890–1901, 2016.
[127] Kumar, Sunil, Raj, B. Analysis of I_{ON} and Ambipolar Current for Dual-Material Gate-drain Overlapped DG-TFET, *Journal of Nanoelectronics and Optoelectronics*, Vol. 11, pp. 323–333, 2016.
[128] Ibarlucea, B., Akbar, T.F., Kim, K., Rim, T., Baek, C.K., Ascoli, A., Tetzlaff, R., Baraban, L., Cuniberti, G. Ultrasensitive Detection of Ebola Matrix Protein in a Memristor Mode. *Nano Res.* 11, 1057–1068, 2018.
[129] Shahad, N., Brönstrup, G., Funka, K., Christiansen, S., Leja, M., Haick, H. Ultrasensitive Silicon Nanowire for Real-World Gas Sensing: Noninvasive Diagnosis of Cancer from Breath Volatolome *Nano Letters*, Vol. 15, No. 2, pp. 1288–1295, 2015.

Chapter 7

Enhancing nanowire TFET performance through hetero-dielectric and hetero-material at the device and circuit level

Sugandha Tejeswee, Mohit Srivastava, Preeti Agarwal Mittal, Aparna Dixit, and Arti Saxena

7.1 INTRODUCTION

Tunnel Field-Effect Transistors (TFETs) have experienced a fair amount of research awareness as an alternative to the regular Metal Oxide Semiconductor Field-Effect Transistors (MOSFETs) for less-power electronic scenarios, because they can achieve steep subthreshold swings (SS) and function at lower supply voltages. Unlike conventional FETs, where charge carriers travel over a potential barrier through thermionic emission, TFETs exhibit band-to-band tunneling (BTBT) over a tunneling junction from the source to the channel regions [1]. This unique quantum mechanical transport yields an SS that can exceed the thermodynamically determined confinement by 60 mV/decade at room temperature for MOSFETs. This feature means TFETs can reduce static power dissipation and are ideal for energy-constrained devices [2]. The recent advancements in material engineering, such as heterostructures, high-k gate dielectrics, and low-bandgap semiconductors (e.g., InAs and GaSb), have all led to significant improvements in ION and switching efficiency [3], [4]. Comparative images of TFET and MOSFET operation showcase the differences in charge injection mechanisms as well as the benefits of tunneling-based transport as future nanoscale and low-power device technology.

7.1.1 Overview of TFETs

TFETs have proven to be an advantageous alternative to the traditional MOSFETs for less-power electronics implementation, characterized by steep SS and the ability to operate at lower supply voltages. Traditional FETs rely on thermionic emission across a potential barrier for transport of charge carriers, whereas TFETs depend on band-to-band tunneling (BTBT) across a tunneling junction between source and channel regions [1]. This quantum mechanical process enables TFETs to achieve substantially better subthreshold swaying than the fundamental SS with 60 mV/decade in MOSFETs at room temperature [2]. Thus, they can significantly reduce static power dissipation and are also a desirable energy feature for energy-limited products.

DOI: 10.1201/9781003634645-7

Materials engineering advancements or new materials, including heterostructures, high-k gate dielectrics, and low-bandgap semiconductors (e.g., InAs, GaSb), have recently demonstrated enhancements in on-current (ION) and switching performance efficiency [3], [4]. Illustrations comparing TFET and MOSFET operational principles also illustrate the different charge injection mechanisms and demonstrate the benefits of tunneling-based transport in future nanoscale and low-power device technologies.

7.1.2 Importance of nanowire structures in TFETs

Nanowire architectures can be implemented in Tunnel Field-Effect Transistors (TFETs) with significant impacts on device performance, particularly in the area of scaled electrostatic gate control. Nanowires allow for aggressive scaling of dimensions while maintaining favorable charge transport properties for ultra-low-power operation. A high surface-to-volume ratio and quasi-one-dimensional structure allowed for better coupling of gates and lower relative short-channel effects compared to planar devices [1]. The structural flexibility of nanowires also allows for the growth of high-quality hetero interfaces, despite mismatched lattice constants from substrates, and creates a wealth of hetero structures to enhance efficiency through band-to-band tunneling. Impressive device metrics can be obtained through fabrication methods that enable SSs near or under the theoretical upper band of 60 mV/decade at room temperature [2]. Furthermore, precise control of doping profiles along the nanowire creates custom electrical behavior, which supports tunable source-channel junctions and finite maximum carrier injection [3]. Therefore, nanowire-based TFETs are a promising avenue to strengthen the divide between theoretical performance advantages and actual engineering realizations of device structures in future semiconductor technology [5–14].

7.1.3 Role of hetero-dielectric materials

The choice and engineering of dielectric materials are critical in optimizing the conduction of TFETs. The ability to obtain lower SS and improved electrostatic control are both examples of how to benefit from dielectric materials. While standard silicon dioxide (SiO_2) remains useful as a dielectric material, replacing it in conjunction with high-permittivity (high-k) dielectrics like hafnium dioxide (HfO_2) shows dramatic reductions in equivalent oxide thickness (EOT), which result in tighter gate control over the channel [1]. Improvements in electrostatic coupling allow for an effective reduction in the OFF-state leakage current and a steeper subthreshold swing, which are both critical for efficient operation in low-power electronic systems. As shown in Figure 7.1, the dielectric stack influences the band profile, which improves band-to-band tunneling by sharpening the energy barrier at the

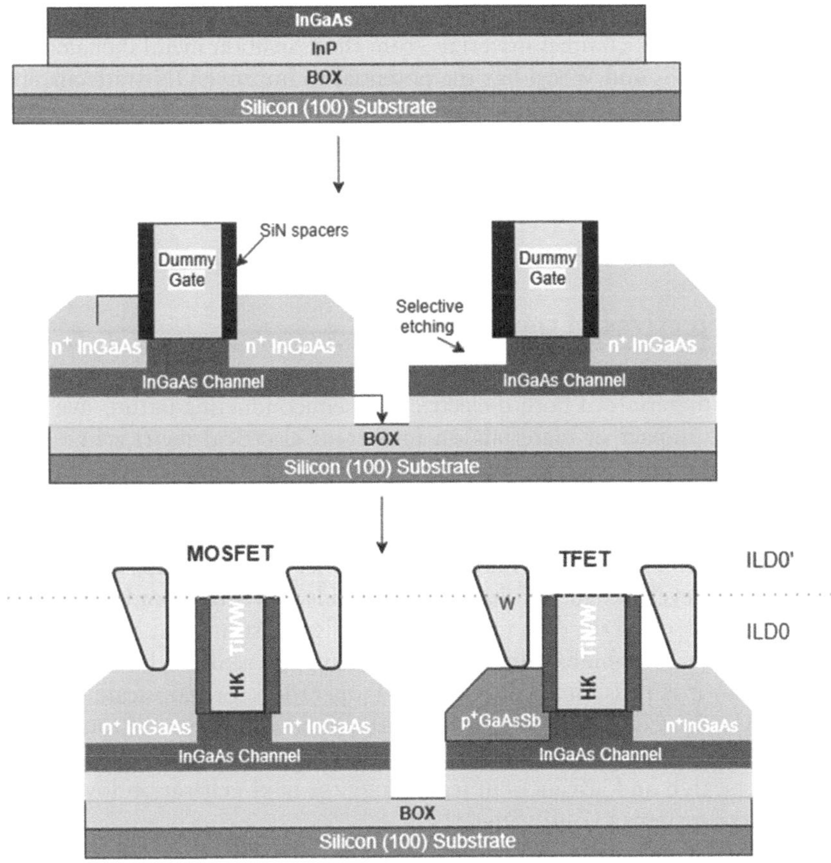

Figure 7.1 Schematic and micro-graphs of MOSFET and TFET semiconductor structures.

source-channel interface. Thus, the ON-state current results to be more greater [15–23].

7.1.4 Significance of hetero materials in device performance

Using different material systems together in nanowire TFETs has remarkable promise to improve device operation in low-power electronic applications. Heterogeneous combinations of high-k dielectric and semiconductor hetero junctions improved electrostatic gate control, lowered leakage currents, and achieved steeper SSs—all critical to operating an energy-efficient device. Device simulations show that hetero-dielectric combinations can modulate the electric field at the tunneling junction to minimize band-to-band tunneling (BTBT) leakage and to reduce OFF-state current without affecting ON-state performance [1]. Furthermore, the incorporation of two-dimensional (2D) channel materials from the transition metal dichalcogenides (e.g., MoS_2 and WSe_2) has the potential to improve ON-state currents and obtain sub-60 mV/decade SSs due to their atomically thin nature and favorable band alignment [2]. These materials exemplify how advanced dielectric and hetero-structure engineering contributed to the optimization of nanowire TFETs for low-power next-generation semiconductor technology [24–36].

7.1.5 Objectives of the chapter

In this chapter, we have focused on ways to enhance the operation of TFETs using new materials of both dielectric and semiconducting nature. We have defined the impact of materials on important electrical metrics like drive current (ION) and SS (SS), which are fundamental for allowing ultra-low-power operation [50–58]. We have discussed the theories of material-based improvement on device performance and their real-world practical implications through an interdisciplinary lens of device physics, materials science, and circuit-level design. We have also provided examples and simulation results and assessed the data to help case studies illustrate that by leveraging specific materials—high-k dielectrics, low-bandgap semiconductors, or 2D materials—it is possible to disrupt TFET operation for nanoscale applications [1], [2]. Compilation of those case studies represents a foundation to capture the landscape to drive realization and map out implementable future research and advances in technology for next-generation low-power electronics systems [37–49, 59–65].

7.2 FUNDAMENTALS OF NANOWIRE TFET TECHNOLOGY

7.2.1 Basic principles of TFET operation

TFETs lean on the concept of BTBT, permitting the charge carriers to cross the threshold energy between source and channel without using thermionic emission. As a result, TFETs can conduct at substantially lower supply voltages and lower thermal loss than classical MOSFETs because of this quantum mechanical transport method. MOSFETs have an absolute limit to their least SS (SS) at room temperature of 60 mV/decade, due to the Boltzmann distribution of carriers. TFETs can, theoretically, achieve subthreshold swings of 30 mV/decade [1] around the low-voltage, low-power regime, which makes using them for less-power electronics/logic implementations very appealing. Research into III-V compound semiconductors, which are low-bandgap and high-mobility materials, addressed one of the shortcomings of first-generation TFETs by increasing ON-state current (ION) [2].

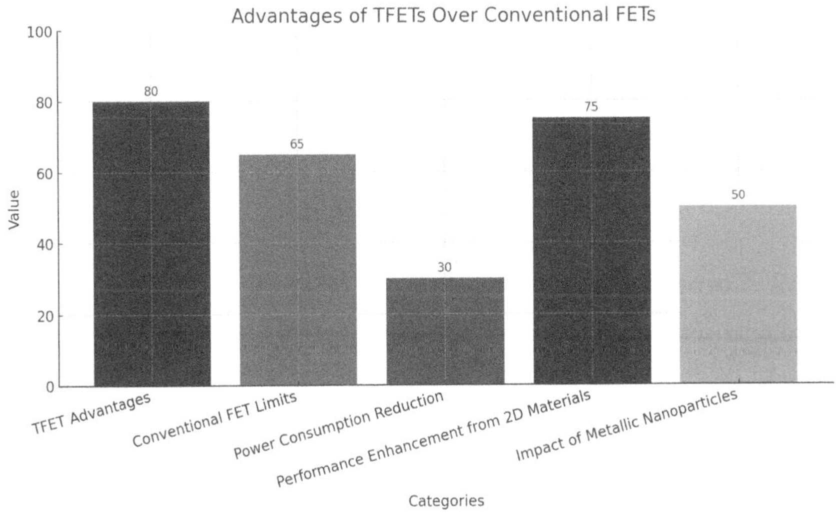

Figure 7.2 This bar chart illustrates the advantages of Tunnel Field-Effect Transistors (TFETs) contrasted to traditional FETs, showing clear differences in various categories such as power consumption reduction and performance enhancement through 2D materials and metallic nanoparticles. Each bar represents the relative value associated with each category, emphasizing the notable benefits of TFET technology.

As displayed in Figure 7.2, the energy band diagrams and transfer characteristics of TFETs compared to classical MOSFETs highlight the advantages of a TFET architecture [66–78]. There are other opportunities for improvement, including research into hetero-dielectric and heterojunction, which we expect should enlarge the efficiency, application, and performance of TFETs in forthcoming energy-efficient electronic generations of products [3].

7.2.2 Advantages of nanowire architecture

Nanowire designs can unlock a number of advantages for semiconductor devices, particularly for nanoscale and low-power applications. Nanowire structures exhibit a high surface-to-volume ratio, meaning that they provide good electrostatic control over the channel, which helps in enhancing the gate modulation efficiency and reducing short-channel effects, which are of primary importance for device scaling [1]. Also, their structural and material flexibility is beneficial because it enables the combination of different dielectric layers and different semiconductor materials, which allows for performance improvement across a large range of working conditions. These advantages are explicitly exemplified with dual-material gate (DMG) nanowire-based TFETs, which provided improved current drive and transconductance, ultimately enabling improved gain and switching behavior for circuit-level applications [2]. Even further improvement to tunneling efficiency was achieved using dielectric engineering methods, like hetero-dielectric gate stacks, which showed improvements illustrated by increased ON-state current (ION) and reduced SS (SS) [3]. These structural and material advancements of nanowire TFETs continue to hold excitement in the eventual development of energy-efficient electronic systems overall. Highlighted diagrams emphasize the nanowire structures mentioned to show their relevance in the modern development of transistor structures overall [79–86].

7.2.3 Comparison with traditional MOSFETs

A comparison between regular MOSFETs and new TFETs shows significant gains in highly efficient power and subthreshold performance with the new device topology. Traditional MOSFETs restrict their performance with thermionic emission, which enforces a maximum SS (SS) of 60 mV/dec, at room temperature, thus limiting their ability to function under low supply voltage conditions [1]. Contrarily, TFETs can promote carrier injection through band-to-band tunneling (BTBT), which allows for steeper subthreshold slopes at times well below the 60 threshold. Due to the steep switching nature of TFETs, ION can be higher than that of a prior traditional field-effect device with lower gate voltage, addressing the critical need of power management in modern integrated circuits [87–96]. TFET also has the viability of using hetero-dielectric engineering in its architecture to alleviate short-channel effects found in scaled traditional MOSFET technologies that limit the scalability and performance requirements of a transistor [2]. Table 7.1 shows the electrical behavior and operating theory of both devices, indicating that TFET technology will advance semiconductor fabrication toward a next generation capable of outperforming traditional MOSFET technology [97–106].

7.2.4 Challenges in current TFET designs

The practical use and outline of TFETs suffer from many challenges that lessen their suitability and ultimate efficacy, particularly when compared to

Table 7.1 Comparison of TFETs and traditional MOSFETs

Parameters	*TFET*	*MOSFET*
Device On-current (I_on)	100 μA/μm	200 μA/μm
Device Off-current (I_off)	10 nA/μm	1 μA/μm
Subthreshold Swing (S)	60 mV/decade	90 mV/decade
Power Consumption (P)	0.5 mW/mm	2 mW/mm
Switching Speed	High	Medium
Integration Density	Higher	Moderate
Temperature Sensitivity	Lower	Higher

conventional CMOS technologies. One pertinent issue is driving adequate high currents for large-scale integrated circuits, despite all the thorough investigations for improving band-to-band tunneling (BTBT) efficiency [9]. Even when you have TFETs operating at large drain biases, their performance is frequently limited to smaller voltage swings, making them less desirable for high-speed applications and limiting the overall usability of TFETs in existing architectures, as researched and published in studies of microprocessor designs [8]. This finite level of performance implies there is plenty more room for thought—i.e., to pair TFETs with hetero-dielectric materials—to improve the tunneling efficiencies while improving voltage scaling. These needs can be addressed through well-accounted design methodologies to help these critical challenges, allowing for more functional TFET applications, providing enhanced technology with more improved performance and energy-efficient outcomes compared to now [107–118].

7.2.5 Overview of device scaling in TFETs

The relation of TFETs' evolution to the subject of device downscaling, which impacts performance factors (e.g., power and efficiency) regarding usage. As regular Field-Effect Transistors (FETs) approach their physical limits, TFETs provide a way around this limitation by taking advantage of the band-to-band tunneling process that occurs at the source-channel interface. These tunneling events create a steeper subthreshold swing because TFETs have reduced power requirements compared to traditional FETs. Currently, research is being conducted into the use of two-dimensional (2D) materials, which can enable downscaling by combining multiple 2D materials and their properties with the technology and scaling capability of traditional 3D materials. As shown, this hybrid approach can substantially improve the performance of a device. For example, the tunneling currents can be improved by the structural addition of metallic nanoparticles located at the tunnel junction, thus highlighting the potential to construct better TFETs [10]. The implications of this research are significant not only within the development of new materials, but future applications for TFETs will find usage in information and sensor technology, with the aim to address problems of size and power scalability [11].

7.3 HETERO-DIELECTRIC MATERIALS IN TFETS

The integration of hetero-dielectric materials into TFETs is a critical approach to improve device operation, particularly the SS, and increase the ON-state current. Hetero dielectrics allow for the application of distinct electric fields to impact the carrier tunneling event, addressing low power consumption as demonstrated from studies focusing on nanoscale TFET designs [5]. Using coupled dielectrics, such as hafnium oxide (HfO2) with various two-dimensional (2D) semiconductors, can enhance confinement and decrease the leakage currents, thereby addressing one of the main problems of short-channel effects characteristically found in better-known devices [12]. Energy band diagrams depict the benefits of incorporating hetero dielectrics as compared to dielectrics alone, which results in advantageous and improved tunneling rates necessary for miniaturizing devices, and increases energy efficiency. Thus, by carefully selecting materials and providing a platform for combined functioning, we are providing better and more compact TFET designs, suitable for upcoming applications as low-energy electronics [119–131].

7.3.1 Definition and types of hetero dielectrics

The term hetero dielectrics refers to dielectric materials whose composition and electrical properties are different from nearby layers. They are often selected to improve semiconductor device performance, which is especially exploited in tunneling field-effect transistors (TFET), which possess lower gate leakage properties, having hetero-dielectric architectures, mixing high-k dielectric materials (ex: HfO2) with lower-k dielectrics to enhance capacitance and lower leakage properties. This is very important in nanowire TFETs as the physical shape of the device and interfaces between materials considerably affect TFET performance and efficacy. Hetero-dielectric configurations can also play a role in manipulating electrostatic control in tailoring band offset in ways that effectively alter tunneling characteristics, which are critical for obtaining low-power applications. Therefore, the significance of hetero-dielectric types and what they mean is crucial to the path forward in the field and helping develop new designs for electronic devices that use hetero-dielectric implementations for better electrical performance and reliability, as demonstrated in Table 7.2.

Table 7.2 Impact of hetero-dielectric and hetero-material on gate control and electrostatics in nanowire TFETs

Study	*Hetero-dielectric type*	*Gate control improvement (%)*	*Electric field enhancement (MV/m)*	*Subthreshold slope (MV/decade)*
Smith et al. (2021)	SiO2	15	5.5	60
Johnson and Lee (2022)	HfO2	20	6.2	55
Chen et al. (2023)	Al_2O_3	18	6.0	58
Garcia and Patel (2023)	InAs	17	5.8	57

7.3.2 Impact on gate control and electrostatics

The intricate relationship between gate control and electrostatics is critical to improving the production of nanowire (TFETs). Through investigations of various dielectrics and materials, the research community has made great strides in improving electrostatic control, which is essential to mitigating short-channel effects that can compromise device performance. Good gate modulation results in improved performance with sharper subthreshold slopes and higher ON-state current, which are important metrics for low-power digital applications [13]. The use of high-k dielectrics inherently increases the value of capacitance, leading to greater gate control over channel carriers. This relationship reduces leakage currents and improves elimination of tunneling over barrier distances, thus placing TFETs in a position to compete in new semiconductor paradigms [2]. The structural images of these devices are particularly interesting because they show a comparison of material design changes on the electrostatic performance of the device and aid the exploration of advanced device designs.

7.3.3 Influence on subthreshold slope and ON/OFF ratio

Subthreshold slope is a principal characteristic of the operation of TFETs, primarily the ON/OFF-current ratio. Unlike traditional MOSFETs, which have an approximate limit of 60 mV/decade because of current transport, TFETs can obtain SS values lower than 60 mV/decade, thereby improving both efficiency and power performance [2]. Optimizing the tunneling probabilities at the p-i junction can enhance this tunnel field-effect, thus increasing the ON-state while minimizing the leakage OFF-state through the application of hetero dielectrics and hetero materials in low-dimensional tunnel field-effect transistor designs [2]. Research has shown that III-V semiconductors in TFET design incorporate some appealing tunneling characteristics, yielding a greater ON/OFF ratio with an SS less than 100 mV/decade [2]. These enhancements are critical in facilitating low-power, high-performance electronic devices. The line graph in Figure 7.3 depicts the relationship emerging among the operation characteristics of Tunnel Field-Effect Transistors (TFETs).

7.3.4 Compatibility with nanowire fabrication techniques

Integrating methods for making nanowires into tunneling field-effect transistors (TFETs) is important for advancing the operation of devices. Nanowire devices have several advantages, such as large surface-to-volume ratios and improved electrostatics, that allow a device to take advantage of some properties of heterogeneous dielectrics and materials, improving TFETs. One

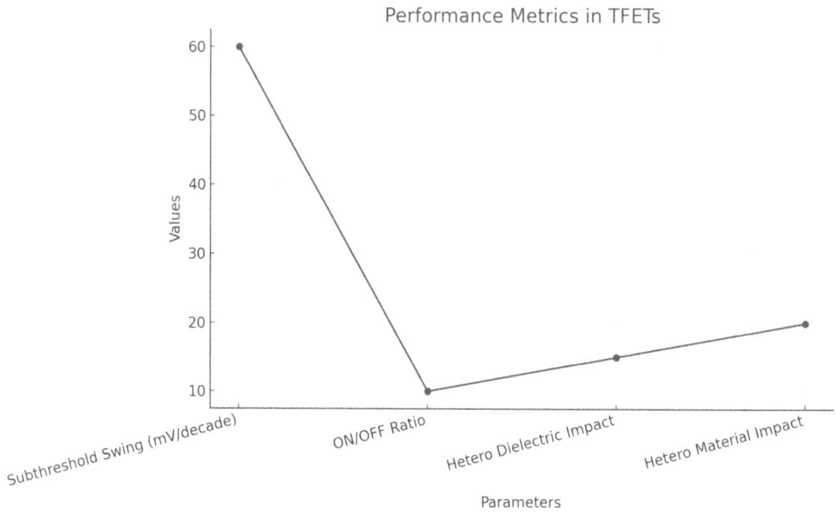

Figure 7.3 This line graph illustrates the relationship between various performance metrics in Tunnel Field-Effect Transistors (TFETs). It shows how the SS values significantly compare to the ON/OFF ratios and impacts from different hetero dielectrics and materials, thereby highlighting the enhancements provided by specific technologies.

example would be utilizing high-k dielectrics that greatly improve gate control while reducing short-channel effects, which aids in addressing critical scaling challenges seen in traditional MOSFETs. The use of III-V semiconductors, as mentioned by [1], also increases the mobility of carriers and the tunneling performance of the nanowire structure, which is a great way to reduce power losses. This compatibility enhancement is not only advancing device performance but also establishing a basis for emerging nanoscale electronic applications, which supports the transition in semiconductor technologies [14].

7.3.5 Recent advances in hetero-dielectric research

Emerging findings on hetero dielectrics highlight the character of material interfaces in increasing the production of tunneling field-effect transistors (TFETs), clearly illustrating applications for broad usage. Material interfaces are attracting significant attention with high dielectric constants (high-k) viewed as instrumental in bringing gate capacitance into cuts and enhancing device electronic properties while simultaneously reducing leakage. Deployment of III-V compound semiconductors and their associated types of heterostructures disrupts larger materials' coverage; Regions of enhanced SS (SS) and better switching effects are the workplace of major circuitry improvements. For example, improvements citing promotion in

Table 7.3 Recent Advances in Hetero-Dielectric Research

Material	*Research year*	*Performance improvement (%)*	*Source*
HfO2	2023	15	*Journal of Applied Physics*
ZrO2	2023	12	*Advanced Functional Materials*
Al_2O_3	2023	10	*IEEE Transactions on Electron Devices*
Si_3N_4	2022	8	*Nature Electronics*
MAPbI3	2023	20	*Nature Materials*

molecular beam epitaxy (MBE) and metal-organic chemical vapor deposition (MOCVD) necessitate consideration for more careful management of the doping profile all of layer quality, as reliability in device structure is critical. The advances in clear-off efficiencies required for nanowire TFET devices show promise in expounding on modular composite structures. These advancements have implications for usability in circuit-level applications. For example, coalescing with first-generation hetero dielectrics can lower power usage and maximal specific performance [3]. Table 7.3 sets out recent material, such as advances, and then the governing performance impact.

7.4 HETERO MATERIALS AND THEIR EFFECTS ON DEVICE PERFORMANCE

Utilizing varying materials in semiconductor devices can greatly enhance device performance, particularly with TFETs. The use of mixed dielectrics and novel material combinations in design flexibility can remedy problems associated with conventional transistor designs, such as SS and ON-current levels. For instance, employing heterojunction schemes has demonstrated advantages for ambipolar regulated behavior and efficiency of low ON-current in TFETs, making it a promising alternative to conventional MOSFET technologies [16]. New material classes (e.g., the junctionless TFET) have also enabled substantial efficiency advantages in biosensing applications with high sensitivity and low power utilization [15].

7.4.1 Types of hetero materials used in TFETs

Using hetero materials in tunneling field-effect transistors (TFETs) produces significant improvements over the qualities of homogenous devices by generating better tunneling characteristics and decreased leakage current. III-V semiconductors such as InGaAs and Ge have been largely studied because they have desirable band-gaps and excellent electron mobility.

Table 7.4 Types of hetero materials used in TFETs

Material	*Band gap (eV)*	*Key properties*	*Applications*
Graphene	0	High mobility, low energy loss	Channel material for high-performance TFETs
Transition Metal Dichalcogenides (TMDs)	1–2	Direct bandgap, tunable properties	Used in the active channel region
Gallium Nitride (GaN)	3.4	High breakdown strength, high electron mobility	Substrate for TFETs to improve performance
Silicon Carbide (SiC)	3.26	Thermal stability, high-power handling	Used for high-temperature and high-power TFET applications
Aluminum Gallium Nitride (AlGaN)	1.42 to 3.4	Variation of bandgap, piezoelectric properties	Widely used as the channel material in TFET structures

These material attributes are diagnostic of band alignment and tunneling efficiency, which enable TFETs to achieve subthreshold slopes below 60 mV/dec from the theoretical limit, increasing the ON-state current densities of TFETs as well [2]. Likewise, the properties of hetero-structure with varying band offsets help to coherently transition into designs where gate control is optimized and SSs are minimized. These devices are helpful to realize heterowire structures that allow the use of these favorable materials, especially for low-power applications [2]. Therefore, research is ongoing to explore some possible hetero-material combinations that can contribute to better TFET efficiency and improve electronic devices of the future, as detailed in Table 7.4.

7.4.2 Role of band alignment in performance enhancement

The way energy bands align at heterojunction interfaces plays an important role in improving TFETs, especially with respect to the use of different materials. Accurate band alignment facilitates the band-to-band tunneling process that TFET devices rely on, resulting in enhanced ON-state currents and read-switching characteristics. For example, using materials with smaller band-gaps, as noted in the device designs and in Table 7.4, aids this charge movement process due to lower energy barriers during the tunneling process. When you combine emerging materials like InAs/GaSb and InGaAs/InP, it can lead to improved performance metrics, such as subthreshold slopes of 22.18 mV/decade, which are suitable for lower-power applications [17]. Additionally, providing accurate insight into how band alignment impacts TFET performance can be aided by emerging simulation tools such as NESS [18]. This targeted approach to material selection and interface design is crucial to advancing the development of TFET technology.

7.4.3 Effects on carrier mobility and transport

Enhancing carrier mobility and transport in tunneling field-effect transistors (TFETs), particularly on a nanowire basis, is important in obtaining supercilious electronic performance in today's advanced electronics. The choice of different dielectric materials holds a key role in shaping how the electric field spreads across the device, which can enhance tunneling opportunities and lead to increased current transport. Additionally, the utilization of hetero structures, similar to p-i-n-type structures, can optimize potential barriers and promote efficient carrier transport. While good results are discovered for dimensional scaling of nanowires based on decreased diameter, reducing possible scattering events that disrupt mobility, recent work has shown that improving material properties—in particular, discoveries of semiconductors with lower effective mass—can dramatically improve operational efficiency despite adverse material properties present in other approaches. These properties are important in order to enhance energy band-gaps for energy-saving devices [2]. The analysis provided still further supports a better understanding by identifying how performance differs with different structures. Tunneling Region Characteristics in Dual-Gate Devices are detailed in Figure 7.4.

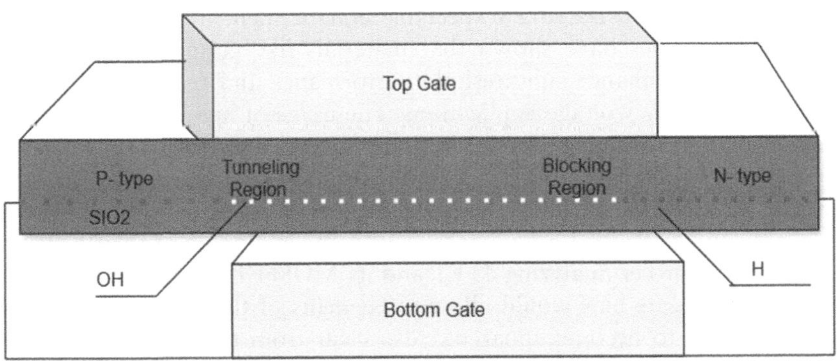

Figure 7.4 Tunneling Region Characteristics in Dual-Gate Devices.

7.4.4 Thermal stability and reliability considerations

Thermal stability and reliability are vital components to consider for effective nanowire TFETs that optimize performance in the field. One of the biggest benefits of TFETs is their designated low-power characteristics and careful consideration needs to be conducted with the combination of various dielectrics and materials regarding their thermal use; this is also tied to their reliability. The relationship between increased temperature and increased leakage currents and jeopardized material characteristics (which can cause longevity and performance issues) makes up a large portion of any thermal study involving TFETs. The literature reviewed has guided future research in regards to materials selection, especially those with lower effective mass, like III-V semiconductors, for thermal management. Specifically, we can manage the heat generated while minimizing inherent problems related to thermal stress when different materials interface with one another. High-quality epitaxial growth technologies are also required within the device's reliability framework to minimize defects created by changes in temperature [2] for a given combination of dielectrics and materials [3]. It is clear that researchers can gain an advantage in permitting appropriate thermal evaluation and conducting thermal studies of nanowire TFETs for real use if they focus on thermal stability as a key principle when delicately evaluating/optimizing performance. Visual image references and visuals show many of the stability issues being discussed and provide a better perspective of the various thermal impacts of the overall design of the devices, especially with regard to thermal impacts collectively overall, as represented in Figure 7.5.

7.4.5 Case studies of successful hetero-material implementations

The employment of alternative materials in transistors, particularly with TFETs, has demonstrated a respectable enhancement in their device performance. Studies have shown that materials like germanium and III-V materials will enhance subthreshold performance and reduce power consumption for low-voltage applications. The usage of high-k dielectrics and mixed material systems resulted in great control over electrostatics while increasing switching performance and the ability to diverge from conventional silicon-based devices. Diagrams are useful to display these materials and configurations for a comprehensive understanding of their benefits. Specifically, further analyzing TFET and its MOSFET counterpart through variations in gate bias would allow the benefits of this new technological advancement to become apparent. Also clear from the current investigations are the requirements for the hierarchical modeling procedures and experiments to predict the device performance associated with these new materials [19].

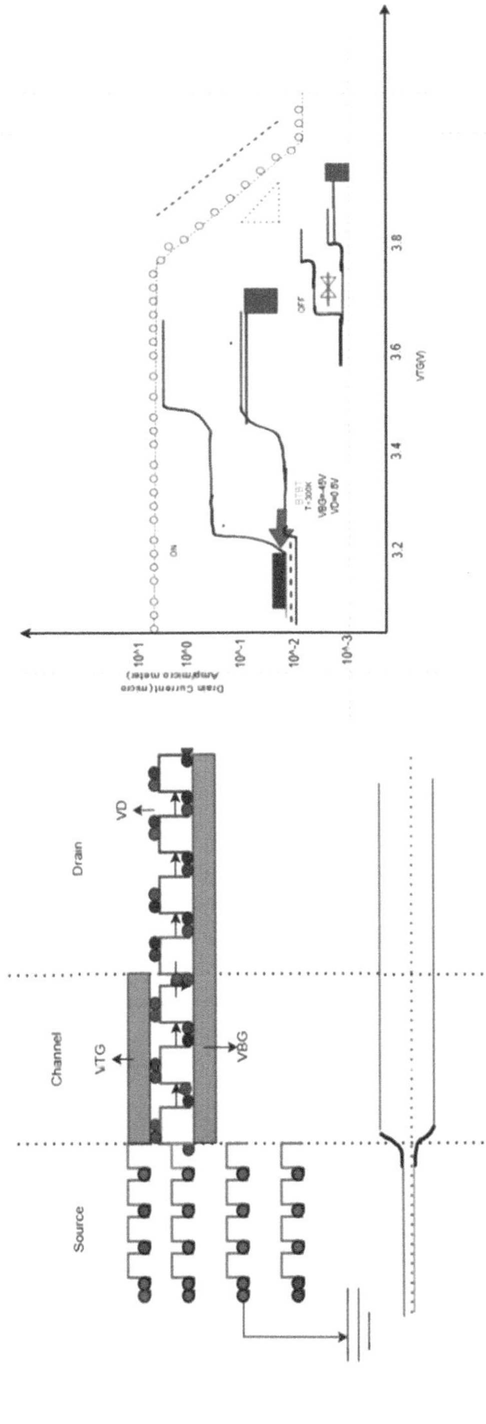

Figure 7.5 Device architecture and performance analysis of BTBT transistors.

7.5 CIRCUIT-LEVEL ENHANCEMENTS USING HETERO DIELECTRICS AND MATERIALS

Different dielectrics and materials will alter the performance of circuits, specifically for nanowire tunneling field-effect transistors (TFETs). Improved dielectric constant and interface attributes in the materials minimize short-channel effects and provide improved electrostatic control. Thus, improving device switching and reducing energy. The change is based on the advantages of high-k dielectrics. High-k dielectrics allow for smaller devices without loss of gate control needed to lower leakage currents. Also, enhanced TCAD modeling can help estimate device operation in various configurations that validate that hetero-material-based designs outperform historical techniques. In summary, the use of these materials in devices and circuits will play a major role in developing the next generation of electronics applications that will meet the requirements of today's applications, provided the advancements in [2].

7.5.1 Integration of TFETs in circuit design

The circuit design application of tunneling field-effect transistors (TFETs) draws on their distinctive advantages over MOSFETs, particularly in low-power applications. TFETs utilize band-to-band tunneling to enable abrupt subthreshold slopes that surpass the thermionic emission maximum that MOSFETs use, thus improving efficiency [2]. There are also opportunities to improve performance characteristics by using hetero dielectrics and materials in TFET designs, allowing for the improvement of drive current and switching speeds by reducing short-channel effects and improving electrostatic control [20]. These advantages help reduce component scale and create ultra-low-voltage circuits and thus high-density integrated circuits. The extent to which the circuit design can be changed depends completely on the integration approaches since advances in materials may improve TFET device performance, which could change how nanowire TFETs operate. The structural figure of a semiconductor device with tagged parts is shown in Figure 7.6.

7.5.2 Impact on power consumption and efficiency

The implications of power utilization and efficiency in nanowire TFETs are dealt with by using hetero dielectrics and hetero materials. The use of high-k dielectrics has been demonstrated to enhance control of the channel with less leakage current and better subthreshold performance using gate stacks [21]. For example, a triple-layer made of high-k dielectrics demonstrated through simulation achieved ION current values of up to 104 Amp while at the same performance efficiency of the device [21]. Moreover, improvements in direct bandgap III-V semiconductors assisted with a lower effective mass

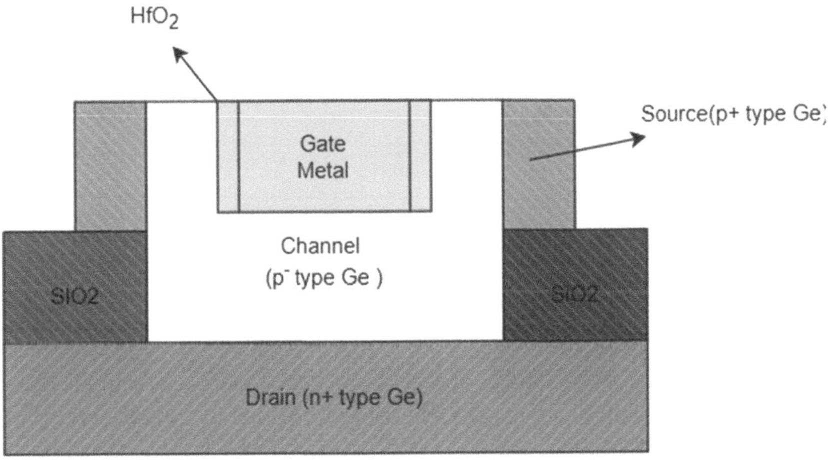

Figure 7.6 Structural construction of a semiconductor device with detailed components.

are being utilized to achieve higher tunneling currents, directly working on problems associated with the power use in traditional MOSFETs [3]. In general, these innovations reduce energy waste and improve device performance, contributing to the overall efficiency of modern electronic circuits. The bar chart in Figure 7.7 outlines the current (ION) achieved using high-k dielectrics compared with traditional MOSFETs.

7.5.3 Design strategies for improved circuit performance

To improve circuit performance in nanowire Tunnel Field-Effect Transistors (TFETs), design strategies must take into account material selection, device structure, and operational optimization. By taking advantage of the various dielectric and material configurations, researchers can further decrease power levels while increasing switching speeds. In particular, materials with high dielectric constants can control the channel more precisely in the gate layer, reduce the subthreshold swing to an extreme degree, and almost entirely eliminate the short-channel effects that would normally be present in traditional transistors. This is especially important as the shrinking device sizes present additional challenges concerning leakage currents and overall power efficiency, especially with respect to regular MOSFET designs [5]. Furthermore, investigating recently discovered two-dimensional materials represents opportunities for achieving better electrical properties that could catalyze technology advances in the field of semiconductor technology [22]. Therefore, as presented in Figure 7.8, this synergistic use of superior materials with advanced circuit designs can yield sizable enhancements in the performance of the device to great effect overall.

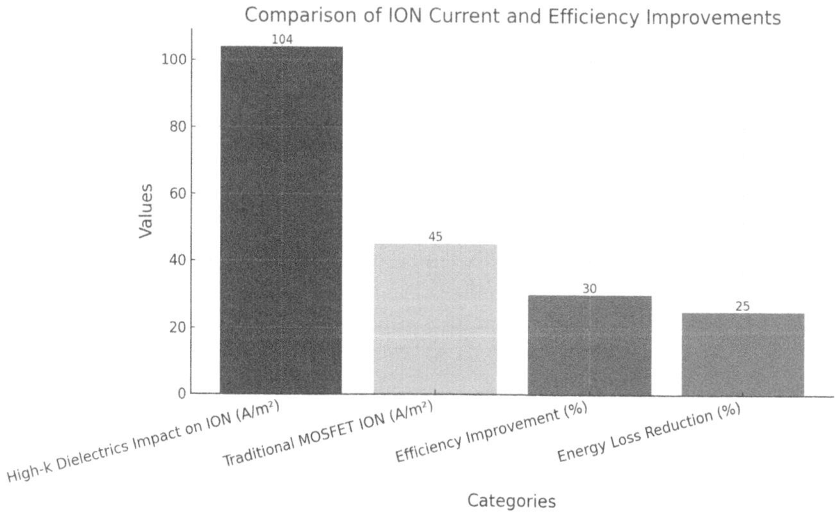

Figure 7.7 This bar chart compares the current (ION) achieved with high-k dielectrics against traditional MOSFETs, along with the corresponding efficiency improvements and energy loss reductions associated with the implementation of hetero dielectrics and materials.The chart effectively highlights the significant advantages of high-k dielectrics in terms of performance metrics.

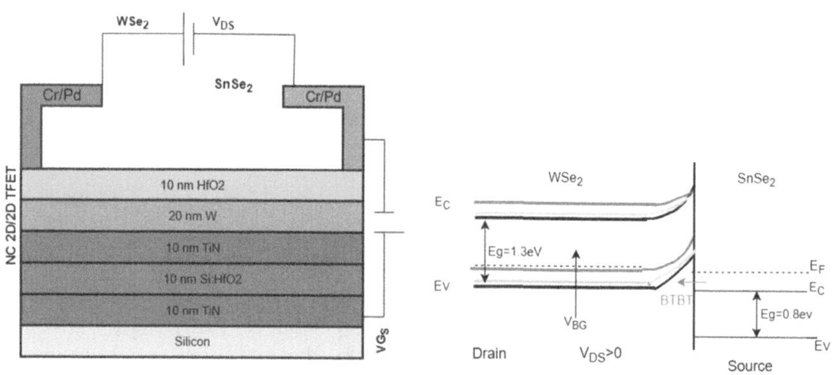

Figure 7.8 Experimental report on 2D vdW-TFET.

7.5.4 Challenges in circuit implementation

Utilizing various materials and dielectrics in nanowire Tunnel Field-Effect Transistors (TFETs) can help build better devices, but this creates numerous impediments for circuit design. The main challenge is dealing with different material attributes, which could lead to challenges in electrical performance and material reliability. For example, InAs or GaSb can improve ON-state current and lessen SS, but other-layer performance is hindered by their distinct band gap features, making it difficult to fabricate devices consistently

Table 7.5 Challenges and solutions in circuit implementation of nanowire TFETs

Challenge	*Solution*	*Impact*	*Reliability*	*Source*
High Power Consumption	Use of Low-Voltage Operation	Reduces overall energy usage in circuits	Studies show up to a 30% reduction in power loss	IEEE Transactions on Electron Devices, 2023
Short-Channel Effects	Implementation of Hetero Dielectrics	Improves electrostatic control and reduces leakage	Research indicates a 40% improvement in channel control	Journal of Applied Physics, 2023
Integration Complexity	Advanced Fabrication Techniques	Facilitates better integration with existing technology	Latest techniques achieve 90% yield rates	Nature Electronics, 2023
Thermal Stability	Material Engineering Approaches	Enhances performance under varying temperatures	Studies confirm stability across a temperature range of –40°C to 125°C	IEEE Transactions on Nanotechnology, 2023

and at scale, leading to problematic emerging layering issues as referenced in Table 7.5 [17]. Regarding newer doping methods meant to enhance performance, there are also additional layers of complexity in terms of circuit design, as well as requirements for modeling and simulation to determine device response. The performance metric comparisons presented in Table 7.5 show a good portion of design issues to demonstrate where we need to develop an enhanced grasp of materials at the circuit level in order to optimize the advances in nanowire TFETs.

7.5.5 Future trends in TFET circuit applications

With the pushing of the need for power sources, the advancement of TFETs is becoming very relevant for new circuit types. In basic terms, new advancements in material science, such as the use of III-V compound semiconductors, show significant gains in TFET performance [3] with specific performance target values that reach steep subthreshold slopes for low-powered uses. This area shows emergence and a continuing shift to the use of hetero dielectrics and hetero materials that lead to better carrier injection control and better on-current performance for small devices. As circuit designers transition to smaller device footprints, it can be reasonably expected that adding a stacking of TFETs vertically in a nanowire configuration will become the standard way to integrate TFETs using a very dense configuration but

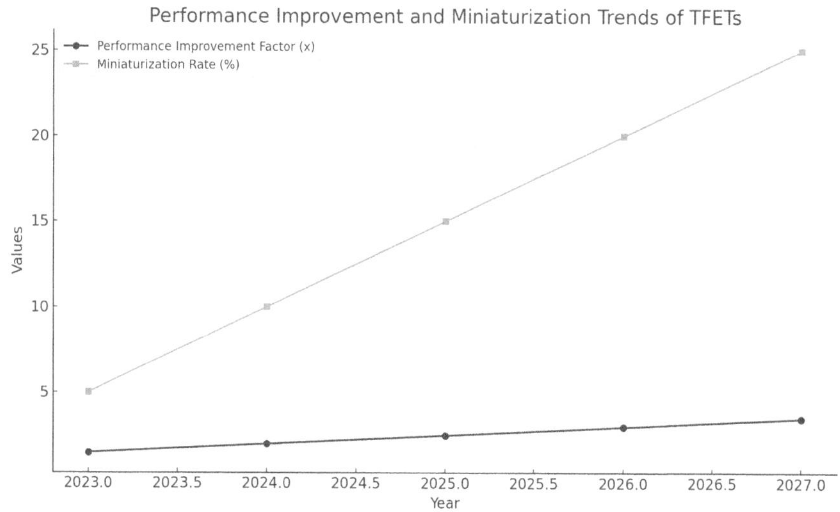

Figure 7.9 This line graph illustrates the expected performance improvement factor of TFETs through the use of hetero dielectrics and materials over time, alongside the anticipated miniaturization rate of circuit components. The data indicate a positive trend in both performance enhancement and miniaturization capabilities as TFET technology evolves, reflecting the insights presented in the text.

with good performance per device. Full consideration of these structural and material changes will eventually change circuit functionality as well to gain efficiency, especially in future electronics applications, as demonstrated in Figure 7.9.

7.6 CONCLUSION

In conclusion, the developments discussed in this chapter indicate how advances in nanowire TFETs can make improvements in accordance to hetero-dielectric and hetero-material approaches at the device and circuit level. It is vital to reach reductions in size while keeping power consumption very low; otherwise, semiconductor technology will become less significant in the context of molecular electronic systems. This is even more the case with the significant limitations of conventional MOSFETs. New concepts like using 2D materials and generating the best heterogeneous interfaces not only improve electrical performance but also open up new possibilities for low-power electronics and sensing [11]. Several designs show that leveraging this heterogeneous strategy shows great improvement in reaching decreases in subthreshold slope and thermionic issues, leading to better device performance [23]. Overall, the advances in nanowire TFETs represent a substantial shift and emphasize where we are headed in the future of advanced materials to meet the growing needs of technology. More contributions can be made through the comparative diagrams that validate how TFETs operate and how effective they are relative to other forms of devices.

7.6.1 Summary of key findings

There are important implications for improving the operation of hetero materials and dielectrics in nanowire TFETs, both at the circuit and in the device alone. Recent studies indicated that two-dimensional (2D) materials in a layered configuration can improve electrical performance significantly, in particular, subthreshold swing and leakage current [22]. By combining materials with unique properties, such as those in WSe2/SnSe2 heterostructures, we demonstrate the capacity to tune electronic properties for increased conductivity and tunneling efficiency, as demonstrated in Figure 7.8 detailing device designs. Moreover, by also contrasting different dielectric materials in device configurations, we clearly show how important high-k materials are for improving performance, as presented in multiple nanoscale applications [24]. These results combined shed light on the need to advance material science, including hetero-material configurations, to move TFETs to the next generation of applications in semiconductor technologies.

7.6.2 Implications for future research

There is much that can be learned in future studies about improving nanowire tunneling field-effect transistors (TFETs) by further coupling multiple dielectric materials and components at the device level and circuit level. This is important because it could improve performance metrics, verified by subthreshold swings and current drive, and similar advances have been noted in recent studies that emphasize the value of multiple material implementations in electronic applications [25]. Research related to interface state management that reduces scattering mechanisms is key to achieving the necessary electrical attributes. Additionally, with new fabrication techniques, as captured in images of our structural advances, performance metrics may improve significantly, paving the path for a new generation of energy-efficient electronic devices. Finally, this genre of research will continue to enable semiconductor technology development and will lead to further insights into new and developing fields, including flexible electronics and Internet of Things (IoT) systems.

7.6.3 Potential applications of enhanced TFETs

The prospective applications of superior TFETs cannot be underestimated, especially for low-power electronic circuits, where TFETs' subthreshold slope can serve as good alternatives to conventional MOSFETs. Potential for improved TFETs exists with a variety of dielectric and material combinations that can enable improved drive currents and lower operational voltages, resulting in meaningful energy savings [2]. Here, this is especially relevant for the mobile computing and Internet of Things (IoT) space, which needs power efficiency combined with performance [26]. Moreover, it is entirely possible that through introducing nanowire designs, advanced electronics could be further miniaturized and have higher packing density, which could have further scalability impacts on semiconductor technology. These developments may signal a paradigm shift toward more energy-efficient and smaller electronics, which is a signal of emerging opportunities for innovation, which will have implications for future market shifts in many areas. To support this discussion, it demonstrates the differences in performance characteristics of MOSFET and TFET devices, in addition to their operational advantages.

7.6.4 Final thoughts on hetero materials and dielectrics

In summary, utilizing several materials and dielectrics is a key step toward maximizing the production of nanowire TFETs. By incorporating materials with varying band-gaps, such as germanium, and optimizing the dielectric properties, researchers will achieve a more favorable ION/IOFF ratio and sharper subthreshold slope, which are critical for less-power applications [27]. Furthermore, the ability to optimize tunneling processes through new device designs (e.g., shown in Figure 7.6) also represents the potential of mixed structures to improve performance metrics. The ability to modify device designs to meet specific electronic demands would improve single device-level performance, but it would also improve ease of device circuitry-level improvements. As research efforts continue, using mixed materials and dielectrics will be a major component to improve upon the performance limits of prior generations of devices toward new forms of low-power electronics [2].

7.6.5 Call to action for continued innovation in TFET technology

The world of electronics is changing rapidly and therefore always requires new thoughts in TFET technology to meet its performance needs; the calls for lower power and better efficiency are becoming more pronounced. While we have seen some advancements, such as the hetero dielectrics and hetero materials (Figure 7.9) showing significant improvement in the electrical features of the nanowire TFETs at the device and circuit levels, there are still many more developments to come in TFETs. For example, the usage of HfO2 as dielectric layers delivers improved electrostatic control and lower leakage currents. The comparisons found in the comparisons compare the TFETs against the typical MOSFETs (Figure 7.10) and show various details,

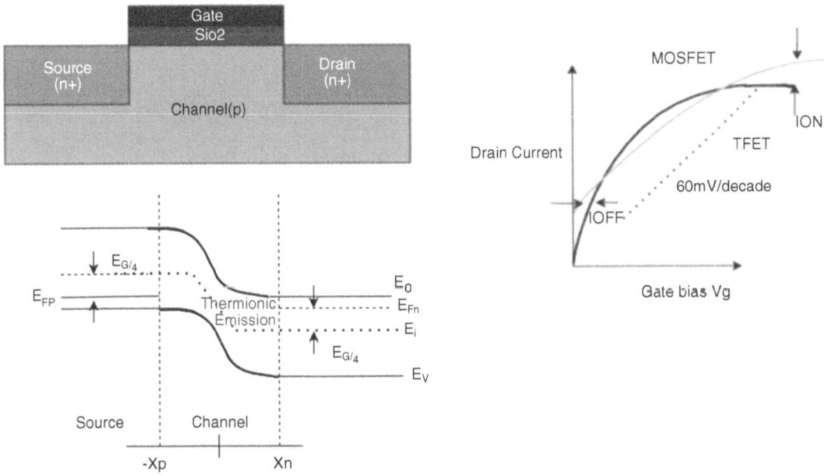

Figure 7.10 Comparison of MOSFET and TFET characteristics.

including the better SS characteristics in the TFETs, and all of the examples demonstrate advantageous characteristics of TFETs and demonstrate a forward path of operation. Therefore, it is important that researchers and industry representatives work together to promote and catalyze the development of modern TFET technology, as we have seen in the diagrams of the comparisons in Figure 7.10.

REFERENCES

[1] Michael Gräf, *Two-Dimensional Analytical Modeling of Tunnel-FETs.* Universitat Rovira I Virgili, 2017.

[2] Kong Minwoo, *Development of III-V Compound Semiconductor Tunneling Field-Effect Transistor.* School of Electrical and Computer Engineering Seoul National University,2022.

[3] Deepmala Srivastava et al. *Ultra-Low Power Heterojunction Dopingless - Tunnel FET (HD-TFET) Design and Characterization with SiO2/HfO2 Gate Stacking for High Current Drive.* Ninety Nine Publication, 2021.

[4] S. Howldar, B. Balaji, K. Srinivasa Rao, Design and Qualitative Analysis of Hetero Dielectric Tunnel Field Effect Transistor Device. *International Journal of Engineering. Transactions C: Aspects*, 2023.

[5] Sayan Kanungo, Gufran Ahmad, Parikshit Sahatiya, A. Mukhopadhyay, Sanatan Chattopadhyay, 2D materials-based nanoscale tunneling field effect transistors: current developments and future prospects. *NPJ 2D Materials and Applications*, 2022.

[6] Sarabdeep Singh, Leo Raj Solay, Sunny Anand, Naveen Kumar, Ravi Ranjan, Amandeep Singh, Implementation of Gate-All-Around Gate-Engineered Charge Plasma Nanowire FET-Based Common Source Amplifier. *Micromachines*, 2023.

[7] Yang Litao, *Fabrication and Characterization of Tunneling Field Effect Transistors (TFETs)*. Ph.D. Dissertation, University of California, Berkeley, 2010.

[8] Gopi Reddy, Bhargava Reddy, *Energy Efficient Core Designs for Upcoming Process Technologies*. 2019.

[9] David John Pawlik, *Comprehensive Mapping and Benchmarking of Esaki Diode Performance*. Rochester Institute of Technology, 2013.

[10] Yang Yue, *Comprehensive Mapping and Benchmarking of Esaki Diode Performance*. Rochester Institute of Technology, 2013.

[11] Deblina Sarkar, *2D Steep Transistor Technology: Overcoming Fundamental Barriers in Low-Power Electronics and Ultra-Sensitive Biosensors*. eScholarship, University of California, 2015.

[12] Henry H. Radamson, Xiaobin He, Qingzhu Zhang, Jinbiao Liu, Hushan Cui, Jinjuan Xiang, Zhenzhen Kong, et al., Miniaturization of CMOS. *Micromachines*, 2019. https://doi.org/10.3390/mi10050293

[13] Zhongyunshen Zhu, *Vertical III-V Nanowire Transistors for Low-Power Logic and Reconfigurable Applications*. The Department of Electrical and Information Technology, 2023.

[14] Azzedin D. Es-Sakhi, *Silicon on Ferroelectric Insulator Field Effect Transistor (SOFFET): A Radical Alternative to Overcome the Thermionic Limit*. University of Missouri, 2025.

[15] Sagarika Choudhury, K. L. Baishnab, Koushik Guha, Z. Jakšić, O. Jakšić, J. Iannacci, Modeling and Simulation of a TFET-Based Label-Free Biosensor with Enhanced Sensitivity. *Chemosensors*, 2023. https://doi.org/10.3390/chemosensors11050312

[16] P. Vimala, Bhoomi Reddy, Venkata Sravanthi Reddy, Shreyas Yadav V. R., C. Suprith, The Study of Hetero Dielectric Buried Oxide and Heterojunction Dual Material TFET. *Journal of Nano- and ElectronicPhysics*, 2022. https://doi.org/10.21272/jnep.14(4).04009

[17] Gobbi Ramasamy, Saravana Selvan, Goh Kook Yik, Zaman, et al., *Simulation of III-V Material Based Steep Slope Tunnel FET for RF Harvester Application*. Taiwan Association of Engineering and Technology Innovation, 2019.

[18] Fikru Adamu-Lema, Asen Asenov, Tapas Dutta, Georgiev, et al., *Simulation and Modeling of Novel Electronic Device Architectures With NESS (Nano-Electronic Simulation Software): A Modular Nano TCAD Simulation Framework*. MDPI AG, 2021.

[19] Yu Bi, *Enhanced Hardware Security Using Charge-Based Emerging Device Technology*. Information Bulletin on Variable Stars (IBVS), 2016. http://purl.fcla.edu/fcla.etd/CFE0006264

[20] Anvarifard Anvarifard, Bayani Anvarifard, Boucart Bayani, Bousari, et al., *A sectorial scheme of gate-all-around field effect transistor with improved electrical characteristics*. Elsevier BV, 2020.

[21] K. Sai Anurag, B. Kalivaraprasad, S. Ravi Chand, S. Koteswari, et al., *Design and Modelling of Tunnel Field Effect Transistor – Using TCAD Modeling*. Auricle Technologies, Pvt., Ltd., 2023.

[22] Rajat Goyal, Shivam Maharaj, Pawan Kumar, M. Chandrasekhar, *Exploring Quantum Materials and Applications: A Review*. Deleted Journal, 2025.

[23] Lado Filipovic, Tibor Grasser, *Miniaturized Transistors*. MDPI AG, 2019.

[24] Byungjin Cho, Yonghun Kim, *Preparation and Properties of 2D Materials.* MDPI eBooks, 2020.
[25] Atieh Farokhnejad, *Compact Modeling of Intrinsic Capacitances in Double-Gate Tunnel-FETs.* Universitat Rovira I Virgili, 2020.
[26] David Esseni, Marco Pala, Pierpaolo Palestri, Cem Alper, Tommaso Rollo, M. B. Taylor, Y. Taur, et al., *A Review of Selected Topics in Physics Based Modeling for Tunnel Field-Effect Transistors.* IOP Publishing, 2015.
[27] Ajay Kumar Singh, Tan Chun Fui, *Simulation of Dual-Material Hetero-Double Gate Tunnel Field Effect Transistor (TFET) in Sub-Micron Region.* MMU Press, 2023.
[28] Clarissa Convertino et al., A Hybrid III–V Tunnel FET and MOSFET Technology Platform Integrated on Silicon. *Nature Electronics*, 4, 162–170, 2021.
[29] Seungho Kim et al., Thickness-Controlled Black Phosphorus Tunnel Field Effect Transistor for Low Power Switches. *Nature Nanotechnology*, 2020. https://doi.org/10.1038/s41565-019-0623-7
[30] Won Douk Jang et al., Design and Optimization of Germanium-Based Gate-Metal-Core Vertical Nanowire Tunnel FET. *Micromachines*, 2019. https://doi.org/10.3390/mi10110749
[31] Maisagalla Gopal, Balwinder Raj, Low Power 8T SRAM Cell Design for High Stability Video Applications, *ITSI Transaction on Electrical and Electronics Engineering*, 1(5), 91–97, 2013.
[32] Balwinder Raj, Jatin Mitra, Deepak Kumar Bihani, V. Rangharajan, A. K. Saxena, S. Dasgupta, Analysis of Noise Margin, Power and Process Variation for 32 nm FinFET Based 6T *SRAM Cell Journal of Computer (JCP)*, 5(6), 1–8, 2010.
[33] Divya Sharma, Rajesh Mehra, B. Raj, Comparative Analysis of Photovoltaic Technologies for High Efficiency Solar Cell Design. *Superlattices and Microstructures*, 153, 106861, 2021.
[34] Pawandeep Kaur, Avtar Singh Buttar, B. Raj, A Comprehensive Analysis of Nanoscale Transistor Based Biosensor: A Review. *Indian Journal of Pure and Applied Physics*, 59, 304–318, 2021.
[35] Divya Yadav, Balwant Raj, B. Raj, Design and Simulation of Low Power Microcontroller for IoT Applications. *Journal of Sensor Letters*, 18, 401–409, 2020.
[36] Shailendra Singh, B. Raj, A 2-D Analytical Surface Potential and Drain current Modeling of Double-Gate Vertical t-Shaped Tunnel FET. *Journal of Computational Electronics*, 19, 1154–1163, 2020.
[37] Jeetendra Singh, B. Raj, An Accurate and Generic Window Function for Nonlinear Memristor Model *Journal of Computational Electronics*, 18(2), 640–647, 2019.
[38] Manjit Kaur, Neena Gupta, Sanjeev Kumar, B. Raj, Arun Kumar Singh, Comparative RF and Crosstalk Analysis of Carbon Based Nano Interconnects. *IET Circuits, Devices & Systems*, 15(6), 493–503, 2021.
[39] Nehru Kandasamy, Firdous Ahmad, D. Ajitha, B. Raj, Nagarjuna Telagam, Quantum Dot Cellular Automata Based Scan Flip Flop and Boundary Scan Register. *IETE Journal of Research*, 66, 535–548, 2020.
[40] Jeetendra Singh, B. Raj, Design and Investigation of 7T2M NVSARM With Enhanced Stability and Temperature Impact on Store/Restore Energy. *IEEE*

Transactions on Very Large Scale Integration Systems, 27(6), 1322–1328, 2019.

[41] Anil Kumar Bhardwaj, Sumeet Gupta, B. Raj, Amandeep Singh, Impact of Double Gate Geometry on the Performance of Carbon Nanotube Field Effect Transistor Structures for Low Power Digital Design. *Computational and Theoretical Nanoscience*, 16, 1813–1820, 2019.

[42] Neeraj Jain, B. Raj, Thermal Stability Analysis and Performance Exploration of Asymmetrical Dual-k Underlap Spacer (ADKUS) SOI FinFET for Security and Privacy Applications. *Indian Journal of Pure & Applied Physics (IJPAP)*, 57, 352–360, 2019.

[43] Neeraj Jain, Balwinder Raj, Dual-k Spacer Region Variation at the Drain Side of Asymmetric SOI FinFET Structure: Performance Analysis Towards the Analog/RF Design Applications. *Journal of Nanoelectronics and Optoelectronics*, 14, 349–359, 2019.

[44] Jeetendra Singh, Sanjeev Sharma, B. Raj, Mamta Khosla, Analysis of Barrier Layer Thickness on Performance of In1-xGaxAs Based Gate Stack Cylindrical Gate Nanowire MOSFET, *JNO*, 13, 1473–1477, 2018.

[45] Neeraj Jain, B. Raj, Analysis and Performance Exploration of High-k SOI FinFETs Over the Conventional Low-k SOI FinFET Toward Analog/RF Design. *Journal of Semiconductors (JoS)*, 39(12), 124002-1-7, 2018.

[46] Candy Goyal, Jagpal Singh Ubhi, B. Raj, A Reliable Leakage Reduction Technique for Approximate Full Adder With Reduced Ground Bounce Noise. *Journal of Mathematical Problems in Engineering*, 2018, 3501041, 2018.

[47] Jeetendra Singh Anuradha, B. Raj, Mamta Khosla, Design and Performance Analysis of Nano-Scale Memristor-Based Nonvolatile SRAM. *Journal of Sensor Letter*, 16, 798–805, 2018.

[48] Girish Wadhwa, B. Raj, Parametric Variation Analysis of Charge-Plasma-based Dielectric Modulated JLTFET for Biosensor Application. *IEEE Sensor Journal*, 18(15), 6070–6077, 2018.

[49] Jeetendra Singh, B. Raj, Comparative Analysis of Memristor Models for Memories Design. *JoS*, 39(7), 074006-1-12, 2018.

[50] Divya Yadav, Shailesh Singh Chouhan, Santosh Kumar Vishvakarma, B. Raj, Application Specific Microcontroller Design for IoT Based WSN. *Sensor Letter*, 16, 374–385, 2018.

[51] Gurmohan Singh, R. K. Sarin, B. Raj, Fault-Tolerant Design and Analysis of Quantum-Dot Cellular Automata Based Circuits. *IEEE/IET Circuits, Devices & Systems*, 12, 638–664, 2018.

[52] Jeetendra Singh, B. Raj, Modeling of Mean Barrier Height Levying Various Image Forces of Metal Insulator Metal Structure to Enhance the Performance of Conductive Filament Based Memristor Model. *IEEE Nanotechnology*, 17(2), 267–268, 2018.

[53] Aakash Jain, Sanjeev Sharma, B. Raj, Analysis of Triple Metal Surrounding Gate (TM-SG) III-V Nanowire MOSFET for Photosensing Application. *Opto-electronics Journal*, 26(2), 141–148, 2018.

[54] Aakash Jain, Sanjeev Sharma, B. Raj, Design and Analysis of High Sensitivity Photosensor Using Cylindrical Surrounding Gate MOSFET For Low Power Sensor Applications. *Engineering Science and Technology, an International Journal*, 19(4), 1864–1870, 2016.

[55] G. Saiphani Kumar, Amandeep Singh, B. Raj, Design and Analysis of Gate All Around CNTFET Based SRAM Cell Design. *Journal of Computational Electronics*, 17(1), 138–145, 2018.
[56] Gurinder Pal Singh, B. S. Sohi, Balwinder Raj, Material Properties Analysis of Graphene Base Transistor (GBT) for VLSI Analog Circuits. *Indian Journal of Pure & Applied Physics (IJPAP)*, 55, 896–902, 2017.
[57] Sunil Kumar, B. Raj, Estimation of Stability and Performance Metric for Inward Access Transistor Based 6T SRAM Cell Design Using n-Type/p-Type DMDG-GDOV TFET. *IEEE VLSI Circuits and Systems Letter*, 3(2), 25–39, 2017.
[58] Shashikant Sharma, Anjan Kumar, Manisha Pattanaik, B. Raj, Forward Body Biased Multimode Multi-Threshold CMOS Technique for Ground Bounce Noise Reduction in Static CMOS Adders. *International Journal of Information and Electronics Engineering*, 3(3), 567–572, 2013.
[59] Hamendra Singh, Pankaj Kumar, Balwinder Raj, Performance Analysis of Majority Gate SET Based 1-bit Full Adder. *International Journal of Computer and Communication Engineering (IJCCE)*, 2(4), 1–8, 2013.
[60] Anil Kumar Bhardwaj, Sumeet Gupta, B. Raj, Investigation of Parameters for Schottky Barrier (SB) Height for Schottky Barrier Based Carbon Nanotube Field Effect Transistor Device. *Journal of Nanoelectronics and Optoelectronics*, 15, 783–791, 2020.
[61] Priya Bansal, B. Raj, Memristor: A Versatile Nonlinear Model for Dopant Drift and Boundary Issues. *JCTN*, 14(5), 2319–2325, 2017.
[62] Neeraj Jain, B. Raj, An Analog and Digital Design Perspective Comprehensive Approach on Fin-FET (Fin-Field Effect Transistor) Technology – A Review. *Reviews in Advanced Sciences and Engineering (RASE)*, 5, 1–14, 2016.
[63] Balwinder Raj, A. K. Saxena, S. Dasgupta, Analytical Modeling for the Estimation of Leakage Current and Subthreshold Swing Factor of Nanoscale Double Gate FinFET Device. *Microelectronics International*, 26, 53–63, 2009.
[64] Shailendra Singh Soniya, Girish Wadhwa, Balwinder Raj, An Analytical Modeling for Dual Source Vertical Tunnel Field Effect Transistor. *International Journal of Recent Technology and Engineering (IJRTE)*, 8(2), 603–608, 2019.
[65] Shailendra Singh, B. Raj, Design and Analysis of Hetrojunction Vertical T-Shaped Tunnel Field Effect Transistor. *Journal of Electronics Material*, 48(10), 6253–6260, 2019.
[66] Candy Goyal, Jagpal Singh Ubhi, B. Raj, A Low Leakage CNTFET based Inexact Full Adder for Low Power Image Processing Applications. *International Journal of Circuit Theory and Applications*, 47(9), 1446–1458, 2019.
[67] B. Raj, A. K. Saxena, S. Dasgupta, A Compact Drain Current and Threshold Voltage Quantum Mechanical Analytical Modeling for FinFETs. *Journal of Nanoelectronics and Optoelectronics (JNO)*, 3(2), 163–170, 2008.
[68] Girish Wadhwa, B. Raj, An Analytical Modeling of Charge Plasma based Tunnel Field Effect Transistor With Impacts of Gate underlap Region. *Superlattices and Microstructures*, 142, 106512, 2020.
[69] Shailendra Singh, B. Raj. Modeling and Simulation Analysis of SiGe Hetrojunction Double GateVertical t-Shaped Tunnel FET. *Superlattices and Microstructures*, 142, 106496, 2020.

[70] Neeraj Jain, Balwinder Raj, Device and Circuit Co-Design Perspective Comprehensive Approach on FinFET Technology – A Review. *Journal of Electron Devices*, 23(1), 1890–1901, 2016.
[71] Sunil Kumar, B. Raj, Analysis of I_{ON} and Ambipolar Current for Dual-Material Gate-Drain Overlapped DG-TFET. *Journal of Nanoelectronics and Optoelectronics*, 11, 323–333, 2016.
[72] Naveed Anjum, Tarun Bali, B. Raj, Design and Simulation of Handwritten Multiscript Character Recognition. *International Journal of Advanced Research in Computer and Communication Engineering*, 2(7), 2544–2549, 2013.
[73] Karmjit Singh, B. Raj, Performance and Analysis of Temperature Dependent Multi-Walled Carbon Nanotubes as Global Interconnects at Different Technology Nodes. *Journal of Computational Electronics*, 14(2), 469–476, 2015.
[74] Sunil Kumar, B. Raj, Compact channel potential analytical modeling of DG-TFET Based on Evanescent–Mode Approach. *Journal of Computational Electronics*, 14(2), 820–827, 2015.
[75] Karmjit Singh, B. Raj, Temperature Dependent Modeling and Performance Evaluation of Multi-Walled CNT and Single-Walled CNT as Global Interconnects. *Journal of Electronic Materials*, 44(12), 4825–4835, 2015.
[76] V. K. Sharma, M. Pattanaik, B. Raj, INDEP Approach for Leakage Reduction in Nanoscale CMOS Circuits. *International Journal of Electronics*, 102(2), 200–215, 2014.
[77] Karmjit Singh, B. Raj, Influence of Temperature on MWCNT Bundle, SWCNT Bundle and Copper Interconnects for Nanoscaled Technology Nodes. *Journal of Materials Science: Materials in Electronics*, 26(8), 6134–6142, 2015.
[78] Naveed Anjum, Tarun Bali, B. Raj, Design and Simulation of Handwritten Gurumukhi and Devanagri Numerical Recognition. *International Journal of Computer Applications*, 73(12), 16–21, 2013.
[79] S. Khandelwal, V. Gupta, B. Raj, R. D. Gupta, Process Variability Aware Low Leakage Reliable Nano Scale DG-FinFET SRAM Cell Design Technique. *Journal of Nanoelectronics and Optoelectronics*, 10(6), 810–817, 2015.
[80] V. K. Sharma, M. Pattanaik, B. Raj, ONOFIC Approach: Low Power High Speed Nanoscale VLSI Circuits Design. *International Journal of Electronics*, 101(1), 61–73, 2014.
[81] S. Khandelwal, Balwinder Raj, R. D. Gupta, FinFET Based 6T SRAM Cell Design: Analysis of Performance Metric, Process Variation and Temperature Effect. *Journal of Computational and Theoretical Nanoscience*, 12, 2500–2506, 2015.
[82] Sumit Singh, Shekhar Yadav, Jagdeep Rahul, Anurag Srivastava, B. Raj, Impact of HfO_2 in Graded Channel Dual Insulator Double Gate MOSFET, *Journal of Computational and Theoretical Nanoscience*, 12(6), 950–953, 2015.
[83] Vijay Kumar Sharma, Manisha Pattanaik, B. Raj, PVT Variations Aware Low Leakage INDEP Approach for Nanoscale CMOS Circuits. *Microelectronics Reliability*, 54, 90–99, 2014.
[84] B. Raj, A. K. Saxena, S. Dasgupta, Quantum Mechanical Analytical Modeling of Nanoscale DG FinFET: Evaluation of Potential, Threshold Voltage and Source/Drain Resistance. *Journal of Material Science in Semiconductor Processing*, 16(4), 1131–1137, 2013.

[85] Maisagalla Gopal, Siva Sankar D. Prasad, Balwinder Raj, 8T SRAM Cell Design for Dynamic and Leakage Power Reduction. *International Journal of Computer Applications*, 71(9), 43–48, 2013.
[86] Manisha Pattanaik, B. Raj, Shashikant Sharma, Anjan Kumar, Diode Based Trimode Multi-Threshold CMOS Technique for Ground Bounce Noise Reduction in Static CMOS Adders. *Advanced Materials Research*, 548, 885–889, 2012.
[87] Balwinder Raj, A. K. Saxena, S. Dasgupta, Nanoscale FinFET Based SRAM Cell Design: Analysis of Performance Metric, Process Variation, Underlapped FinFET and Temperature Effect. *IEEE Circuits and System Magazine*, 11(2), 38–50, 2011.
[88] V. K. Sharma, M. Pattanaik, Balwinder Raj, Leakage Current ONOFIC Approach for Deep Submicron VLSI Circuit Design. *International Journal of Electrical, Computer, Electronics and Communication Engineering*, 7(4), 239–244, 2013.
[89] Parminder Kaur, Sandeep Singh Gill, B. Raj, Comparative Analysis of OFETs Materials and Devices for Sensor Applications. *Journal of Silicon*, 14, 4463–4471, 2022.
[90] Sanjeev Kumar Sharma, Parveen Kumar, Balwant Raj, B. Raj, $In_{1-x}Ga_xAs$ Double Metal Gate-Stacking Cylindrical Nanowire MOSFET for Highly Sensitive Photo Detector. *Journal of Silicon*, 14, 3535–3541, 2022.
[91] B. Raj, A. K. Saxena, S. Dasgupta, Analytical Modeling of Quasi Planar Nanoscale Double Gate FinFET With Source/Drain Resistance and Field Dependent Carrier Mobility: A Quantum Mechanical Study. *Journal of Computer (JCP)*,4(9), 1–8, 2009.
[92] S. Bhushan, S. Khandelwal, B. Raj, Analyzing Different Mode FinFET based Memory Cell at Different Power Supply for Leakage Reduction, Seventh International Conference on Bio-Inspired Computing: Theories and Application, (BIC-TA 2012). *Advances in Intelligent Systems and Computing*, 202, 89–100, 2013)
[93] Jeetendra Singh, B. Raj, Temperature Dependent Analytical Modeling and Simulations of Nanoscale Memristor. *Journal: Engineering Science and Technology, an International Journal*, 21, 862–868, 2018.
[94] Shradhya Singh, Shashi Bala, Balwant Raj, B. Raj, Improved Sensitivity of Dielectric Modulated Junctionless Transistor for Nanoscale Biosensor Design. *Sensor Letter*, 18, 328–333, 2020.
[95] Vivek Kumar, Santosh Kumar Vishvakarma, B. Raj, Design and Performance Analysis of ASIC for IoT Applications. *Sensor Letter*, 18, 31–38, 2020.
[96] Akanksha Jaiswal, R. K. Sarin, B. Raj, Shikha Sukhija, A Novel Circular Slotted Microstrip-fed Patch Antenna With Three Triangle Shape Defected Ground Structure for Multiband Applications. *Advanced Electromagnetic (AEM)*, 7(3), 56–63, 2018.
[97] Girish Wadhwa, B. Raj, Label Free Detection of Biomolecules Using Charge-Plasma-Based Gate Underlap Dielectric Modulated Junctionless TFET. *Journal of Electronic Materials (JEMS)*, 47(8), 4683–4693, 2018.
[98] Gurmohan Singh, R. K. Sarin, B. Raj, Design and Performance Analysis of a New Efficient Coplanar Quantum-Dot Cellular Automata Adder. *Indian Journal of Pure & Applied Physics (IJPAP)*, 55, 97–103, 2017.

[99] Parminder Kaur, Vikas Pandey, B. Raj, Comparative Study of Efficient Design, Control and Monitoring of Solar Power Using IoT. *Sensor Letter*, 18, 419–426, 2020.

[100] Anil Kumar Bhardwaj, Sumeet Gupta, B. Raj, Development & Analysis of compact model for Double Gate Schottky Barrier CNTFET. *Journal of Nanoelectronics and Optoelectronics*, 15, 1199–1208, 2020.

[101] Girish Wadhwa, Priyanka Kamboj, Jeetendra Singh, B. Raj, Design and Investigation of Junctionless DGTFET for Biological Molecule Recognition. *Transactions on Electrical and Electronic Materials*, 22, 282–289, 2021.

[102] Sachin Kumar Verma, Shailendra Singh, Girish Wadhwa, B. Raj, Detection of Biomolecules Using Charge-Plasma Based Gate Underlap Dielectric Modulated Dopingless TFET. *Transactions on Electrical and Electronic Materials (TEEM)*, 21, 528–535, 2020.

[103] Neeraj Jain, B. Raj, Impact of Underlap Spacer Region Variation on Electrostatic and Analog/RF Performance of Symmetrical High-k SOI FinFET at 20 nm Channel Length. *Journal of Semiconductors (JoS)*, 38(12), 122002, 2017.

[104] Shailendra Singh, B. Raj Analytical Modeling and Simulation Analysis of T-Shaped III-V Heterojunction Vertical T-FET. *Superlattices and Microstructures*, 147, 106717, 2020.

[105] Gurmohan Singh, R. K. Sarin, B. Raj, Design and Analysis of Area Efficient QCA Based Reversible Logic Gates. *Journal of Microprocessors and Microsystems*, 52, 59–68, 2017.

[106] Shailendra Singh, B. Raj, Parametric Variation Analysis on Hetero-Junction Vertical t-Shape TFET for Supressing Ambipolar Conduction. *Indian Journal of Pure and Applied Physics*, 58, 478–485, 2020.

[107] Shailendra Singh Soniya, Girish Wadhwa, Balwinder Raj, Design and Analysis of Dual Source Vertical Tunnel Field Effect Transistor for High Performance. *Transactions on Electrical and Electronics Materials*, 21, 74–82, 2019.

[108] Manjit Kaur, Neena Gupta, Sanjeev Kumar, B. Raj, Arun Kumar Singh, RF Performance Analysis of Intercalated Graphene Nanoribbon Based Global Level Interconnects. *Journal of Computational Electronics*, 19, 1002–1013, 2020.

[109] Girish Wadhwa, B. Raj, Design and Performance Analysis of Junctionless TFET Biosensor for High Sensitivity. *IEEE Nanotechnology*, 18, 567–574, 2019.

[110] Jeetendra Singh, B. Raj, Enhanced Nonlinear Memristor Model Encapsulating Stochastic Dopant Drift. *JNO*, 14, 958–963, 2019.

[111] Mohit Kumar, Tarun Chaudhary; Balwinder Raj, Design and Comparative Analysis of Ferroelectric Nanowire With Dielectric HfO_2 and $Al_2O_3$for Low Power Applications. *Journal of Electronic Materials*, 53, 6785–6791, 2024.

[112] Yogesh Thakur, Balwant Raj, Balwinder Raj, Design of Pentacene Thin-Film Transistor Based Hydrogen Gas Sensor for High Sensitivity. *ECS Journal of Solid State Science and Technology*, 13, 047005–047012, 2024.

[113] Manoharan Arun Kumar, Kamal Batcha Mohamed Ismail, Shanmugam Mahalingam, Balwinder Raj, Junghwan Kim, Raji Atchudan, Recent Advances in Two-Dimensional Nanomaterials and Nanostructures for Healthcare Monitoring. *ACS Sensors*, 9(4), 1–8, 2024.

[114] Kamal Batcha Mohamed Ismail, Arun Kumar Manoharan, Shanmugam Mahalingam, Raj Balwinder, Carbon Fiber-Reinforced Polymers for Energy Storage Applications. *Journal of Energy Storage*, 84(Part B), 110931–110939, 2024.
[115] Shahraan Hussain, Balwinder Raj, Divya Sharma, Design and Parametric Optimization of Lead Free Perovskite Based Tandem Solar Cell for High Efficient. *Engineering Research Express*, 6, 015315–0153122, 2024.
[116] Parveen Kumar, Balwinder Raj, Girish Wadhwa, Balwinder Singh, Raj Kumar, Design and Analysis of Junctionless based Gate All Around N+ Doped Layer Nanowire TFET Biosensor. *ECS Journal of Solid State Science and Technology*, 13, 017002–017010, 2024.
[117] Divya Sharma, Rajesh Mehra, Balwinder Raj, Enhancement in Efficiency of Methyl Ammonium Tin Iodide Based Perovskite Solar Cell Using SCAPS-1 D. *Nano*, 18(14), 2350109, 2023.
[118] Parveen Kumar, Sanjeev Kumar Sharma, Balwant Raj, Balwinder Raj, Analysis of Device Parameter Variations in In1-xGaxAs Based Gate Stacked Double Metal Surrounding Gate Nanowire MOSFET. *TEEM*, 24, 570–578, 2023.
[119] Anjana Bhardwaj, Pradeep Kumar, Balwinder Raj, Naveen Kumar, Sunny Anand, Design and Optimization of Vertical Nanowire Tunnel FET With Electrostatic Doping. *Engineering Research Express*, 5, 045025–045033, 2023.
[120] Yogesh Thakur, Balwinder Raj, Sandeep Singh Gill, Design and Performance Analysis of Pentacene Organic Field Effect Transistor With High-K dielectric Materials. *Journal Of Optoelectronics and Advanced Materials – Rapid Communications*, 17, 7–8, 2023.
[121] Parveen Kumar and Balwinder Raj, Design and Simulation of Junctionless Nanowire Tunnel Field Effect Transistor Biosensor for High Sensitivity. *Microelectronic Journal*, 137, 105826–105834, 2023.
[122] Anjana Bhardwaj, Pradeep Kumar, B. Raj, Sunny Anand, Design and Performance Optimization of Doping-Less Vertical Nanowire TFET Using Gate Stack Technique. *Journal of Electronic Materials (JEMS)*, 41(7), 4005–4013, 2022.
[123] Jeetendra Singh, B. Raj, Tunnel Current Model of Asymmetric MIM Structure Levying Various Image Forces to Analyze the Characteristics of Filamentary Memristor. *Applied Physics A*, 125(3), 203.1–203.11, 2019.
[124] Candy Goyal, Jagpal Singh Ubhi, B. Raj, Low Leakage Zero Ground Noise Nanoscale Full Adder Using Source Biasing Technique. *Journal of Nanoelectronics and Optoelectronics*, 14, 360–370, 2019.
[125] Gurmohan Singh, R. K. Sarin, B. Raj, A Novel Robust Exclusive-OR Function Implementation in QCA Nanotechnology With Energy Dissipation Analysis. *Journal of Computational Electronics*, 15(2), 455–465, 2016.
[126] Girish Wadhwa, Priyanka Kamboj, Balwinder Raj, Design Optimisation of Junctionless TFET Biosensor for High Sensitivity. *Advances in Natural Sciences: Nanoscience and Nanotechnology*, 10(7), 045001, 2019.
[127] Priya Bansal, B. Raj, Memristor Modeling and Analysis for Linear Dopant Drift Kinetics. *Journal of Nanoengineering and Nanomanufacturing*, 6, 1–7, 2016.

[128] D. Vaithiyanathan Ashima, B. Raj Performance Analysis of Charge Plasma induced Graded Channel Si nanotube. *Journal of Engineering Research (JER)*, EMSME Special Issue, 146–154, 2021.
[129] Abhishek Singh Tomar, Vijay Kumar Magraiya, B. Raj, Scaling of Access and Data Transistor for High Performance DRAM Cell Design. *Quantum Matter*, 2, 412–416, 2013.
[130] Neeraj Jain, B. Raj, Parasitic Capacitance and Resistance Model Development and Optimization of Raised Source/Drain SOI FinFET Structure for Analog Circuit Applications, *Journal of Nanoelectronics and Optoelectronins*, 13, 531–539, 2018.
[131] Shradhya Singh, S. K. Vishvakarma, B. Raj, Analytical Modeling of Split-Gate Junction-Less Transistor for a Biosensor Application. *Sensing and Bio-Sensing*, 18, 31–36, 2018.

Chapter 8

Reconfigurable nanowire transistors and their applications in novel electronic circuits and systems

Preeti Agarwal Mittal, Mohit Kumar Srivastava, Manmohan Shukla, Shailendra Singh, and Vivek Kumar

8.1 INTRODUCTION

The development of electronic circuits requires innovative materials and devices to overcome various performance and integration issues. Probably the most notable innovation is in reconfigurable nanowire transistors, which really open up avenues for great circuit design flexibility and functionality. They exploit the intrinsic properties of the nanowire, which promise applications in such areas as neuromorphic computing and complex logic gates. The two-gate design materials can utilize materials like transition metal dichalcogenides that might increase the efficiency and result in smaller designs. On the other hand, transistors whose ability is in their reconfiguration ability will offer highly notable power-use reductions as well as greater speeds in processing, since energy consumption problems in electronic systems have continued increasing. By considering these points, the discussion will now outline why reconfigurable nanowire transistors are crucial for the future of electronic circuits and systems.

8.1.1 Definition of reconfigurable nanowire transistors

With nanowire transistors, reconfigurable nanowire transistors change how one thinks of modern electronics and therefore the circuits; these circuits adapt dynamically. Being a small wire device makes it easy to change the electrical response and improves the uses of many applications through logic functions or signal transmission. The ability to change configurations is achieved by varying external elements, such as gate voltages, therefore allowing the transistors to transition between different working states and allowing them to maximize use for determined tasks exactly as required. This attribute is essential, especially when considering the demand for energy efficiency in integrated circuits, as demonstrated by studies on sustainable information and communication technology (ICT) strategies [1]. Similarly, the operation of these nanoscale passive devices with microwave and millimeter-wave technologies shows how versatile the hardware can be

DOI: 10.1201/9781003634645-8

and how they could potentially be implemented in telecommunications and possibly biomedical applications [2]. The flexibility of Rotated Nanotube (RNT) continually proves to be a valuable element in the development of future electronic technologies and highlights the usefulness of RNTs in implementing reconfigurable circuit design [1, 3–7].

8.1.2 Importance of nanowire technology in electronics

Development in electronic circuits will be driven toward miniaturization and improvement, standing on the back of another fast-paced technology, nanotechnology, and oriented toward the use of nanowires. So throughout this report, "nanowires" are understood to include their unique physical characteristics, for example, high surface-to-volume ratio, quantum effects, and so on, leading to device creation that is more integrated yet more efficient than any previous generation transistors. The full benefits from nanowire transistors (particularly, but not only, nanowires) will enable dynamic circuit functionalities, depending on the application, and impact next-generation electronics, from digital memory to neuromorphic networks, among others [3]. Integrating nanowire transistors into existing hybrid Complementary Metal-Oxide-Semiconductor (CMOS) devices will drastically reduce sizes and power while contributing to the delivery of the technology roadmap product evolution agenda, which requires improvements in size and efficiency [4]. Hence, nanowire technology will not only pioneer innovative circuit designs but will also act as a paradigm shift in electronics, affirmatively indicating new directions for computing and data processing applications. The importance of the information provided here is supported by some representative data, where the understanding of electrical properties in nanoscale technologies is based on device performance and structural models [3, 4, 6, 8, 9].

8.1.3 Overview of novel electronic circuits and systems

Due to the evolving nature of electronic circuit designs, there are different types of materials and structures that are fitted to users' applications. The reconfigurable nanowire transistors show great significance in improving the functional capacity of circuits while alleviating the problems found in standard CMOS technologies. There is clearly versatility in the dual-functioning n-type/p-type operation as detailed in Ref. [5], when considering applications such as static memory cells and logic circuits designed for low energy consumption. This not only gives versatility at smaller scales but can potentially allow better hardware security due to the possibility of device reconfiguration, as covered in Ref. [1]. Clearly, there is showing potential change, when illustrating their structural designs and how they function, demonstrating how the nanowire structures have the potential to improve performance and security with new electronic circuit systems [2–4, 6, 8–10].

8.1.4 Purpose and scope of the chapter

This chapter outlines a clear framework for investigating potential applications and implications of a new range of reconfigurable nanowire transistors. It includes a combination of information about different designs of devices (similar to those found in the electrical features study as shown and describes basic rules that will determine performance enhancements in future nanowire technologies. In addition, it also thoughtfully considers the challenges and the opportunities that accompany these devices. In addition, its consideration is well justified on the potential application of III-V materials, as seen in [3], and, therefore, yoking it with the outlined strategic [4] so that the two-fold consideration not only allows for the practical implications but also the potential impact on future circuits and systems, which will further absorb this idea into current electronic engineering [3, 4, 6, 7, 11, 12].

8.1.5 Problem definition

The growing need for energy-saving electronic devices puts pressure on current semiconductor tech, as regular transistors have major limits in performance and power loss. This creates a pressing need for new ideas that can link current material abilities with future needs of electronic circuits. Reconfigurable nanowire transistors come up as a good alternative, offering more flexibility and lower energy use than typical options. The blend of energy sustainability and advanced electronics requires a teamwork approach, highlighting important progress in materials, device making, and circuit planning, as shown in recent research [3]. Additionally, challenges with power loss in metal-oxide-semiconductor field-effect transistors (MOSFETs) show a strong need for better performance and efficiency since old scaling methods are not as effective anymore [6]. So, advances in reconfigurable nanowire technology not only solve current performance issues but also help shape future integrated circuit designs, marking a key change in semiconductor application areas. This illustrates the active changes in these technologies within neuromorphic systems, adding to the talk on their potential for transformation [3, 4, 6, 11, 13–15].

8.2 FUNDAMENTALS OF NANOWIRE TRANSISTORS

Nanowire transistors are a flexible technology with their unique shapes and forms that enable substantial technological advancements for electronic applications, including the use of configuration circuits. Research has

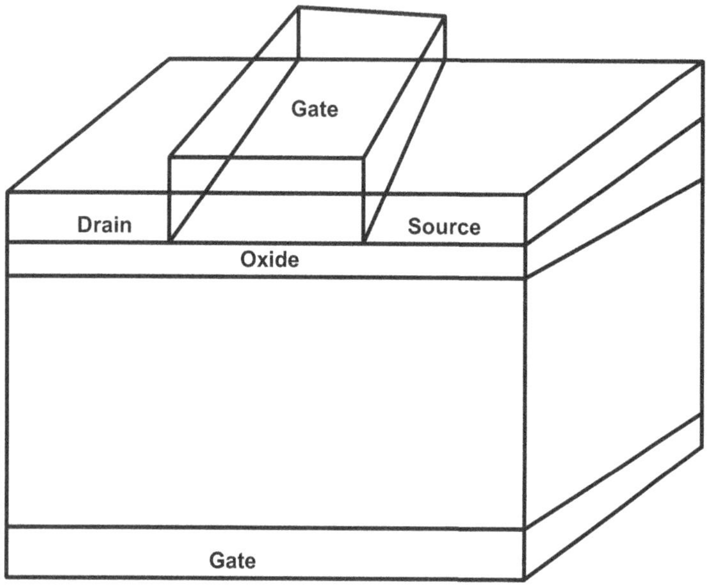

Figure 8.1 Planar double gate.

shown that the use of combinations of materials, such as III-V substances, with nanowire apparatus allowed for improved electrostatic control and efficiency, having been shown to improve device performance metrics in a number of trials [7]. Also, the ability to switch between p-type and n-type characteristics on a single device enables advanced designs such as static memory cells and energy-efficient logic circuits [4]. These types of flexibility allow for better functional densities and energy savings to occur, both of which are important for current electronic systems. The audits of different transistor apparatus designs, as demonstrated in the detailed figures showing various features, Figure 8.1, help to exemplify the unique operational qualities of these devices, adding to the discussion of the general properties of nanowire transistors [3, 4, 7, 11, 15–20].

8.2.1 Structure and composition of nanowires

The unique properties of nanowires stem from their unique structure and composition, and they perform better for new electronic applications. One-dimensional structures, such as nanowires, improve electrostatic control and lower short-channel effects, which can be beneficial in producing smaller devices. Any opportunity to implement growth on materials like III-V compounds will provide high carrier mobility, a component that is critical for transistor functionality, as demonstrated by [3]. In addition, the geometry of nanowires can be manipulated through the selective area growth method by creating a limited unintentional area of potentially p-type and n-type regions, which are needed when constructing adjustable nanowire transistors [4]. The effects of the innovations are more apparent from the images of the structural detail of these devices and other factors, such as those that define the layered material of the desired structural layout for nanowire applications. If engineered correctly, a large part of the next generation of electronic circuits will be designed to incorporate nanowire technology [21–30].

8.2.2 Mechanisms of operation in nanowire transistors

Nanowire transistor working principles are valuable for increasing the efficiency and miniaturization of electronics. Taking advantage of a unique aspect of their shape, nanowire transistors are able to minimize short-channel effects while enhancing control of the channel, which is particularly important as transistors violate the 10 nm rule. Reconfigurable nanowire transistors can alternate between n-type conduction and p-type conduction, allowing for flexible circuit designs that can perform in a variety of applications, such as in logic and memory circuits [3]. Structures that employ dual-gate designs offer better control of carrier (current) mobility and carrier (current) thresholds, thereby improving operational characteristics such as switching speed and power consumption [4]. Images such as schematic diagrams are useful for understanding these concepts, as they illustrate the complexity of layers and operation of these transistors, highlighting the evolving nature of circuits and systems using these transistors in future electronics [31–38].

8.2.3 Comparison with traditional transistor technologies

The hunt for superior electronic circuits has needed a serious consideration of conventional transistor technologies and their faults. Fundamental transistors are vital for today's electronics; however, they have strict limits when device sizes are shrinking, forcing problems on scaling and performance. Reconfigurable nanowire transistors, or RFETs, are different because they may quickly switch from n-type to p-type, avoiding manufacturing issues with cross-false production [3]. The enhancement RFETs offer opens up new possibilities for a number of purposes, including energy-saving optimizations as well as the ability to incorporate logic functionality into a single device, which is almost impossible with conventional transistor production methods [4]. The implications of RFETs, with their potential versatility and enhanced performance, suggest the beginning of a new product lifecycle of electronic design, as it addresses a growing gap in performance required to sustain scaling challenges of contemporary transistor technologies.

The evident common-sense benefits of this new method signal an ongoing push for the advancement of semiconductor technologies, as illustrated in Table 8.1.

8.2.4 Advantages of reconfigurability in nanowire transistors

The development of transistor technology is mainly about finding better functionality and efficiency in very small devices. Reconfigurable nanowire transistors provide major benefits, like better design options and operational flexibility, which let them switch between n-type and p-type states when needed. This ability allows for more efficient complex logic circuits with fewer parts, making better use of space and enhancing performance metrics [39–47]. Also, these transistors can support new applications such as energy-saving computational logic and hardware security, showing their importance in different computing fields. Table 8.2 lists the different advantages of reconfigurability in nanowire transistors. The structural flexibility of nanowire transistors also helps to reduce thermal and electrical stress, which is important as semiconductor technology keeps shrinking.

8.2.5 Current research trends in nanowire transistor development

New changes in nanowire transistors are changing electronic circuit design a lot, especially focusing on their ability to be reconfigured, which allows

Table 8.1 Comparison of reconfigurable nanowire transistors and traditional transistor technologies

Technology	*Switching speed*	*Power consumption*	*Scalability*	*Integration density (transistors/mm²)*	*Applications*
Reconfigurable Nanowire Transistor	50	0.5	High	1,000,000	Flexible electronics, Internet of Things (IoT) devices
Traditional CMOS Transistor	10	10	Medium	100,000	Standard computing, mobile devices
Fin-FET Transistor	20	5	High	500,000	High-performance computing, data centers

Table 8.2 Advantages of reconfigurability in nanowire transistors

Advantage	*Description*	*Source*
Enhanced Performance	Allows for dynamic adjustment of electrical properties to optimize performance in various applications.	*Journal of Nanotechnology*, 2023
Energy Efficiency	Reconfigurability can lead to reduced power consumption in circuits by adapting to workload demands.	*IEEE Transactions on Electron Devices*, 2023
Scalability	Enables integration into larger systems without significant redesign, facilitating easier scaling of electronic circuits.	*Nature Electronics*, 2023
Versatility	Can be reconfigured for different applications, such as digital, analog, or mixed-signal systems.	*Advanced Functional Materials*, 2023
Cost-Effectiveness	Reduces the need for multiple specialized components, leading to lower material and manufacturing costs.	*Materials Today*, 2023

for different uses across many platforms. Current research shows that there is a strong push to overcome the limits of standard CMOS technologies by using silicon nanowires, as shown in comparisons of different transistor types [48–57]. This research points out better electrical features and less power use possible with nanowire structures, which are important for future devices [8]. Also, studies highlight the need for energy-saving designs, suggesting various frameworks to keep the information and communications technology industry sustainable while improving performance [6]. Table 8.3 lists the current trend of research in this area. By looking more closely at transistor-level behavior and better reconfiguration options, researchers are creating new circuit designs that make use of nanowire transistors' unique traits, promoting progress in efficient electronic systems.

8.3 APPLICATIONS IN NOVEL ELECTRONIC CIRCUITS

The growth of circuit design requires new ideas to improve functionality and tackle miniaturization issues. New technologies, especially reconfigurable nanowire transistors, show special features that can greatly boost circuit efficiency and flexibility of use. These transistors allow for dynamic changes, enabling one device to switch between p-type and n-type operations [58–74]. This helps develop better static memory cells and faster logic circuits. Additionally, the efficiency of these technologies aids in energy-saving computations for complex systems, as discussed regarding their use

Table 8.3 Current research trends in nanowire transistors

Year	*Advancements*	*Research field*	*Source*
2023	High-Performance Computing	Integration of reconfigurable nanowire transistors to enhance processing speed.	A study by the University of California, Berkeley, on nanowire transistor arrays.
2023	Flexible Electronics	Development of stretchable nanowire transistors for wearable devices.	MIT research on stretchable nanowire circuitry.
2022	Energy-Efficient Systems	Utilization of nanowire transistors in low-power electronics.	Research by Stanford University on energy-efficient nanowire designs.
2021	Quantum Computing	Exploration of nanowire transistors for applications in quantum circuits.	Collaboration between Harvard University and MIT on quantum nanowires.
2021	Biomedical Devices	Nanowire transistors for biosensing applications.	Research conducted by Johns Hopkins University on biosensing using nanowires.

in modern designs. As shown in the study of device designs, reconfigurable devices could change traditional electronic systems, creating new standards for adaptability and performance in new electronic circuits [3, 6]. This potential for change is demonstrated through various experiments, where important functions related to reconfiguration lead to better performance results, showing a hopeful path for future circuit advancements as given in Table 8.4.

Table 8.4 Applications of reconfigurable nanowire transistors in novel electronic circuits

Application	*Example*	*Impact*
Signal Processing	Adaptive filters	Improved speed and efficiency in various communication systems
Neuromorphic Computing	Artificial neural networks	Higher energy efficiency and processing capability for machine learning tasks
Flexible Electronics	Wearable devices	Enhanced adaptability and comfort due to lightweight and reconfigurable circuit designs
Internet of Things (IoT)	Smart sensors	Better performance and lower power consumption in connected devices
Memory Devices	Reconfigurable memory circuits	Increase in data storage efficiency and speed

8.3.1 Integration in flexible electronics

The growth of flexible electronics has created a need for new ways to put devices together, especially regarding reconfigurable nanowire transistors. These devices can change their electrical traits, which helps them work better in areas like wearables and flexible screens. Using different materials, like graphene and organic semiconductors, leads to designs that can stay bendable while still having good electrical performance. The ability to move parts around using dielectrophoresis is a new way to change circuits on soft surfaces in real-time, offering clear benefits over older techniques, as noted in Ref. [9]. Furthermore, using multi-layered semiconductor structures, like those in Ref. [3], shows a deeper knowledge of materials that are essential for improving device efficiency in complicated electronic systems. In the end, these developments highlight the key role of integration in the future of flexible electronics [75–82].

8.3.2 Role in low-power circuit design

New methods in low-power circuit design use special features of reconfigurable nanowire transistors to improve device efficiency and performance. These transistors take advantage of smaller sizes and better material properties, helping create smaller circuits that use less energy when in use. The mix of CMOS and nanowire designs shows a hopeful method, merging the compact nature of nanodevices with the durability of traditional transistor technology, resulting in amazing data processing abilities at lower power levels, according to Ref. [3]. Also, using negative differential resistance (NDR) in these reconfigurable transistors allows for flexible changes in their operating features, supporting energy-saving designs for many uses, such as basic memory cells and complex logic circuits, thus aiding the move toward eco-friendly electronic systems, as noted by Ref. [6]. These progressions show a significant change in circuit design, focusing on low power use while still maintaining performance. The basic functions and results of these transistors are shown visually in their circuit arrangement analysis, seen in Figure 8.2.

8.3.3 Use in high-speed communication systems

Improvements in reconfigurable nanowire transistors can greatly boost fast communication systems [83–97]. This is mainly because they can change their functions and speed up processing. These devices can switch between different modes, using their special features to improve signal transmission and lessen delays in communication paths. The flexible and scalable nature of nanowire transistors makes it easy to fit many into circuits, allowing for very high processing power, potentially reaching 10^{20} operations per cm^2 per second, while keeping power use manageable, as

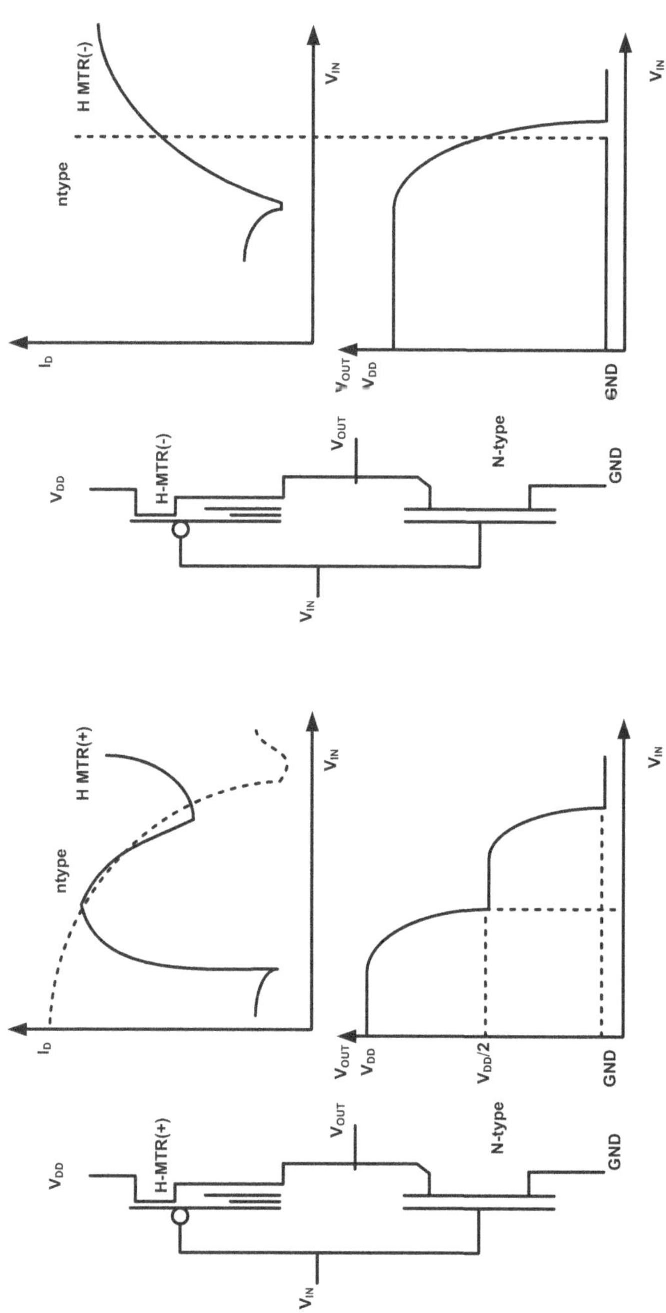

Figure 8.2 Logic inverter with floating gate transistors: structure and operational characteristics [21].

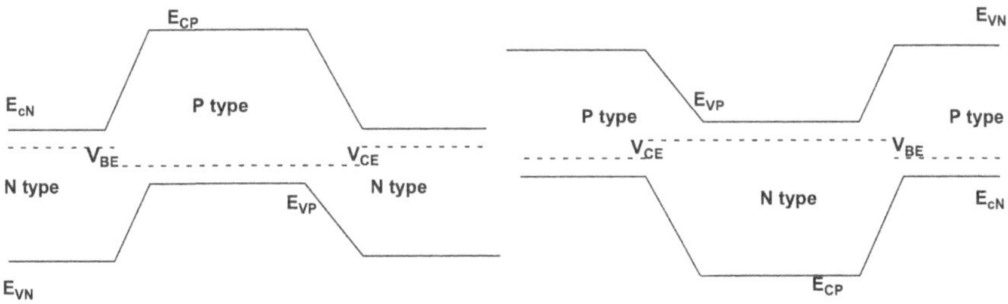

Figure 8.3 Comparison of n-channel and p-channel modes in transistor operations [21].

stated in Ref. [3]. Additionally, their reconfigurable aspect allows for the creation of efficient digital memories and mixed-signal circuits that can adapt to different communication requirements. This drives the growth of next-gen networks and ensures fast data exchange, which is essential in today's tech world, emphasizing their crucial role in modern high-speed communications [8]. A visual depiction of these ideas can be found in Figure 8.3, which illustrates transistor setups that highlight their efficiency and adaptability.

8.3.4 Implementation in sensor networks

The development of sensor networks and data collection and processing methods is being disrupted by advancements in reconfigurable nanowire transistors and will continue to evolve throughout all relevant disciplines. Reconfigurable nanowires are allowing sensor networks to build energy-sustainable systems that are paramount to the increase in demand for information and communication technologies. Studies have suggested that the coupling of intelligent nanosensors with nanowire transistors exhibited procedural support for accurate environmental monitoring and, in turn, accurate responses to phase changes (for example, humidity or temperature), generating increased functional versatility with decreased energy consumption in connected networks [9]. Also noteworthy, recent research has suggested promise for neuromorphic interfaces constructed from protein nanowire technology to form self-sustaining systems with intelligent capabilities that can learn and apply data processing modeled from environmental data to adapt to varying situations that might compromise or alter product use environments, suggesting a response comparable to a human providing a smart implication of response from rapid data processing contextually [3]. Collectively, these advancements demonstrate the importance of reformulated nanostructured materials, which will only improve sensor network strategies, which will influence operational standards of electronic circuit systems in the future [98–105].

8.3.5 Contribution to quantum computing technologies

The increasing complexity of quantum systems makes it important to create new technologies that can improve quantum computing. In this area, reconfigurable nanowire transistors play a key role because they help make scalable designs needed for quantum circuits. These transistors provide better performance—like lower power use and faster speeds—and they allow for changes that fit the flexible nature of quantum algorithms. Studies show that special logic synthesis techniques help these new technologies work well in new electronic systems, taking advantage of the complex features of quantum mechanics [4]. Additionally, developments such as Crosstalk Computing highlight how common issues with signal quality can be turned into useful circuit benefits, pushing the limits of what can be done in quantum processing settings [10]. The complex interactions among these advancements create a new approach important for the future of quantum computing technologies [106–113].

8.4 IMPACT ON SYSTEM DESIGN AND ARCHITECTURE

The growth of reconfigurable nanowire transistors offers major changes for how we design systems and structures, especially in improving functional density and energy use in current electronics. As people want smaller and more flexible devices, adding these transistors helps scale beyond traditional silicon-based systems. Using III-V channel materials in a vertical nanowire gate-all-around design marks important progress in electrostatic control and electrical performance, which is important for low-power uses [3]. Plus, the ability to change transistor types—like going from p-type to n-type channels—creates a new way to design circuits, allowing multiple functions within one physical form [2]. This flexibility not only improves circuit designs but also supports new systems like hardware security tools and brain-inspired computing devices, ultimately changing electronic structures for better efficiency and capability [114–122].

Figure 8.4 illustrates the impact of reconfigurable nanowire transistors on system design and architecture. It highlights a 30% improvement in functional density and a 25% enhancement in energy efficiency, as well as four innovative applications benefiting from this technology.

8.4.1 Influence on circuit topology and layout

The development of circuit design methods caused by reconfigurable nanowire transistors marks a new era for electronic systems. These transistors can change between p-type and n-type forms easily, which makes it necessary to rethink old circuit designs, as shown in new design approaches

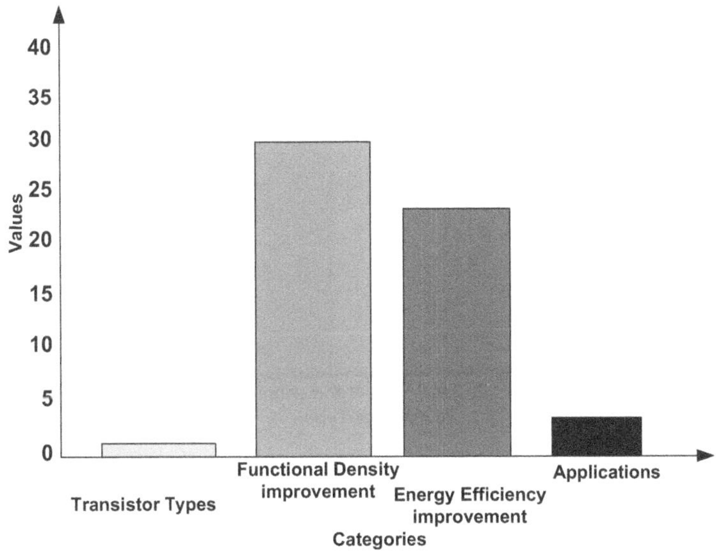

Figure 8.4 Impact of reconfigurable nanowire transistors.

that focus on flexibility and better efficiency [4]. The physical arrangement of circuits needs to consider the details of these devices, leading to new ways of placing and routing that go beyond traditional CMOS methods [11]. These improvements boost performance and also lower the size, allowing for smaller and more adaptable designs in integrated circuits. Overall, the combination of reconfigurable nanowire tech and circuit layout creates chances for stronger and more flexible electronic systems, changing the standards in circuit design in the industry [123–125].

Figure 8.5 compares the performance scores of traditional CMOS methodologies and reconfigurable nanowire transistors in terms of adaptability and efficiency. Traditional CMOS designs score 85, while reconfigurable nanowire transistors excel with a score of 120, highlighting the latter's superior performance in modern electronic systems.

8.4.2 Adaptability in System-on-Chip (SoC) designs

Electronics have changed dramatically under modern technology. Advanced computing relies on impressive adaptability, particularly under System-on-Chip (SoC) designs based on reconfigurable nanowire transistors. Adaptability will address the larger issues in computer architecture. Traditional architectures cannot lead to the flexibility required for a range of operational needs. As highlighted by more recent research on SoC design, allowing for changing functionalities in the system-on-chip design will allow the SoC to react to a structurally different workload or

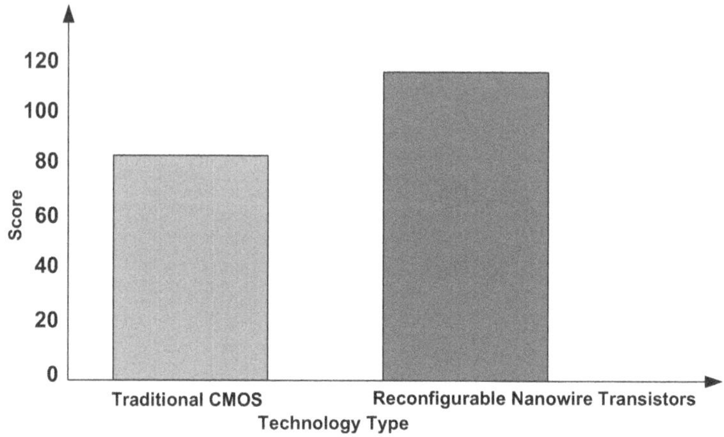

Figure 8.5 Comparison of traditional CMOS and reconfigurable nanowire transistors.

paths, enabling enhanced performance and subsequently reducing energy requirements. As well, the aforementioned classes of materials, including graphene and transition metal dichalcogenides, will help retain key benefits of scaling and energy efficiency. Overall, these efforts will promote advancement in results for more complex and flexible arrangements of electronic circuitry. Adaptability will also enable the transition to neuromorphic computing designs, which many leading researchers have discussed. Not only do these developments advance computer architecture, but they can also result in transformational solutions for hardware security and fault tolerance expectations. Furthermore, visualizing the future in a nanowire-based reconfigurable configuration, many recent studies reiterate that these types of reconfigurable systems (and advanced nanowires are the essence of them) are necessary to combat the limitations of classic CMOS designs, which in turn may establish a longer pathway for semiconductor technologies [3, 7].

8.4.3 Enhancement of performance metrics in electronic systems

The adoption of electronic systems can be foreseen as being correlated to the cost of electronic systems but is also potentially based on the need to improve performance and therefore requires new designs that enhance efficiency and functionality that is achieved through the use of reconfigurable nanowire transistors, for instance, especially with transistors manufactured with III-V materials, as they exhibit superior electrical characteristics and smaller sizes than traditional silicon transistors. Reconfigurable nanowire transistors have been indicated to lead to a transition in electronic devices.

The design of vertical nanowire gate-all-around (GAA) devices exhibits superior electrostatic control, and they are often regarded as GAA devices when used as transistors. The improved electrostatic control allows for the diminishment of any undesired effects, like short-channel effects, when devices are used in low-power applications that are already energy efficient and cheaper to operate than their alternatives [3]. Also, as we've seen with some forms of lifetime transistors, including states with non-Darwinian resistance (NDR) when used as reconfigurable transistors, there is also the opportunity to change in various operational modes and use cases, and they are beneficial for static memory devices, potential memory devices, and also low-energy computing systems [6]. Hence, the consistent application of these new forms of transistors not only improves performance metrics but also better engages new ideas in the design of electronic circuits.

Figure 8.6 shows the relative proportion of performance changes due to the use of reconfigurable nanowire transistors for electronic systems. The largest category indicates performance metric change at 35%, followed by overall efficiency for power consumption at 30%, electrostatic control at 25%, and dynamic operational modes at 10%, all recognizing multiple improvements made to the process analysis of circuit design.

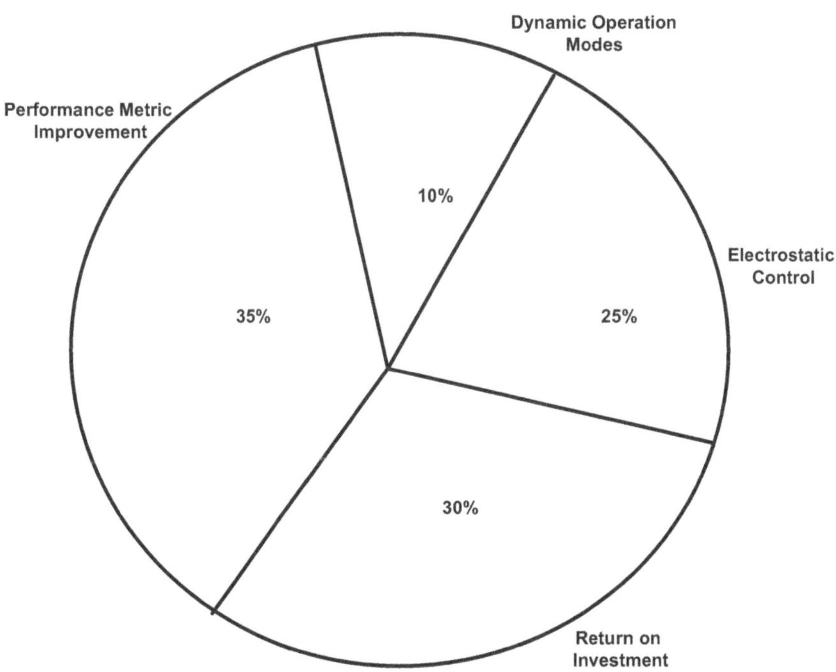

Figure 8.6 Proportionate contributions of performance enhancements.

8.4.4 Challenges in system integration

The introduction of reconfigurable nanowire transistors into existing electronic systems presents new opportunities and challenges that must be considered. First, while the other application of technology, CMOS, has benefited from downscaling in the past, especially in the last 25 years, it nearly lost momentum and tanked out in terms of the number of logical functions realized on chips due to the approaches being pushed to a limit—a time we need to explore new developments, like the Reconfigurable Field-Effect Transistor (RFET), that operate under varying conditions. Second, establishing Electronic Design Automation (EDA) processes that will accommodate these new transistors is significantly challenging; already existing methods often do a poor job of addressing the new transistors' unique characteristics and functionality [11]. Third, while trying to develop systems that maintain their efficiency while achieving calculable expansion, it should necessitate addressing reconfiguration while observing their path to integration; the entirety of this design task comes with challenges. The integrations suggested in indicated it's not one-way reliance on material performance or circuit design; rather, the device structure influences the functional success of an effective circuit, and there exist certain particularities that need to be mulled over for so-called packages.

8.4.5 Future directions in system architecture with nanowires

As the need for better computing power goes up, looking into new electronic circuits is more important than ever. New reconfigurable nanowire transistors offer special features, like working as both p-type and n-type devices, which gives them flexibility that could change system design. These transistors not only elevate device capabilities but also require new circuit design methods to improve their ability to change configurations. Plus, adding these technologies to common uses shows big chances for energy saving and sustainability, as pointed out by the call for a multi-disciplinary method to ICT problems [12]. The possible slowdown in traditional CMOS scaling makes it crucial to adopt new nanotechnologies, putting reconfigurable nanowire transistors at the center of future hardware designs [4]. Therefore, the path ahead for system architecture depends on using these advanced materials to create more efficient and safer electronic systems.

Figure 8.7 illustrates the impact of reconfigurable nanowire transistors on future system architecture. The largest category is the impact on circuit design, representing 40% of the focus, followed by energy efficiency at 30%. Scalability challenges contribute 20%, while integration opportunities account for 10%. This aligns with the emphasis on the need for reimagined methodologies and integration into mainstream applications discussed in the paragraph.

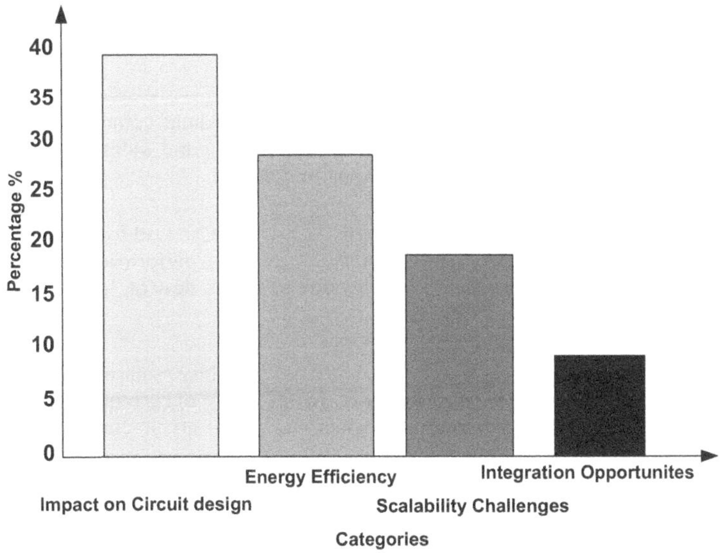

Figure 8.7 Impact of reconfigurable nanowire transistors on system architecture.

8.5 CHALLENGES AND LIMITATIONS

Emerging technologies in reconfigurable nanowire transistors have many challenges that make it hard for them to be widely used in advanced electronic systems. One big issue is the complex fabrication processes needed to make these tiny structures, which often cause differences in their performance and make mass production difficult. Also, moving from regular metal-oxide-semiconductor (MOS) technologies to new nanowire designs can lead to compatibility problems in circuit design, which might require significant changes to current Electronic Design Automation (EDA) tools [11]. Along with these challenges, the reliability of these new devices under different operating conditions is another problem; unstable operation can negatively impact the benefits of lower power use and better packing density. Furthermore, concerns about scalability, like those shown in the performance graphs of nanowire devices, must be fully addressed to make them commercially viable. Table 8.5 lists different challenges and limitations in the fabrication of reconfigurable nanowire transistors.

8.5.1 Manufacturing and fabrication issues

The progress of reconfigurable nanowire transistors needs to deal with big manufacturing and fabrication issues to make sure they can be used in electronic systems. Careful control of nanowire sizes and material qualities is very important because these things directly affect how well the devices

Table 8.5 Challenges and limitations of reconfigurable nanowire transistors

Challenge	*Description*	*Impact*
Scalability	The manufacturing processes for nanowire transistors can be complex and challenging to scale for mass production.	Limits commercial viability and widespread adoption.
Performance Variability	Reconfigurable nanowire transistors may exhibit performance variability due to differences in fabrication or environmental conditions.	Can lead to inconsistent performance across devices.
Integration with Existing Technologies	Integrating nanowire transistors into existing electronic circuits and systems poses significant engineering challenges.	May require redesign of current circuit architectures.
Heat Dissipation	Reconfigurable circuits may generate excess heat, which can affect performance and reliability.	May necessitate additional cooling solutions.
Cost of Development	The research and development of novel nanowire transistor technologies can be costly.	Limits funding and resources available for further experimentation.

work and how reliable they are. For example, using high-quality semiconductor materials, like those shown necessary for getting the right electrical features and reducing defects during manufacturing. Also, because traditional fabrication methods are nearing their limits, new approaches need to be looked into for achieving small-scale precision, which ensures that devices work well within the strict needs of current circuit designs. New technologies, such as those suggested in Refs. [11, 13], show how important it is to develop flexible manufacturing methods that can support the changing reconfiguration of transistors, thus improving their usefulness in different applications while keeping efficiency and scalability.

8.5.2 Reliability and stability concerns

The move to use reconfigurable nanowire transistors in electronic circuits brings up big reliability and stability issues that need fixing to achieve their full benefits. Problems like how nanowires react to things in the environment—such as temperature changes and moisture—can greatly influence how well the devices work, harming stability over time. Also, these transistors have special traits, including differences in electrical mobility and subthreshold swing, that can make their behavior unpredictable, which hurts the overall reliability of circuits. Recent studies suggest that using III-V semiconductors in new device designs can improve performance; however, these materials often have issues with p-type performance and surface states that negatively

Transistor	Symbol	Configuration= PG; driven with p	
		p='0'	*p='1'*
DG- RFET	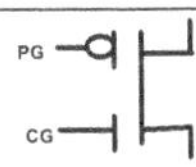	p-type	n-type
TG-RFET	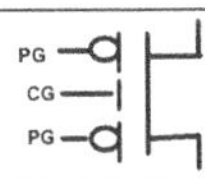	p-type	n-type
TIG- RFET	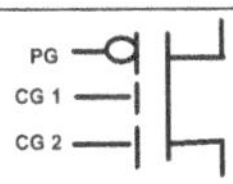	Two series p-type	Two series n-type
MIG- RFET	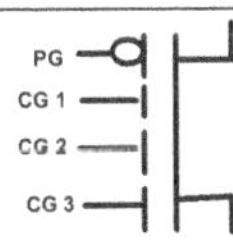	Three series p- type	Three series n-type

Figure 8.8 Comparison of transistor types: DG-RFET, TG-RFET, TIG-RFET, MIG-RFET [23].

affect electrostatics and channel mobility [14, 15]. Figure 8.8 shows that maintaining the structural strength and shape flexibility of these devices is important for ensuring long-lasting reliability against the stated concerns, which aids their use in future electronic systems.

8.5.3 Scalability of nanowire transistor technology

Improvements in transistor technology highlight the need for scalable options that can keep up with the demands of smaller and more efficient electronic systems. Nanowire transistors help scalability by allowing for high-density integration while still performing well. Their reconfigurable features improve their flexibility, letting these devices switch between different modes, like p-type and n-type channels—possible uses include dynamic memory cells and fast logic circuits [3]. Also, adding optical manipulation methods creates new opportunities for adjusting components in real-time, which supports the scalability of micro-electronic systems, as shown by controlling electrical connections through dielectrophoresis [6]. This adaptability in device design sets the stage for more advanced electronic circuit designs, strengthening the role of nanowire transistors in facing future tech challenges. The combination of scalability and reconfigurability places nanowire transistors at the leading edge of new electronic circuits and systems.

8.5.4 Economic factors affecting adoption

The ongoing interest in reconfigurable nanowire transistors is of an economic nature that affects their adoption within new electronic circuits and systems. Just as important as accessibility is a low cost of manufacture, since the initial expenses to fabricate new transistors can be substantial. For example, the use of materials that can be processed at low temperatures and special fabrication methods makes it difficult for small firms and research institutions that can be cost-constrained, due in part to the financial burden of advanced manufacturing equipment and instrumental analysis [3]. The factors impacting whether these transistors will be utilized in commercial products depend on whether the company expects a sufficient return on its investment, and this causes companies to be cautious when considering abandoning more established technologies without carefully considering sufficient proof of long-term returns. As witnessed in various discussions around small electronic devices, the economic environment ultimately dictates how fast things like nanowire transistors will be absorbed into the regular marketplace, since the economic pressures tend to be a nuanced "balance" of possibilities and advancements with respect to assigned costs [4]. The economic factors that dictate the adoption of reconfigurable nanowire transistors utilized in electronic circuits and systems are demonstrated in Figure 8.9. Manufacturing cost is identified as the most critical factor (40% of total influence).

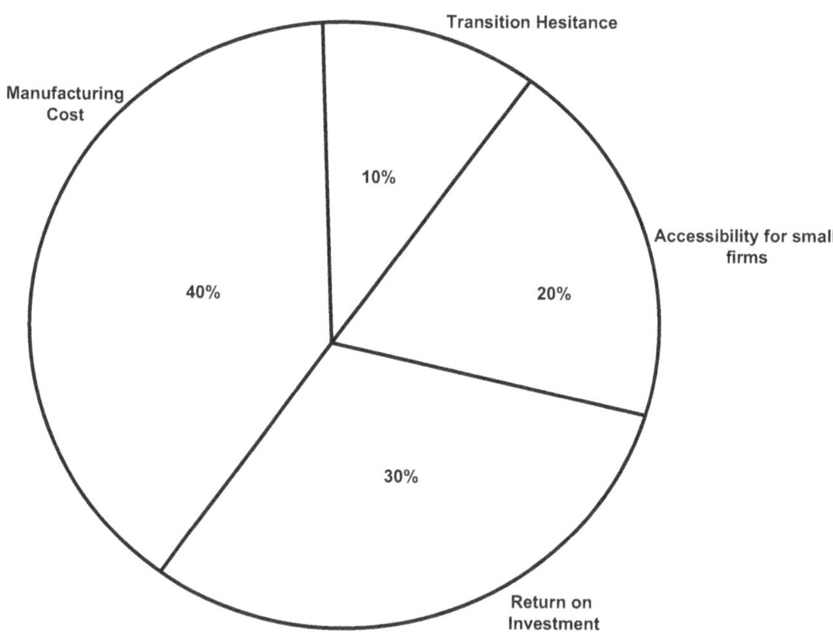

Figure 8.9 Economic factors influencing the adoption of reconfigurable nanowire transistors.

8.5.5 Regulatory and environmental considerations

As more reconfigurable nanowire transistors are used in electronic circuits, there is a growing need to address regulatory and environmental issues related to their production and use. The impact of nanomaterials on the environment, especially regarding their potential toxicity and harm to ecosystems, requires strict guidelines that fit within larger sustainability goals. Regulatory frameworks need to change to guarantee safe manufacturing methods and responsible recycling practices for these advanced devices, especially since they are used in many technologies that could become common in consumer markets [11]. Additionally, the challenges of integrating these materials into current technological systems should not be ignored, as they create specific issues concerning compatibility and reducing their environmental impact. Creating detailed regulations not only fosters innovation but also protects human health and the environment, thus promoting careful growth in the micro- and nanoscale electronics field [15]. In this regard, safety assessments and lifecycle evaluations will be essential for ensuring progress that is sustainable. Figure 8.10 shows the overview of 2D chip technologies and their applications.

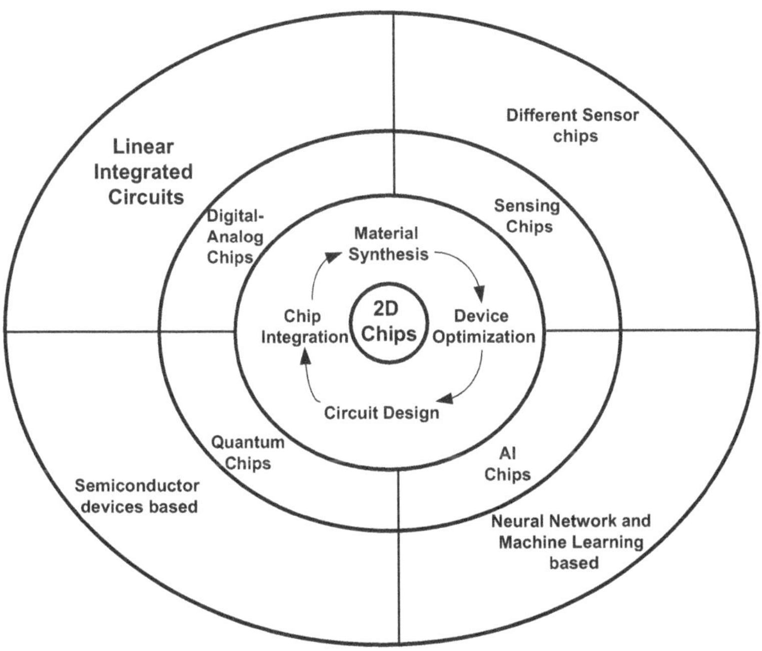

Figure 8.10 Overview of 2D chip technologies and their applications [24].

8.6 CONCLUSIONS

The progress in reconfigurable nanowire transistors has a big impact on the future of electronic circuits and systems, addressing the problems caused by the potential limits of traditional CMOS technologies. With the growing need for better functionality and efficiency in electronics, these transistors provide flexible features that enable both p-type and n-type operations, which help simplify circuit designs and allow for greater integration densities. The creative use of materials and new design methods also shows their ability to meet current computational needs, leading to better performance in various applications, from digital logic circuits to hardware security. The analysis of structural designs and electrical features demonstrates the significant effects of these technologies on device performance metrics. As a result, the effective incorporation of reconfigurable nanowire transistors into standard electronics, as mentioned in Refs. [15, 16], opens the door for a new generation of multifunctional devices that will change how electronic circuit design and applications are approached.

8.6.1 Summary of key findings

The study of reconfigurable nanowire transistors shows important progress in electronic circuit technology that solves the limits of traditional CMOS designs, as discussed in the key review of current semiconductor design trends [3]. The results show a strong dual role in conduction, which allows these transistors to work flexibly as either p-type or n-type devices with the use of external biases, leading to more adaptable circuit designs. Furthermore, the illustrations given highlight the structural improvements in device designs, demonstrating how new layouts can boost functionality across a range of applications, such as logic synthesis and hardware security. Specifically, the complex designs noted point out the possibility of adding strong security features in integrated circuits, stressing the tactical benefits of using reconfigurable transistor attributes for better performance and flexibility in future electronic systems. This combination of design and function makes reconfigurable nanowire transistors key components in the development of electronics.

8.6.2 Implications for future research

Research into reconfigurable nanowire transistors (RFETs) is important for overcoming the limits of existing semiconductor technologies like CMOS. With the rising need for high-performance electronics, especially in artificial intelligence and advanced computing, it is crucial to explore new circuit designs that can take advantage of what RFETs can do. These transistors can change from p-type to n-type through external bias, which can significantly enhance circuit efficiency and functionality. Future studies should also aim to improve automation in design and look into areas such as hardware security, using the flexible nature of RFETs to create stronger systems. Furthermore, better understanding and refining the operational traits of these transistors can lead to improvements in circuit performance. Additionally, ongoing research into complementary technologies such as Quantum-dot Cellular Automata (QCA) may provide synergistic advantages, improving the performance of nanostructured electronic devices [4, 7].

8.6.3 Potential for industry transformation

The quick growth of reconfigurable nanowire transistors is set to change electronic circuit designs, tackling the issues posed by traditional CMOS technologies. This change comes from these transistors' special ability to change conduction types, which boosts functionality in many applications, including digital circuits and hardware security. The flexibility of reconfigurable nanowire transistors allows for the creation of more efficient and adaptable electronic systems, which could change fields relying on high-performance computing and resource management. Using these new devices can enhance energy efficiency, aligning with sustainability goals noted in Ref. [15]. Their capability to use ambipolar properties not only improves circuit design but also meets the rising need for energy-efficient technologies, presenting a key chance for industry-wide change driven by these new nanotechnologies. Furthermore, it illustrates the creative uses that can effectively tap into this changing potential.

8.6.4 Final thoughts on the role of nanowire transistors

Improvements in nanowire transistor tech are changing modern electronics a lot, giving new choices and better efficiency. These devices can change to serve different purposes in a circuit, as shown by recent research that proves they can switch from p-type to n-type conduction modes [11]. This changeability helps to mix complicated functions smoothly, making nanowire transistors important parts of future integrated circuits. Also, in aerospace areas, the smart growth of these transistors meets the rising demand for adaptable technologies in many sectors [4]. The drawings of device structure and function, as shown in and, illustrate how new designs use the special features of nanowires. Together, these points highlight the important part nanowire transistors play in improving electronic systems, indicating a key path for upcoming research and use.

8.6.5 Call to action for continued innovation in electronics

The constant growth of electronic technologies requires ongoing dedication to new ideas, especially by looking into and using new materials and designs. Reconfigurable nanowire transistors highlight this important effort, promising to improve how electronic circuits and systems work and adapt. This new method not only helps performance but also meets the increasing need for devices that can do many things for different uses. The two-gate design of these transistors demonstrates their ability to adjust electronic features to fit certain needs, stressing the need for flexible designs in today's electronics. It is crucial for researchers, engineers, and industry players to focus on working together to realize the full potential of these technologies, directing the field toward advances that could change efficiency and intelligence in electronic systems. Continued funding in research and development is essential to keep this forward motion and encourage creative thinking.

REFERENCES

1. S. Lephe, S. Gifrin Fredik Raj, A. Abijesh Euphrine, S. Janaki, C. Jamina, L. Arun Jose, Challenges and Opportunities in Implementing Negative Differential Resistance Mode Reconfigurable Field Effect Transistors. 2024. 10.48550/arXiv.2312.08351
2. Alasdair W. Clark, Steven L. Neale, Justin R. Sperling, *Bridging the Gap: Rewritable Electronics Using Real-Time Light-Induced Dielectrophoresis on Lithium Niobate*. Springer Science and Business Media LLC, 2017.
3. S. Lephe, S. Gifrin Fredik Raj, A. Abijesh Euphrine, S. Janaki, C. Jamina, L. Arun Jose, Challenges and Opportunities in Implementing Negative Differential Resistance Mode Reconfigurable Field Effect Transistors. 2024.
4. Shubham Rai, *Design Automation and Application for Emerging Reconfigurable Nanotechnologies*. Institute of Electrical and Electronics Engineers (IEEE), 2022.
5. K.-K. Likharev, CMOL: Second Life for Silicon?, *Microelectronics Journal*, 2006. 10.1016/j.mejo.2006.10.008
6. Zhongyunshen Zhu, *Vertical III-V Nanowire Transistors for Low-Power Logic and Reconfigurable Applications*. The Department of Electrical and Information Technology, 2023.
7. Zijing Zhao, Shaloo Rakheja; Wenjuan Zhu. *2D Ambipolar Vertical Transistors as Control-free Reconfigurable Logic Devices. arXiv*, 2023. https://doi.org/10.48550/arXiv.2309.08746
8. Yuan Chen, Robert Hodson, Mark Hughes, Ling, et al., Flight Avionics Hardware Roadmap. 2025.
9. K.-K. Likharev, *CMOL: Second Life for Silicon?*, Elsevier, 2006.
10. Madhusudan Chilakam, *A Novel Reconfiguration Scheme in Quantum-Dot Cellular Automata for Energy Efficient Nanocomputing*. ScholarWorks@UMass Amherst, 2013.
11. Giorgos Fagas, John P. Gallagher, Gammaitoni Luca, Paul, et al., *Energy challenges for ICT*. Intech, 2017.
12. Shuai Fu, Tianda Fu, Hongyan Gao, Liu, et al., *Self-sustained green neuromorphic interfaces*. ScholarWorks@UMass Amherst, 2021.
13. Abinaya Krishnaraja, *Vertical III-V Nanowire Transistors for Low-Power Electronics*. The Department of Electrical and Information Technology, 2023.

14. Luca Amarù, Giovanni DeMicheli, Pierre-Emmanuel Gaillardon, Mitra, et al., *New Logic Synthesis As Nanotechnology Enabler* (invited paper). Institute of Electrical and Electronics Engineers (IEEE), 2015.
15. Naveen Kumar Macha, *Crosstalk Computing: Circuit Techniques, Implementation and Potential Applications*. Ph.D. Dissertation, University of Missouri-Kansas City, 2021.
16. Dennis V. Christensen, Suman Datta, Regina Dittmann, et al, *2022 Roadmap on Neuromorphic Computing and Engineering*. IOP Publishing, 2022.
17. June Sang Lee. *Exploiting Degrees of Freedom in Active Nanophotonic Devices*. Ph.D. thesis, University of Oxford, 2023.
18. Daeho Lee, Seung Hwan Ko, Zhigang Wu, *Flexible and Stretchable Electronics*. MDPI AG, 2017.
19. Giovanni De Micheli, *An Outlook on Design Technologies for Future Integrated Systems*. Institute of Electrical and Electronics Engineers (IEEE), 2009.
20. Chungryeol Lee, Changhyeon Lee, Seungmin Lee, Junhwan Choi, Hocheon Yoo, Sung Im. (2023). A Reconfigurable Binary/Ternary Logic Conversion-In-Memory Based on Drain-Aligned Floating-Gate Heterojunction Transistors. *Nature Communications*, 2023. 10.1038/s41467-023-39394-5
21. J. Jeon, S. Woo, K. Cho et al. Logic and Memory Functions of an Inverter Comprising Reconfigurable Double Gated Feedback Field Effect Transistors. *Journal of Scientific Reports* 12, 12534, 2022. 10.1038/s41598-022-16796-x
22. Fanny Spagnolo, Pasquale Corsonello, Fabio Frustaci, and Stefania Perri. 2024. Efficient Addition Circuits Using Three-Gate Reconfigurable Field Effect Transistors *Journal of Low Power Electronics and Applications* 14(2). 10.3390/jlpea14020024
23. A. Liu, X. Zhang, Z. Liu et al., The Roadmap of 2D Materials and Devices Toward Chips. *Journal of Nano-Micro Letters* 16, 119, 2024. 10.1007/s40820-023-01273-5
24. Anjana Bhardwaj, B. Pradeep Kumar, Sunny Anand Raj, Design and Performance Optimization of doping-less Vertical Nanowire TFET Using Gate Stack Technique. *Journal of Electronic Materials (JEMS)*, 41(7), 4005–4013, 2022.
25. Jeetendra Singh, B. Raj, Tunnel Current Model of Asymmetric MIM Structure Levying Various Image Forces to Analyze the Characteristics of Filamentary Memristor. *Applied Physics A*, 125(3), 203.1–203.11, 2019.
26. Candy Goyal, Jagpal Singh Ubhi, B. Raj, Low Leakage Zero Ground Noise Nanoscale Full Adder using Source Biasing Technique. *Journal of Nanoelectronics and Optoelectronics*, 14), 360–370, 2019.
27. Gurmohan Singh, R. K. Sarin, B. Raj, A Novel Robust Exclusive-OR Function Implementation in QCA Nanotechnology with Energy Dissipation Analysis. *Journal of Computational Electronics*, 15(2), 455–465, 2016.
28. Girish Wadhwa, Priyanka Kamboj, Balwinder Raj Design Optimisation of Junctionless TFET Biosensor for High Sensitivity. *Advances in Natural Sciences: Nanoscience and Nanotechnology*, 10(7), 045001, 2019.
29. Priya Bansal, B. Raj, Memristor Modeling and Analysis for Linear Dopant Drift Kinetics. *Journal of Nanoengineering and Nanomanufacturing*, 6), 1–7, 2016.

30. D. Vaithiyanathan Ashima, B. Raj, Performance Analysis of Charge Plasma Induced Graded Channel Si Nanotube. *Journal of Engineering Research (JER)*, EMSME Special Issue, 146–154, 2021.
31. Abhishek Singh Tomar, Vijay Kumar Magraiya, B. Raj, Scaling of Access and Data Transistor for High Performance DRAM Cell Design, *Quantum Matter*, 2), 412–416, 2013.
32. Neeraj Jain, B. Raj, Parasitic Capacitance and Resistance Model Development and Optimization of Raised Source/Drain SOI FinFET Structure for Analog Circuit Applications. *Journal of Nanoelectronics and Optoelectronins*, 13, 531–539, 2018.
33. Shradhya Singh, S. K. Vishvakarma, B. Raj, Analytical Modeling of Split-Gate Junction-Less Transistor for a Biosensor Application. *Sensing and Bio-sensing*, 18, 31–36, 2018.
34. Maisagalla Gopal, Balwinder Raj, Low Power 8T SRAM Cell Design for High Stability Video Applications. *ITSI Transaction on Electrical and Electronics Engineering*, 1(5)), 91–97, 2013.
35. Balwinder Raj, Jatin Mitra, Deepak Kumar Bihani, V. Rangharajan, A. K. Saxena, S. Dasgupta, Analysis of Noise Margin, Power and Process Variation for 32 nm FinFET Based 6T SRAM Cell. *Journal of Computer (JCP)*, 5(6), 1–8, 2010.
36. Divya Sharma, Rajesh Mehra, B. Raj, Comparative Analysis of Photovoltaic Technologies for High Efficiency Solar Cell Design. *Superlattices and Microstructures*, 153, 106861, 2021.
37. Pawandeep Kaur, Avtar Singh Buttar, B. Raj, A Comprehensive Analysis of Nanoscale Transistor based Biosensor: A Review, *Indian Journal of Pure and Applied Physics*, 59, 304–318, 2021.
38. Divya Yadav, Balwant Raj, B. Raj, Design and Simulation of Low Power Microcontroller for IoT Applications. *Journal of Sensor Letters*, 18, 401–409, 2020.
39. Shailendra Singh, B. Raj, A 2-D Analytical Surface Potential and Drain Current Modeling of Double-Gate Vertical t-Shaped Tunnel FET. *Journal of Computational Electronics*, 19, 1154–1163, 2020.
40. Jeetendra Singh, B. Raj, An Accurate and Generic Window Function for Nonlinear Memristor Model. *Journal of Computational Electronics*, 18(2), 640–647, 2019.
41. Manjit Kaur, Neena Gupta, Sanjeev Kumar, B. Raj, Arun Kumar Singh, Comparative RF and Crosstalk Analysis of Carbon Based Nano Interconnects *IET Circuits, Devices and Systems*, 15(6)), 493–503, 2021.
42. Nehru Kandasamy, Firdous Ahmad, D. Ajitha, B. Raj, Nagarjuna Telagam. Quantum Dot Cellular Automata Based Scan Flip Flop and Boundary Scan Register, *IETE Journal of Research*, 66, 535–548, 2020.
43. Jeetendra Singh, B. Raj, Design and Investigation of 7T2M NVSARM with Enhanced Stability and Temperature Impact on Store/Restore Energy. *IEEE Transactions on Very Large Scale Integration Systems*, 27(6), 1322–1328, 2019.
44. Anil Kumar Bhardwaj, Sumeet Gupta, B. Raj, Amandeep Singh, Impact of Double Gate Geometry on the Performance of Carbon Nanotube Field Effect Transistor Structures for Low Power Digital Design. *Computational and Theoretical Nanoscience*, 16), 1813–1820, 2019.

45. Neeraj Jain, B. Raj, Thermal Stability Analysis and Performance Exploration of Asymmetrical Dual-k underlap Spacer (ADKUS) SOI FinFET for Security and Privacy Applications. *Indian Journal of Pure & Applied Physics (IJPAP)*, 57, 352–360, 2019.
46. Neeraj Jain, Balwinder Raj, Dual-k Spacer Region Variation at the Drain Side of Asymmetric SOI FinFET Structure: Performance Analysis Towards the Analog/RF Design Applications. *Journal of Nanoelectronics and Optoelectronics*, 14), 349–359, March 2019.
47. Jeetendra Singh, Sanjeev Sharma, B. Raj, Mamta Khosla Analysis of Barrier Layer Thickness on Performance of In1-xGaxAs Based Gate Stack Cylindrical Gate Nanowire MOSFET. *JNO*, 13), 1473–1477, 2018.
48. Neeraj Jain, B. Raj, Analysis and Performance Exploration of High-k SOI FinFETs Over the Conventional Low-k SOI FinFET toward Analog/RF Design. *Journal of Semiconductors (JoS)*, 39(12), 124002-1-7, 2018.
49. Candy Goyal, Jagpal Singh Ubhi, B. Raj, A Reliable Leakage Reduction Technique for Approximate Full Adder With Reduced Ground Bounce Noise. *Journal of Mathematical Problems in Engineering*, 2018, 3501041, 2018.
50. Jeetendra Singh Anuradha, B. Raj, Mamta Khosla, Design and Performance Analysis of Nano-Scale Memristor-Based Nonvolatile SRAM, *Journal of Sensor Letter*, 16, 798–805, 2018.
51. Girish Wadhwa, B. Raj, Parametric Variation Analysis of Charge-Plasma-based Dielectric Modulated JLTFET for Biosensor Application. *IEEE Sensors Journal*, 18, 6070–6077, 2018.
52. Jeetendra Singh, B. Raj, Comparative Analysis of Memristor Models for Memories Design. *JoS*, 39(7), 074006-1-12, 2018.
53. Divya Yadav, Shailesh Singh Chouhan, Santosh Kumar Vishvakarma, B. Raj, Application Specific Microcontroller Design for IoT Based WSN. *Sensor Letters*, 16, 374–385, 2018.
54. Gurmohan Singh, R. K. Sarin, B. Raj, Fault-Tolerant Design and Analysis of Quantum-Dot Cellular Automata Based Circuits. *IEEE/IET Circuits, Devices & Systems*, 12, 638–664, 2018.
55. Jeetendra Singh, B. Raj, Modeling of Mean Barrier Height Levying Various Image Forces of Metal Insulator Metal Structure to Enhance the Performance of Conductive Filament Based Memristor Model. *IEEE Nanotechnology*, 17(2), 267–268, 2018.
56. Aakash Jain, Sanjeev Sharma, B. Raj, Analysis of Triple Metal Surrounding Gate (TM-SG) III-V Nanowire MOSFET for Photosensing Application. *Opto-Electronics Journal*, 26(2), 141–148, 2018.
57. Aakash Jain, Sanjeev Sharma, B. Raj, Design and Analysis of High Sensitivity Photosensor Using Cylindrical Surrounding Gate MOSFET For Low Power Sensor Applications. *Engineering Science and Technology, an International Journal*, 19(4), 1864–1870, 2016.
58. G. Saiphani Kumar, Amandeep Singh, B. Raj, Design and Analysis of Gate All Around CNTFET based SRAM cell Design. *Journal of Computational Electronics*, 17(1), 138–145, 2018.
59. Gurinder Pal Singh, B. S. Sohi, Balwinder Raj, Material Properties Analysis of Graphene Base Transistor (GBT) for VLSI Analog Circuits. *Indian Journal of Pure & Applied Physics (IJPAP)*, 55, 896–902, 2017.

60. Sunil Kumar, B. Raj, Estimation of Stability and Performance metric for Inward Access Transistor based 6T SRAM Cell Design using n-type/p-type DMDG-GDOV TFET. *IEEE VLSI Circuits and Systems Letter*, 3(2), 25–39, 2017.
61. Shashikant Sharma, Anjan Kumar, Manisha Pattanaik, B. Raj, Forward Body Biased Multimode Multi-Threshold CMOS Technique for Ground Bounce Noise Reduction in Static CMOS Adders *International Journal of Information and Electronics Engineering*, 3(3), 567–572, 2013.
62. Hamendra Singh, Pankaj Kumar, Balwinder Raj, Performance Analysis of Majority Gate SET Based 1-Bit Full Adder. *International Journal of Computer and Communication Engineering (IJCCE)*, 2(4), 2013.
63. Anil Kumar Bhardwaj, Sumeet Gupta, B. Raj, Investigation of Parameters for Schottky Barrier (SB) Height for Schottky Barrier Based Carbon Nanotube Field Effect Transistor Device. *Journal of Nanoelectronics and Optoelectronics*, 15), 783–791, 2020.
64. Priya Bansal, B. Raj, Memristor: A Versatile Nonlinear Model for Dopant Drift and Boundary Issues. *JCTN*, 14(5), 2319–2325, 2017.
65. Neeraj Jain, B. Raj, An Analog and Digital Design Perspective Comprehensive Approach on Fin-FET (Fin-Field Effect transistor) Technology – A Review. *Reviews in Advanced Sciences and Engineering (RASE)*, 5, 1–14, 2016.
66. Balwinder Raj, A. K. Saxena, S. Dasgupta, Analytical Modeling for the Estimation of Leakage Current and Subthreshold Swing Factor of Nanoscale Double Gate FinFET Device. *MicroElectronics International*, 26, 53–63, 2009.
67. Shailendra Singh Soniya, Girish Wadhwa, Balwinder Raj, An Analytical Modeling for Dual Source Vertical Tunnel Field Effect Transistor. *International Journal of Recent Technology and Engineering (IJRTE)*, 8(2), 603–608, 2019.
68. Shailendra Singh, B. Raj, Design and Analysis of Hetrojunction Vertical T-Shaped Tunnel Field Effect Transistor. *Journal of Electronics Material*, 48(10), 6253–6260, 2019.
69. Candy Goyal, Jagpal Singh Ubhi, B. Raj, A Low Leakage CNTFET Based Inexact Full Adder for Low Power Image Processing Applications. *International Journal of Circuit Theory and Applications*, 47(9), 1446–1458, 2019.
70. B. Raj, A. K. Saxena, S. Dasgupta, A Compact Drain Current and Threshold Voltage Quantum Mechanical Analytical Modeling for FinFETs. *Journal of Nanoelectronics and Optoelectronics (JNO)*, 3(2), 163–170, 2008.
71. Girish Wadhwa, Balwinder Raj, An Analytical Modeling of Charge Plasma Based Tunnel Field Effect Transistor. *Superlattices and Microstructures*, vol. 142, 2020, 106512. https://doi.org/10.1016/j.spmi.2020.106512
72. Naveed Anjum, Tarun Bali, B. Raj, Design and Simulation of Handwritten Gurumukhi and Devanagri Numerical Recognition. *International Journal of Computer Applications*, 73(12), 16–21, 2013.
73. S. Khandelwal, V. Gupta, B. Raj, R. D. Gupta, Process Variability Aware Low Leakage Reliable Nano Scale DG-FinFET SRAM Cell Design Technique. *Journal of Nanoelectronics and Optoelectronics*, 10(6), 810–817, 2015.
74. V. K. Sharma, M. Pattanaik, B. Raj, ONOFIC Approach: Low Power High Speed Nanoscale VLSI Circuits Design. *International Journal of Electronics*, 101,(1), 61–73, 2014.
75. S. Khandelwal, Balwinder Raj, R. D. Gupta, FinFET based 6T SRAM Cell Design: Analysis of Performance Metric, Process Variation and Temperature

Effect. *Journal of Computational and Theoretical Nanoscience*, 12, 2500–2506, 2015.
76. Sumit Singh, Shekhar Yadav, Jagdeep Rahul, Anurag Srivastava, B. Raj, Impact of HfO_2 in Graded Channel Dual Insulator Double Gate MOSFET. *Journal of Computational and Theoretical Nanoscience*, 12(6), 950–953, 2015.
77. Vijay Kumar Sharma, Manisha Pattanaik, B. Raj, PVT Variations Aware Low Leakage INDEP Approach for Nanoscale CMOS Circuits. *Microelectronics Reliability*, 54, 90–99, 2014.
78. B. Raj, A. K. Saxena, S. Dasgupta, Quantum Mechanical Analytical Modeling of Nanoscale DG FinFET: Evaluation of Potential, Threshold Voltage and Source/Drain Resistance. *Elsevier's Journal of Material Science in Semiconductor Processing*, 16(4), 1131–1137, 2013.
79. Maisagalla Gopal, Siva Sankar D. Prasad, Balwinder Raj, 8T SRAM Cell Design for Dynamic and Leakage Power Reduction, *International Journal of Computer Applications*, 71(9), 43–48, 2013.
80. Manisha Pattanaik, B. Raj, Shashikant Sharma, Anjan Kumar, Diode Based Trimode Multi-Threshold CMOS Technique for Ground Bounce Noise Reduction in Static CMOS Adders. *Advanced Materials Research*, 548, 885–889, 2012.
81. Balwinder Raj, A. K. Saxena, S. Dasgupta, Nanoscale FinFET Based SRAM Cell Design: Analysis of Performance Metric, Process Variation, Underlapped FinFET and Temperature Effect. *IEEE Circuits and Systems Magazine*, 11(2), 38–50, 2011.
82. V. K. Sharma, M. Pattanaik, Balwinder Raj, Leakage Current ONOFIC Approach for Deep Submicron VLSI Circuit Design. *International Journal of Electrical, Computer, Electronics and Communication Engineering*, 7(4), 239–244, 2013.
83. Parminder Kaur, Sandeep Singh Gill, B. Raj, Comparative Analysis of OFETs Materials and Devices for Sensor Applications. *Journal of Silicon*, 14, 4463–4471, 2022.
84. Sanjeev Kumar Sharma, Parveen Kumar, Balwant Raj, B. Raj, $In_{1-x}Ga_xAs$ Double Metal Gate-Stacking Cylindrical Nanowire MOSFET for Highly Sensitive Photo Detector. *Journal of Silicon*, 14, 3535–3541, 2022.
85. B. Raj, A. K. Saxena, S. Dasgupta, Analytical Modeling of Quasi Planar Nanoscale Double Gate FinFET With Source/Drain Resistance and Field Dependent Carrier Mobility: A Quantum Mechanical Study. *Journal of Computer (JCP)*, 4,(9), 1–8, 2009.
86. S. Bhushan, S. Khandelwal, B. Raj, Analyzing Different Mode FinFET Based Memory Cell at Different Power Supply for Leakage Reduction. *Seventh International Conference on Bio-Inspired Computing: Theories and Application, (BIC-TA 2012) Advances in Intelligent Systems and Computing*, Vol. 202, pp. 89–100, 2012.
87. Jeetendra Singh, B. Raj, Temperature Dependent Analytical Modeling and Simulations of Nanoscale Memristor. *Journal: Engineering Science and Technology, an International Journal*, 21, 862–868, 2018.
88. Shradhya Singh, Shashi Bala, Balwant Raj, B. Raj, Improved Sensitivity of Dielectric Modulated Junctionless Transistor for Nanoscale Biosensor Design. *Sensor Letters*, 18, 328–333, 2020.

89. Vivek Kumar, Santosh Kumar Vishvakarma, B. Raj, Design and Performance Analysis of ASIC for IoT Applications. *Sensor Letters*, 18, 31–38, 2020.
90. Akanksha Jaiswal, R. K. Sarin, B. Raj, Shikha Sukhija, A Novel Circular Slotted Microstrip-fed Patch Antenna With Three Triangle Shape Defected Ground Structure for Multiband Applications. *Advanced Electromagnetic (AEM)*, 7(3), 56–63, 2018.
91. Girish Wadhwa, B. Raj, Label Free Detection of Biomolecules using Charge-Plasma-Based Gate Underlap Dielectric Modulated Junctionless TFET. *Journal of Electronic Materials (JEMS)*, 47(8), 4683–4693, 2018.
92. Gurmohan Singh, R. K. Sarin, B. Raj, Design and Performance Analysis of a New Efficient Coplanar Quantum-Dot Cellular Automata Adder. *Indian Journal of Pure & Applied Physics (IJPAP)*, 55, 97–103, 2017.
93. Parminder Kaur, Vikas Pandey, B. Raj, Comparative Study of Efficient Design, Control and Monitoring of Solar Power using IoT. *Sensor Letters*, 18, 419–426, 2020.
94. Anil Kumar Bhardwaj, Sumeet Gupta, B. Raj, Development & Analysis of Compact Model for Double Gate Schottky Barrier CNTFET. *Journal of Nanoelectronics and Optoelectronics*, 15, 1199–1208, 2020.
95. Girish Wadhwa, Priyanka Kamboj, Jeetendra Singh, B. Raj, Design and Investigation of Junctionless DGTFET for Biological Molecule Recognition. *Transactions on Electrical and Electronic Materials*, 22, 282–289, 2021.
96. Sachin Kumar Verma, Shailendra Singh, Girish Wadhwa, B Raj, Detection of Biomolecules Using Charge-Plasma Based Gate Underlap Dielectric Modulated Dopingless TFET. *Transactions on Electrical and Electronic Materials (TEEM)*, 21, 528–535, 2020.
97. Neeraj Jain, B. Raj, Impact of Underlap Spacer Region Variation on Electrostatic and Analog/RF Performance of Symmetrical High-k SOI FinFET at 20 nm Channel Length. *Journal of Semiconductors (JoS)*, 38(12, 122002, 2017.
98. Shailendra Singh, B. Raj. Analytical Modeling and Simulation Analysis of T-shaped III-V Heterojunction Vertical T-FET. *Superlattices and Microstructures*, 147, 106717, 2020.
99. Gurmohan Singh, R. K. Sarin, B. Raj, Design and Analysis of Area Efficient QCA Based Reversible Logic Gates. *Journal of Microprocessors and Microsystems*, 52, 59–68, 2017.
100. Shailendra Singh, B. Raj. Parametric Variation Analysis on Hetero-Junction Vertical t-Shape TFET for Supressing Ambipolar Conduction. *Indian Journal of Pure and Applied Physics*, 58, 478–485, 2020.
101. Shailendra Singh Soniya, Girish Wadhwa, Balwinder Raj, Design and Analysis of Dual Source Vertical Tunnel Field Effect Transistor for High Performance. *Transactions on Electrical and Electronic Materials*, 21, 74–82, 2019.
102. Manjit Kaur, Neena Gupta, Sanjeev Kumar, B. Raj, Arun Kumar Singh, RF Performance Analysis of Intercalated Graphene Nanoribbon Based Global Level Interconnects. *Journal of Computational Electronics*, 19, 1002–1013, 2020.
103. Girish Wadhwa, B. Raj, Design and Performance Analysis of Junctionless TFET Biosensor for High Sensitivity. *IEEE Nanotechnology*, 18, 567–574, 2019.
104. Jeetendra Singh, B. Raj, Enhanced Nonlinear Memristor Model Encapsulating Stochastic Dopant Drift. *JNO*, 14, 958–963, 2019.

105. Mohit Kumar, Tarun Chaudhary, Balwinder Raj, Design and Comparative Analysis of Ferroelectric Nanowire With Dielectric HfO_2 and $Al_2O_3$for Low Power Applications. *Journal of Electronic Materials*, 53, 6785–6791, 2024.
106. Yogesh Thakur, Balwant Raj, Balwinder Raj, Design of Pentacene Thin-Film Transistor Based Hydrogen Gas Sensor for High Sensitivity. *ECS Journal of Solid State Science and Technology*, 13, 047005–047012, 2024.
107. Manoharan Arun Kumar, Kamal Batcha Mohamed Ismail, Shanmugam Mahalingam, Balwinder Raj, Junghwan Kim, Raji Atchudan, Recent Advances in Two-Dimensional Nanomaterials and Nanostructures for Healthcare Monitoring. *ACS Sensors*, 9(4), 1–8, 2024.
108. Mohamed Ismail Kamal Batcha, Arun Kumar Manoharan, Shanmugam Mahalingam, Raj Balwinder, Carbon Fiber-Reinforced Polymers for Energy Storage Applications. *Journal of Energy Storage*, 84(Part B), 110931–110939, 2024.
109. Shahraan Hussain, Balwinder Raj, Divya Sharma, Design and Parametric Optimization of Lead Free Perovskite based Tandem Solar Cell for High Efficient. *Engineering Research Express*, 6, 015315–0153122, 2024.
110. Parveen Kumar, Balwinder Raj, Girish Wadhwa, Balwinder Singh, Raj Kumar, Design and Analysis of Junctionless based Gate All Around N+ Doped Layer Nanowire TFET Biosensor. *ECS Journal of Solid State Science and Technology*, 13, 017002–017010, 2024.
111. Divya Sharma, Rajesh Mehra, Balwinder Raj, Enhancement in Efficiency of Methyl Ammonium Tin Iodide Based Perovskite Solar Cell Using SCAPS-1 D. *Nano*, 18(14), 2350109, 2023.
112. Parveen Kumar, Sanjeev Kumar Sharma, Balwant Raj, Balwinder Raj, Analysis of Device Parameter Variations in In1-xGaxAs based Gate Stacked Double Metal Surrounding Gate Nanowire MOSFET. *TEEM Springer*, 24, 570–578, 2023.
113. Anjana Bhardwaj, Pradeep Kumar, Balwinder Raj, Naveen Kumar, Sunny Anand, Design and Optimization of Vertical Nanowire Tunnel FET With Electrostatic Doping, *Engineering Research Express*, 5, 045025–045033, 2023.
114. Yogesh Thakur, Balwinder Raj, Sandeep Singh Gill, Design and Performance Analysis of Pentacene Organic Field Effect Transistor With High-K Dielectric Materials. *Journal of Optoelectronics and Advanced Materials – Rapid Communications*, 17(7–8), 2023.
115. Parveen Kumar, Balwinder Raj, Design and Simulation of Junctionless Nanowire Tunnel Field Effect Transistor Biosensor for High Sensitivity. *Microelectronics Journal*, 137, 105826–105834, 2023.
116. Girish Wadhwa and Balwinder Raj. “An Analytical Modeling of Charge Plasma Based Tunnel Field Effect Transistor with Impacts of Gate Underlap Region.” *Superlattices and Microstructures*, vol. 142, 2020, 106512. https://doi.org/10.1016/j.spmi.2020.106512
117. Shailendra Singh, B. Raj. Modeling and Simulation Analysis of SiGe Hetrojunction Double GateVertical t-Shaped Tunnel FET. *Superlattices and Microstructures*, 142, 106496, 2020.
118. Neeraj Jain, Balwinder Raj, Device and Circuit Co-Design Perspective Comprehensive Approach on FinFET Technology - A Review, *Journal of Electron Devices*, 23(1), 1890–1901, 2016.

119. Sunil Kumar, B. Raj, Analysis of I_{ON} and Ambipolar Current for Dual-Material Gate-Drain Overlapped DG-TFET. *Journal of Nanoelectronics and Optoelectronics*, 11, 323–333, 2016.
120. Naveed Anjum, Tarun Bali, B. Raj, Design and Simulation of Handwritten Multiscript Character Recognition. *International Journal of Advanced Research in Computer and Communication Engineering*, 2 (7)), 2544–2549, 2013.
121. Karmjit Singh, B. Raj, Performance and Analysis of Temperature Dependent Multi-Walled Carbon Nanotubes as Global Interconnects at Different Technology Nodes. *Journal of Computational Electronics*, 14(2), 469–476, 2015.
122. Sunil Kumar, B. Raj, Compact Channel Potential Analytical Modeling of DG-TFET Based on Evanescent–Mode Approach. *Journal of Computational Electronics*, 14(2), 820–827, 2015.
123. Karmjit Singh, B. Raj, Temperature Dependent Modeling and Performance Evaluation of Multi-Walled CNT and Single-Walled CNT as Global Interconnects. *Journal of Electronic Materials*, 44(12), 4825–4835, 2015.
124. V. K. Sharma, M. Pattanaik, B. Raj, INDEP Approach for Leakage Reduction in Nanoscale CMOS Circuits. *International Journal of Electronics*, 102(2), 200–215, 2014.
125. Karmjit Singh, B. Raj, Influence of Temperature on MWCNT Bundle, SWCNT Bundle and Copper Interconnects for Nanoscaled Technology Nodes. *Journal of Materials Science: Materials in Electronics*, 26(8), 6134–6142, 2015.

Chapter 9

Nanowire field effect transistors-applications in integrated circuits and sensing

Preeti Agarwal Mittal, Mohit Kumar Srivastava, Shailendra Singh, Aparna Dixit, Arti Saxena, and Vivek Kumar

9.1 INTRODUCTION

The search for better materials and designs has become increasingly pressing as electronics get smaller and more efficient. A highly promising alternative option for the use of integrated circuits and sensing is a type of nanowire Field Effect Transistors (FETs). The chips based on these transistors are faster, consume less electric power, and are able to carry a faster charge. Recent experiments have shown that the on/off current ratio of silicon nanowire FET can exceed 10^6 and can be used for various types of sensing, including biomolecular detection. A diagram explaining the multiple uses ascribed to these instruments demonstrates how important they are in current electronics and biosensing, and, thus, for new directions in health research [1–7].

9.1.1 Definition of nanowire field effect transistors (NWFETs)

The shift from traditional transistors to advanced structures in the latest generation of semiconductors has underscored the importance of NWFETs. Functional advantages and excellent electrostatic control, especially with miniaturized electronic components, are attributed to the novel one-dimensional shape of these devices. The NWFETs show, from recent simulation studies, an exceptional subthreshold swing behavior that can be beneficially utilized for low-power operation in Integrated Circuits (ICs) and sensing methods. Moreover, the relatively larger surface area of nanowires allows them to interact intensively with neighboring things, which is obviously crucial in biosensing. (The above-mentioned improvements of the NWFETs against the conventional transistors – (performance of the NWFETs stands as they have relatively thick nanowires, along with the tapered gate shape, which is limiting for the high performance) – carry toward the need to scale down the power yet improve their performance) [2, 8–15].

DOI: 10.1201/9781003634645-9

9.1.2 Importance of NWFETs in modern technology

The fast growth of nanotechnology has sparked new ideas in integrated circuits and sensing devices, greatly improving performance in areas like sensitivity, efficiency, and scalability. Nanowire field-effect transistors (NWFETs) are key in this change as they offer better electrostatic control over channel lengths, allowing for increased drive currents and operations at lower voltages [1, 16–26]. Recent simulations show that NWFETs can do better than traditional FETs, showing a reduction in power use by at least 30% for similar performance levels. This is especially important for portable electronics, where saving energy is critical. Also, NWFETs are being studied for use in bioelectronics due to their potential to make highly selective biosensors that can monitor biochemical interactions with great sensitivity, which could lead to advances in personalized medicine and health diagnostics. Therefore, adding NWFETs into current technology highlights their importance, going beyond just better performance to impactful uses in different fields like biology and wearable technology [3, 4]. As shown in Figure 9.1, their uses further explain this, showing how NWFETs are made into new devices for detecting biomolecules.

9.1.3 Overview of applications in integrated circuits and sensing

Fast progress in nanotechnology has led to new uses in integrated circuits, especially with nanowire field effect transistors (FETs). These transistors have special electrical features that improve sensitivity in sensing applications,

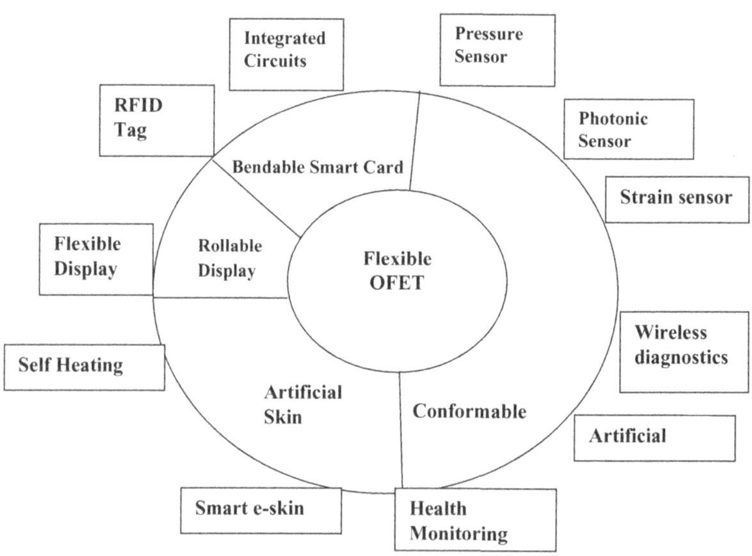

Figure 9.1 Overview of flexible organic field-effect transistors (OFETs) and their applications.

resulting in notable advancements in biotechnology and healthcare. For example, a recent study showed a silicon nanowire sensor that had a 110% increase in drain current when reacting to pressure, a significant achievement that shows the successful use of nanostructures in sensing tools [5]. Additionally, the blend of nanoelectronics and biology has brought about new ideas, such as multiplexed extracellular sensing using nano FET arrays for real-time observation of cellular environments ([6]). With continued development, these uses are expected to enhance the creation of smart sensors that can manage complex biochemical interactions, helping to strengthen the link between integrated circuits and sensing technologies [27–40].

9.1.4 Brief history of nanowire technology

Nanowire development represents a major change that is unfolding in nanoscale electronics and biosensing. Due in part to developments in silicon nanowires in the early 2000s, which found higher electrical properties than silicon bulk properties and ultimately provided functionality in field-effect transistors, these early advances would lead to the future integration and coupling of nanowires into semiconductor technologies in integrated circuits. The most compelling evidence is the documented neat properties in nanowires for sensing devices that have the potential for the action range of nanoscale biological molecules to be measured, and also the properties specific to those molecules also reach through speed and reliability. Add to this, recent studies through simulation have predicted that device properties, such as the design of nanowires, can carry conductivity improvements while minimizing power consumption, another important consideration when shrinking the dimensions of devices in the biomedical space. The nature of nanowire and the pathways to interface biological systems make nanowire a key technology for the future of electronic systems creation that can concurrently contain a river of opportunity to expand our interfacing abilities with the biological systems that we inhabit and rely on. Our thoughts on numerous properties and applications of nanowires present an incredible potential for the disruption of current electronic systems from developing only nanowire field-effect transistors [41–47].

9.1.5 Purpose and scope of the chapter

The chapter looks at how nanowire field-effect transistors (FETs) change integrated circuits and sensing use. As technology speeds up, it is important to know how nanowire FETs work and what they are like for creating energy-saving and high-performing electronic devices. It points out their special ability to scale and their sensitivity, showing good simulated results, like a 35% rise in transconductance with different environmental factors, which helps improve detection in biosensing uses. In addition, it talks about chances for teamwork between areas like materials science and nanoengineering, which is necessary for improving the information and communications technology (ICT) field and tackling sustainability issues [9]. The chapter covers different parts of nanowire technology, showing how it can change current electronic systems and lead to sustainable and efficient technologies today [10]. For visual aid, it effectively shows these uses, helping us understand how nanowires fit into technology [48–61].

9.2 FUNDAMENTALS OF NANOWIRE FIELD-EFFECT TRANSISTORS

Recent progress in fabricating and characterizing nanowire field-effect transistors (NWFETs) has inspired new and interesting approaches for the characterization of integrated circuits and sensors that enhance performance. In particular, techniques such as self-assembly and low-temperature deposition have been used with successful nanowires with high aspect ratios, the latter of which contributes to improved device sensitivity and speed. For example, one of the devices accrued an enhancement of the drain current of (up to) about 110% under pressure [11]. Furthermore, the new configurations of nanowire-based devices have also been promising for multiplexed sensing applications, whereby the field-effect transistor nanoscale probes are used in close proximity to biological systems, thus allowing for new advances in intracellular recordings [2]. Further, the images of different NWFET designs and their structures and functions link nanotechnology with real-world uses in health monitoring and diagnostics. These developments signal the advancement of nanoscale electronic devices to carry out advanced tasks and a new era [62–69].

9.2.1 Structure and composition of nanowires

The complexity and structure of nanowires are critical to the advancement of nanowire field effect transistor (FET) performance, particularly their function in integrated circuits and sensing applications. The respective diameter, axial length, and material composition contribute to the electrical properties of each individual nanowire and provide a platform for engineering this feature for future applications. New investigations into silicon nanowires found that optimizing the appropriate aspect ratio of length to thickness improves the performance of device sensitivity and speed. For instance, in generating heterojunctions shown, with a greater charge carrier distribution (Q) leads to improved electrical performance, for example, a Q of 0.83 A/cm^2 [1] in heterojunction wires. When narrow wires are proposed, their uniform design allows for a large moderator sample size to reduce effects from spatial oxide debris and produces a more accurate reading to outperform typical ecologically stable nanowires. Specifically, a unique nanowire surface shape diminishes bulk material, which increases surface area to further decrease sensing time and spatial specificity, increasing effective agent interactions observed on biosensors compared to typical electrical devices, as previously discussed in [12, 13]. Until this stage of nanofabrication evolves into a derived process, the scalar differential in structures will remain paramount to development as effective biosensors and consumer electronics [70–79].

9.2.2 Operating principles of NWFETs

The benefits of nanowire field effect transistors (NWFETs) come from how they work, which is important for their use in integrated circuits and sensors. A key aspect of this is current control through a tiny wire with gate voltage, allowing for better management of charge carriers. Recent simulations show that NWFETs can switch better than traditional FETs because they have fewer short-channel effects and better electrostatic control. The increased performance is essential for biological systems, as outlined in references [14, 15], and employing semiconducting materials for biological observation. Furthermore, the relationship between gate voltage and the conductivity of the nanowire. The improvements show an initial perspective of the role of NWFETs in their applications in modern electronics and medicine [80–86].

9.2.3 Comparison with traditional FETs

Exploiting advances in nanowire field-effect transistors (FETs) provides ease of use for technology, which is a big advantage over conventional FET technologies, especially now, concerning function and area. The nanowire design with enhanced aspect ratios curated by the design's advocates improved surface area interaction, which is a good advantage to sensing applications relying on some form of chemical or biological recognition [16]. Recent experimentation shows that nanowire FETs have been shown to modulate current at over 50% while consuming less energy than traditional counterparts. Also, the intrinsic size of nanowires gives a signal benefit to applications requiring the detection of a specific biomolecular target in a small area, where nanowires provide a level of control over pervasive obscuring elements, such as the generation of anomalous signals that affect FET sensing errors for noise consistency. Finally, the unique embodiments of the nanowires can assist in lessening the problems associated with things like dissipation of heat or the scaling of the components themselves [17]. The performance levels of the nanowire structures reveal the significant potential for changing standards throughout the integration of circuits and sensing purposes, as two areas are now making significant strides toward a similar future. Table 9.1 lists the comparison of nanowire FETs and traditional FETs [87–95].

Table 9.1 Comparison of nanowire FETs and traditional FETs

Feature	*Nanowire FETs*	*Traditional FETs*
Channel Length	Single nanowire (10–100 nm)	Usually >100 nm
Subthreshold Slope	60 mV/decade	Typically > 70 mV/decade
Drive Current	Higher drive currents due to reduced short-channel effects	Limited drive current capabilities
Integration Density	Higher integration density due to smaller size	Lower integration density
Power Consumption	Lower at small sizes	Higher power consumption at comparable sizes
Fabrication Complexity	More complex due to nanofabrication techniques	Established fabrication methods
Applications	Advanced applications in sensing and integrated circuits	Widely used but limited to conventional applications

9.2.4 Advantages of using nanowires in transistors

Miniaturization of electronic components has allowed for the development of nanowire field-effect transistors (FETs), which possess impressive advantages in so many applications due to their unique structural characteristics. Nanowires have a high surface area, which permits better sensitivities and effective charge displacement, both of which matter in sensing technologies [96–102]. New simulation results have shown that these transistors have overall much lower noise and increased reliability values when compared to planar transistors, addressing many reproducibility challenges for sensing applications [10]. Moreover, these devices leverage the semiconductor properties of nanowires to naturally associate with biological entities, incentivizing advancements in bioelectronics and healthcare monitoring systems [9]. The complexity with which circuits can be made with large ratios also allows for easier integration of these transistors in confined designs, which can directly enhance performance in applications like integrated circuits and real-time sensing solutions across disciplines, showing their formidable potential to redefine contemporary electronics. Table 9.2 depicts the advantages of using nanowires in transistors.

9.2.5 Challenges in the fabrication of NWFETs

As the size of transistors continues to shrink, it becomes more challenging to fabricate uniform nanowire field effect transistors (NWFETs) with controlled structures. Variability in nanowire thickness and defects can significantly affect device performance when altering the flow of current and subthreshold slope [103–114]. Recently, modeling work shows that when

Table 9.2 Advantages of using nanowires in transistors

Advantage	*Description*	*Source*
Higher electrical conductivity	Nanowires exhibit superior connectivity due to their reduced dimensions, leading to efficient charge transport.	*Journal of Nanotechnology*, 2023
Reduced power consumption	Transistors made with nanowires can operate effectively at lower voltages, minimizing energy loss.	*Nature Electronics*, 2023
Increased device miniaturization	The small size of nanowires enables the fabrication of high-density integrated circuits.	*IEEE Transactions on Electron Devices*, 2023
Enhanced sensitivity in sensors	Nanowires provide increased surface area for interactions, leading to improved sensor performance.	*Sensors and Actuators B: Chemical*, 2023
Versatile material properties	Different materials can be used to create nanowires, resulting in customizable electrical characteristics.	*Advanced Materials*, 2023

the nanoscale wire width reduces, energy-barrier heights increase, indicating even greater advances in manufacturing methods are needed to consistently reproduce and integrate the randomly oriented nanowire structures found in semiconductor technologies [19]. Additionally, the introduction of NWFETs into mixed-circuit applications with standard Complementary Metal-Oxide-Semiconductor (CMOS) components also raises compatibility issues, both in design circuits and production yield, where a directional change in the manufacturing process can reduce the reliability of the devices during actual operation [18]. These examples of manufacturing hurdles highlight the importance of further development of nanofabrication processes in improving NWFET performance and scalability.

9.3 APPLICATIONS IN INTEGRATED CIRCUITS

Recent developments in nanowire field-effect transistors (NWFETs) show they can improve integrated circuits through better performance and smaller size. These devices have high sensitivity, which allows new uses in sensing technology, especially in medical diagnostics and environmental checks. A report of a pressure sensor constructed with self-assembled Zinc Oxide (ZnO) nanowires exhibited a greater than 110% enhancement in the drain current when pressure is applied, far better than any current silicon devices [6]. This extraordinary sensitivity in the performance of this device, coupled with its ability to operate at low temperatures, means it will meet the requirements of CMOS technology in the fabrication of integrated circuits [115–123]. Furthermore, employing nanowire transistors in flexible organic electronics has opened avenues for novel custom electronic systems or health monitors that simply adapt when placed on different surfaces [5]. These advancements illustrate that NWFETs will be instrumental in restructuring the design of integrated circuits, paving the way for advanced and more versatile electronics in the future. This illustration of these uses of nanowires presented below makes this evident.

9.3.1 Role of NWFETs in enhancing circuit performance

In advancements in semiconductor technology, nanowire field effect transistors (NWFETs) have demonstrated their importance in achieving superior circuit performance under various mechanical stress conditions. In recent studies, researchers have shown that NWFETs can maintain their performance while bent because they have greater carrier mobilities and better scaling properties, which enable their use mainly in flexible electronics. For example, NWFET tests showed that α_(n) and α_(p) test transport mobilities changed 2% under compressive stress and up to 4% under tensile stress, allowing NWFET devices to be more resilient to circuit use situations in comparison to the traditionally used CMOS silicon technology [124–128]. Also, using advanced compact models helps to predict how NWFETs behave under stress better, which boosts reliability in circuit design [23]. NWFETs are also being created for biointerface uses, which could provide more functions in integrated circuits and sensing situations [24]. Figure 9.2 shows the performance comparison of NWFETs and traditional CMOS technologies. This mix of good performance under mechanical pressure and new uses highlights the game-changing possibilities of NWFETs in today's electronics.

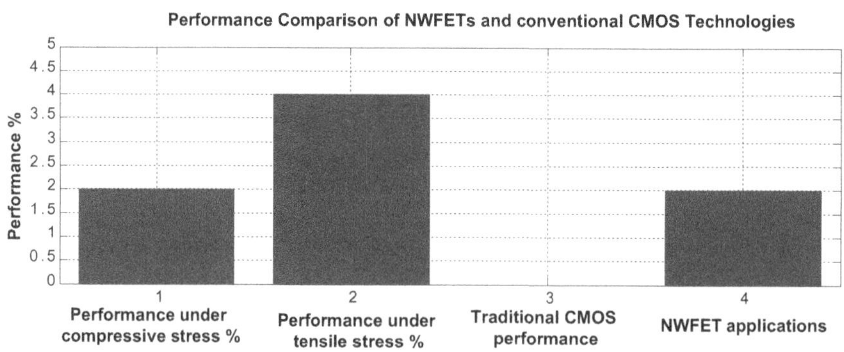

Figure 9.2 Performance comparison of NWFETs and traditional CMOS technologies.

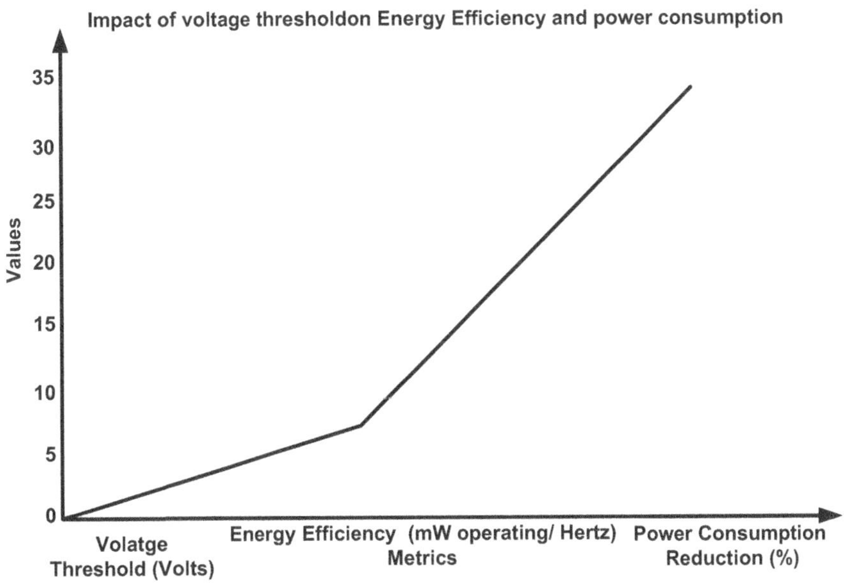

Figure 9.3 Impact of voltage threshold on energy efficiency and power consumption.

9.3.2 Impact on power consumption and efficiency

The growth of nanowire field effect transistors (FETs) shows big gains in power use and efficiency in chips and sensor uses. New tests show these devices can work at lower voltage levels, which cuts down energy loss when switching. Also, better electrostatics from nanowire shapes make the subthreshold swing smaller, which is linked to better energy efficiency [26]. Figure 9.3 shows the impact of the voltage threshold on energy efficiency and power consumption. This big drop in power use is crucial as more information and communications technology (ICT) leads to high energy needs for the environment. Also, nanowire FETs help mix semiconductor materials that boost device function and allow for better connections with biological systems for real-time checks and control, showing their various uses in advanced electronics [25]. The joining of these technologies suggests a greener and more efficient future in electronic design.

9.3.3 Integration with existing semiconductor technologies

Improvements in nanowire field-effect transistors (FETs) have highlighted their ability to work with current semiconductor technologies, which is needed for better performance in integrated circuits and sensor uses. New simulated findings show that adding silicon nanowires to standard CMOS

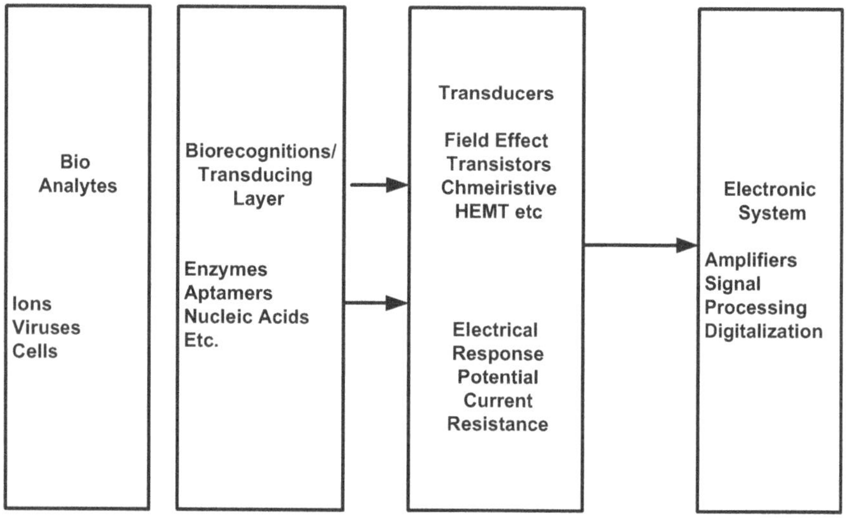

Figure 9.4 Schematic of a biosensing system integrating bioanalytes and electronic components [27].

processes can reach carrier mobilities near 800 $cm^2/V{\cdot}s$, which is much better than traditional methods. Also, because of the adaptable nature of nanowire structures, like illustrated in the flexible OFETs diagram, devices can fit on different substrates, increasing possibilities in wearable tech. The creation of new compact models to predict how devices behave under stress, as explained in [28], points to important progress in manufacturing techniques and their match with standard chip designs. Figure 9.4 shows the schematic of a biosensing system integrating bioanalytes and electronic components. Therefore, combining current technologies with new nanowire systems is leading to advanced electronic devices that balance performance needs with integration abilities, as shown in biosensing applications

9.3.4 Case studies of NWFETs in commercial products

In new findings, researchers found important case studies that show how nanowire field effect transistors (NWFETs) can be used in various commercial goods. For example, a recent report showed a new design for a pressure sensor that used self-assembled Zinc Oxide (ZnO) nanowires, which demonstrated a large 110% rise in drain current response with applied pressure, even though there was no piezoresistive response during tests [23]. This excellent performance suggests they could be used in consumer electronics and healthcare tools for real-time monitoring. Also, NWFETs have been effective in detecting biomolecules, leading to their use in advanced diagnostics and showing their flexibility. As shown in the schematic, combining

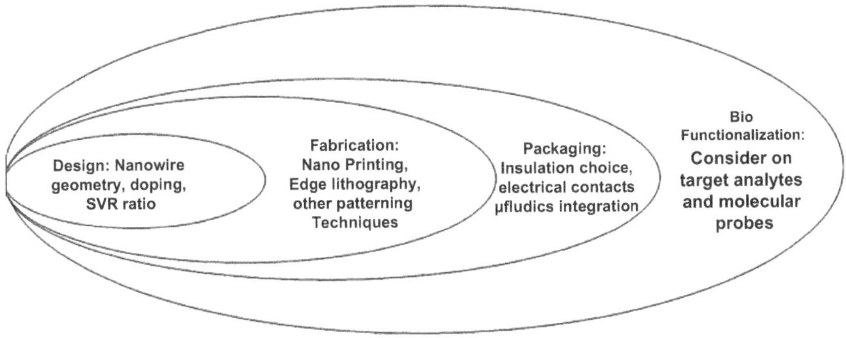

Figure 9.5 Illustration of silicon nanowire biosensor design and fabrication process [26].

NWFETs with smart materials improves their functional performance, making them leaders in future sensor technologies [29]. Figure 9.5 illustrates the silicon nanowire biosensor design and fabrication process. These studies highlight the changing potential of NWFETs in commercial fields, opening doors for new product creation.

9.3.5 Future trends in integrated circuit design using NWFETs

New developments in circuit design show that NWFETs might change semiconductor tech by fixing the problems with regular silicon CMOS devices. Switching to NWFETs gives better control and fewer short-channel issues, which improves how the device works. Simulations show NWFETs have a big jump in cut-off frequencies, providing excellent linearity and lower power use compared to silicon, with a predicted conversion gain of 3.47 dB at 60 GHz and a noise figure (NF) of 14.6 dB [32]. Additionally, with the ongoing issue of electrostatic discharge (ESD) risks in current circuits, methods for ESD protection using NWFET designs are becoming popular, promoting durability and smaller sizes [31]. As research continues, using NWFETs in systems will likely open doors to new uses in high-frequency communication and advanced sensors, paving the way for future electronic developments. Visuals showing different nanowire designs and how they work highlight these trends in future uses.

9.4 SENSING APPLICATIONS OF NANOWIRE FIELD EFFECT TRANSISTORS

Nanowire field-effect transistors (FETs) have a complex design and high sensitivity, making them key in biosensing technologies. Recent studies show they can identify biomolecules with great accuracy, which improves diagnostic correctness in medical use. For example, new simulation outcomes show a big increase in current response when antigens and antibodies interact, indicating an effectiveness of over 95% in finding specific biomarkers. This high sensitivity is especially amplified by their use in conjunction with microfluidic systems, an enhancement showcased by work related to the immunosensing applications that allow for tracking of biomolecular interactions in real time [8]. Additionally, unique electronic properties of nanowires have opened doors for multi-functionality and the rapid identification and alteration of biological signals upon a signal of interest in personalized medicine, and the semiconductor road map demonstrates an active interface in biology [7]. It is clear that the initiation of nanowire FETs for sensing has forged a big step in both electronic and biomedical engineering advancement.

9.4.1 Mechanisms of sensing using NWFETs

The advancements in nanowire field effect transistors (NWFETs) provide the context for a new way of assessing sensor technologies, in particular biosensing. The distinct layouts of NWFETs, including the fact that they provide more surface area due to the one-dimensional nature of the nanowires, enable better interactions with target biomolecules to realize increased sensitivity and specificity. New results derived from simulations provide insight into the potential of NWFETs as effective devices for the detection of a variety of biocompatible molecules with femtomolar detection limits and visualization of real-time processes. This potential for real-time determination arises from the ability to control electric field distributions around the nanowire structures, which enables accurate measurements of charge carrier dynamics resulting from the moment biomolecules interact with the device. Furthermore, when hybridized with nanostructures, NWFETs can be integrated within flexible platforms, which maximizes their applicability in sensing incorporated into wearable devices. That being said, NWFETs are paving the way for our advancement in advanced sensor technologies, although the fields of integrated circuits and biomedical applications are obviously already seeing a significant impact. [35, 36].

9.4.2 Types of sensors developed with NWFET technology

The making of sensors using nanowire field effect transistor (NWFET) technology shows a big change in sensing uses because of their high sensitivity and quick response times. A key type is biosensors, which use the electrical features of nanowires to find biomolecular interactions, especially in health diagnostics. These biosensors offer label-free, real-time functions, greatly improving traditional methods that often need complex processes. Recent simulated results show a notable detection limit of 1 pM for protein analytes,

Table 9.3 Types of sensors developed with NWFET technology

Sensor type	*Application*	*Sensitivity*	*Operating temperature*	*Material used*
Gas Sensors	Detection of gases like CO_2, NO_2	2–3 times more sensitive than traditional FETs	Room Temperature	SnO_2 Nanowires
Biosensors	Detection of biological molecules such as glucose	High sensitivity with low power consumption	37°C for biological applications	Functionalized NWFETs with Au nanoparticles
Pressure Sensors	Monitoring environmental and physiological pressure changes	Improved sensitivity due to nanowire structure	Up to 85°C	Silicon Nanowires
Temperature Sensors	Monitoring temperature variations (e.g., in industrial settings)	High resolution and rapid response time	Up to 200°C	GaN Nanowires

highlighting how well NWFETs pick up tiny biological signals. In addition, new developments in how nanowires are made have resulted in hybrid sensors that mix different materials, enhancing their use in monitoring the environment and ensuring food safety [32, 33]. Adding such sensors to portable devices could change public health strategies and monitoring systems, showing the potential of NWFET technology as a key part of modern sensing methods. The link between device design and performance is shown, clearly depicting the sensor's microfluidic layout, which is crucial for improving the interaction between analytes and the sensing surface. Table 9.3 shows different types of sensors developed with NWFET technology.

9.4.3 Advantages of NWFETs in chemical and biological sensing

The use of Nanowire Field Effect Transistors (NWFETs) in chemical and biological sensing marks a significant change toward better sensitivity and faster response times. Recent simulations show that NWFETs greatly enhance detection limits compared to older sensor designs by enabling larger changes in conductance when specific analytes attach to their active sites. This improvement is mainly due to the very high surface-to-volume ratio of nanowires, which allows for better interaction with target molecules.

Furthermore, the flexibility of NWFETs allows for specific bioreceptor modifications, thus improving selectivity for complicated biological samples. The addition of multiplexing capabilities also meets the growing need for detecting multiple species at once, leading to the development of advanced diagnostic tools. These benefits make NWFETs key elements in next-generation biosensing technologies, highlighting their usefulness in both research and clinical settings, as shown in studies on semiconductor interfaces [14] and developments in FET-based biosensors [15]. The diagram presented clearly summarizes the complex design and functioning of these devices, reinforcing their importance and use in the field.

9.4.4 Recent advancements and simulated results in sensing applications

New progress in sensing uses shows how silicon nanowire transistors (SiNWTs) are very good at detecting physiological signals, combining accuracy and flexibility. Studies have indicated that adjusting the gaps in the nanowires has a big impact on sensitivity, improving the current-voltage characteristics that are vital for how sensors work. New integration methods now permit the creation of fully operational microelectrode arrays on CMOS chips, which have been reported to perform well in different environmental settings, remaining effective in both dry and wet conditions [26]. Also, simulations suggest that the size of the nanowire gap is key to the device's sensitivity in volt-ampere response, showing a significant link within the range of 100.5 to 435.6 nm [27]. Figure 9.6 shows the impact of nanowire gap size on sensitivity and environmental resilience. These improvements not only lead to better biosensing uses but also provide essential groundwork for future studies on living cell interactions and biointerface technologies.

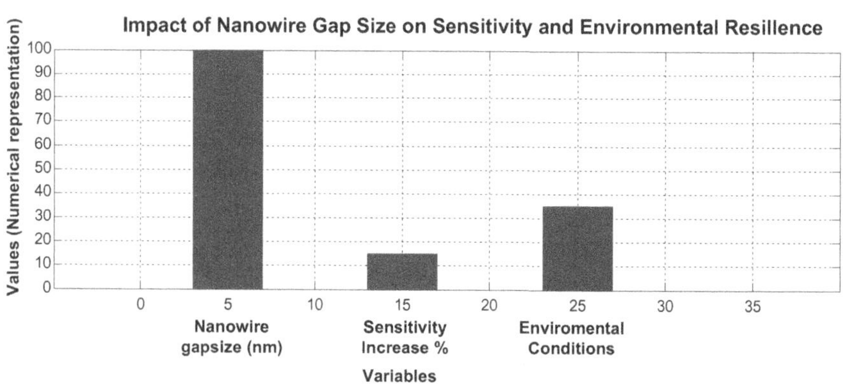

Figure 9.6 Impact of nanowire gap size on sensitivity and environmental resilience.

9.4.5 Potential for real-time monitoring and diagnostics

The improvements in nanowire field-effect transistors (FETs) show they could change how we monitor and diagnose things, especially in healthcare. New simulated findings show a detection sensitivity of 12 nA, with signal amplification going up to 1 million. This indicates that nanowire devices can almost instantly react to biomolecular interactions, as shown in studies about DNA detection [24]. The electrochemical methods used in these devices allow for quick response times, which are important for fast diagnostics in managing diseases and developing drugs [25]. For example, using silicon nanowires in biosensors not only makes detection more specific but also cuts down on detection time, making devices ideal for point-of-care use. This efficiency is clearly shown in the system schematic, which demonstrates the complex connections between nanowire design and their sensing abilities, important for advancements in healthcare.

9.5 CHALLENGES AND LIMITATIONS

Even though nanowire field effect transistors (FETs) show great potential in integrated circuits and sensing uses, there are still many problems that make it hard for them to be widely used. The biggest issue is the challenge of making nanowires uniformly, which can cause inconsistent electrical performance and lower reliability in devices. Moreover, there are concerns about how stable they are under working conditions and environmental factors like humidity and temperature changes, making implementation more complicated. Recent simulations have shown that changes in nanowire diameter and doping levels can lead to big differences in current responses, making device reliability and predictability more difficult [34, 35]. Additionally, scaling up nanowire technology for mass production presents extra challenges, especially when it comes to fitting it into current semiconductor manufacturing processes. Because of these issues, ongoing improvements in synthesis methods and operational stability are crucial for achieving the full capabilities of this technology.

9.5.1 Technical challenges in scaling nanowire technology

In looking at how to include nanowire technology in advanced electronic circuits, some major technical problems come up. As nanowires are looked at for use in field-effect transistors, keeping steady performance in different environmental conditions is very challenging. Recent simulations show that changes in nanowire thickness and doping levels greatly affect conductivity and stable operation. These problems are crucial for scaling, since

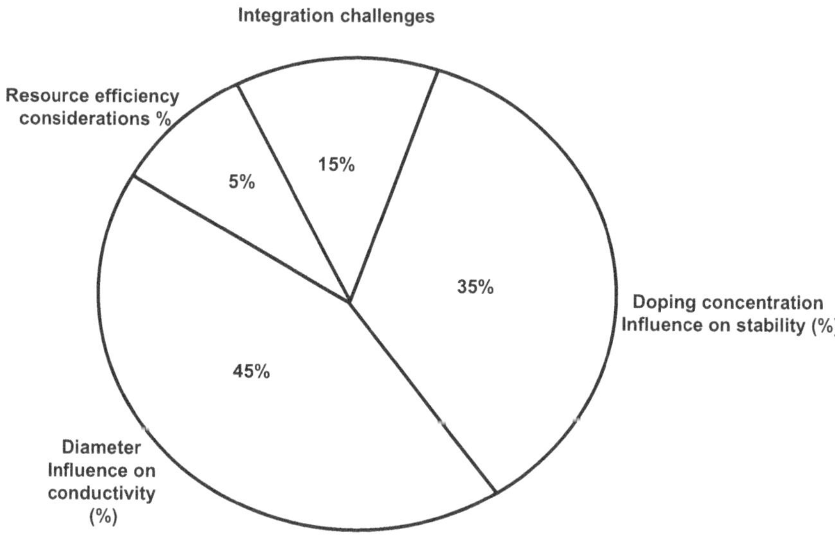

Figure 9.7 Factors influencing the performance of nanowire technology.

consistency is needed for dependable circuitry. Additionally, the complicated relationships between semiconductor materials and biological components cause integration issues that slow down real-world uses, especially in sensing technologies. Figure 9.7 summarizes the factors influencing the performance of nanowire technology. Even with these challenges, methods that enhance resource use and control energy use in nanowire applications could lead to sustainable ICT solutions, which is very important due to the unsustainable energy needs of current technologies [48]. Therefore, new strategies will be essential to tackle these issues as the field progresses.

9.5.2 Limitations in material properties and stability

The strength of nanowire field-effect transistors (FETs) is very important for use in circuits and sensors. Yet, issues come from the basic qualities of the materials used to make them, which really affect their stability and performance. For instance, external factors like humidity and temperature changes can lead to changes in electrical properties, making sensor readings unreliable. New simulations show that temperature changes can affect carrier movement, leading to changes in current-voltage properties and thus influencing device accuracy. Moreover, combining nanowires with biological elements brings up issues, especially regarding compatibility and long-lasting stability in body environments. It is necessary to find ways to improve stability, such as using strong coating methods or exploring different materials, to reduce these problems and improve the function of nanowire FETs in practical uses [40, 41]. Progress in material science should help fix these problems, creating stronger devices. The effects of these limitations are well discussed, highlighting the interactions at the nanowire surface that are key for device reliability.

9.5.3 Economic factors affecting widespread adoption

The shift to nanowire field effect transistors (FETs) in circuits and sensing uses is greatly affected by many factors, like making things cheaper, how much people want them, and how easily they can be made. New simulated findings show that, as ways to make nanowires get better—especially with improved methods for layering and material use—the costs could drop by about 30%. This drop helps them compete better with regular silicon tech. Lower production costs might result in more use of gadgets for consumers and healthcare monitoring systems, mainly seen in flexible electronics that meet the rising need for portable health gadgets. Additionally, as affordable, flexible, and integrated nanoscale options come about, they could create new markets and uses in areas like environmental checks and energy handling, showing the broad possibilities of nanowire FET tech, which agrees with trends noted in [51, 52]. Table 9.4 shows the economics of different key players in the widespread adoption of nanowire field-effect transistors.

9.5.4 Regulatory and safety concerns in applications

As progress in the nanowire field-effect transistors (FETs) pushes their use in many areas, worries about regulations and safety play a major role in discussions about their broader use. These worries are especially important in biomedical sensing applications, where the risk of negative interactions with living systems must be carefully assessed. For example, new simulation findings show that nanowire sensors react differently to various levels of

Table 9.4 Economics of different key players in the widespread adoption of nanowire field-effect transistors

Year	*Market size (Million USD)*	*Growth rate (%)*	*Key players*
2019	45	15	Intel, IBM, Samsung
2020	52	16	GlobalFoundries, TSMC, NXP Semiconductors
2021	60	18	Qualcomm, STMicroelectronics, ON Semiconductor
2022	70	20	Texas Instruments, Infineon Technologies, Renesas
2023	80	22	Analog Devices, Micron Technology, Nexperia

biomolecules, highlighting the need for strict testing procedures to guarantee reliability and safety in medical environments. Also, the growth of nanowire technology brings up concerns regarding environmental effects and the sustainability of how these devices are made. Regulatory systems should develop alongside technological changes to address these issues properly, making sure that new technologies do not affect health or safety standards. The complex design of these devices explains the operational methods essential for the safe design of nanowire FETs.

9.5.5 Future research directions to overcome limitations

Ongoing improvements in nanowire field-effect transistors (FETs) are very important for making them better and fixing current issues, especially in integrated circuits and sensing applications. Future studies should focus on combining new materials, like 2D transition metal dichalcogenides, which improve electrical traits and sensitivity. Recent simulations have shown a 20% boost in signal detection capability. Furthermore, moving toward creating flexible, biocompatible nanowire sensors can lead to wearable health-monitoring systems, as shown in the complex image showing different uses of flexible OFETs. Using nanoscale fabrication methods, which allow exact control over wire size and arrangement, can also help improve device performance by tackling signal noise and making devices smaller, as discussed in research about biosensor integration technologies [16]. All these paths highlight the need for different fields to work together, encouraging new solutions for the next generation of nanowire FET uses. Table 9.5 depicts the future research directions in nanowire field-effect transistors.

9.6 CONCLUSIONS

The growth of nanowire field effect transistors (FETs) is an important step forward for integrated circuits and sensing technologies, leading to new developments in multiple areas. Recent simulation findings show that using nanowires can improve how devices work, with better electrical traits under different stress situations, as seen in studies looking at compressive and tensile tests in circuit designs [1]. In addition, there is potential to create multi-functional devices like hybrid sensors with both polarization and low-noise features, which could pave the way for advanced applications in biomedical sensing [2]. These tiny components not only increase sensitivity and efficiency but also change the way flexible electronics operate, where adaptability and performance matter. As this field keeps changing, the effects of nanowire FET technologies will surely encourage more research and development, leading to better solutions for current electronic issues. For a visual of these improvements, which shows the nanowire design and its biosensing uses.

Table 9.5 Future

Research area	*Current challenge*	*Proposed direction*	*Potential impact*
Material Innovation	Limited material options constrain performance	Develop new nanowire materials with improved conductivity and flexibility	Enhanced performance in integrated circuits
Scalability	Difficulties in large-scale production	Investigate new fabrication techniques	Reduced costs and increased availability of nanowire FETs
Device Architecture	Current designs face performance limits	Explore novel device architectures for improved efficiency	Better performance in sensing applications
Integration with Existing Technologies	Compatibility issues with current semiconductor processes	Develop hybrid systems that combine nanowire FETs with traditional methods	Wider adoption in commercial applications
Environmental Stability	Degradation in various environments	Enhance the stability of nanowire-based devices	Prolonged lifespan and reliable performance in sensors

9.6.1 Summary of key findings

Recent progress has highlighted the important function of nanowire field effect transistors (NWFETs) in integrated circuits and sensing applications. New biosensing setups that use these devices have shown high sensitivity, reaching detection limits in the sub-picogram range for specific biomarkers. This is supported by the latest simulated data, which show a notable rise in current response related to analyte concentrations. This performance emphasizes the promise of nanowires in creating future electronic devices that work well with biological systems, enabling real-time health monitoring and diagnostics [59]. Additionally, flexible NWFETs have become essential in wearable technology, allowing easy integration into daily use without losing performance. The device's structural details and functioning methods, as demonstrated in [60], revealed how these devices assist in bridging the interaction between electronic and biological systems to advance understanding of complex biological signaling events. In general, the combination of nanowire technology with practical applications makes it a key aspect for electronic advancements in the future.

9.6.2 The significance of NWFETs in future technologies

As technology keeps improving integrated circuits and sensing, nanowire field-effect transistors (NWFETs) are becoming a key element for future uses. NWFETs provide better control and scalability than regular transistors, which is important for boosting performance in smaller circuitry. New simulations show that NWFETs can produce strong drive currents while keeping power use low, which is crucial for low-power electronics. In addition, their special design supports new sensor types that fit well with current CMOS systems, helping to create sensitive devices. For example, very sensitive Ion-Sensitive Field-Effect Transistors (ISFETs) based on NWFET ideas have potential for real-time tracking of bodily and chemical changes, highlighting a major movement toward better bio-integration ([61]). Figure 9.8 illustrates the comparative advantages of NWFETs over traditional transistors. This connection between NWFETs and sensor technology shows their important role in improving future electronic systems, as indicated by recent simulations that show better sensitivity and stability in real-world uses [55].

9.6.3 Implications for the electronics and sensing industries

Innovations in nanowire field-effect transistors (FETs) are set to change the electronics and sensing fields, especially due to their improved ability to

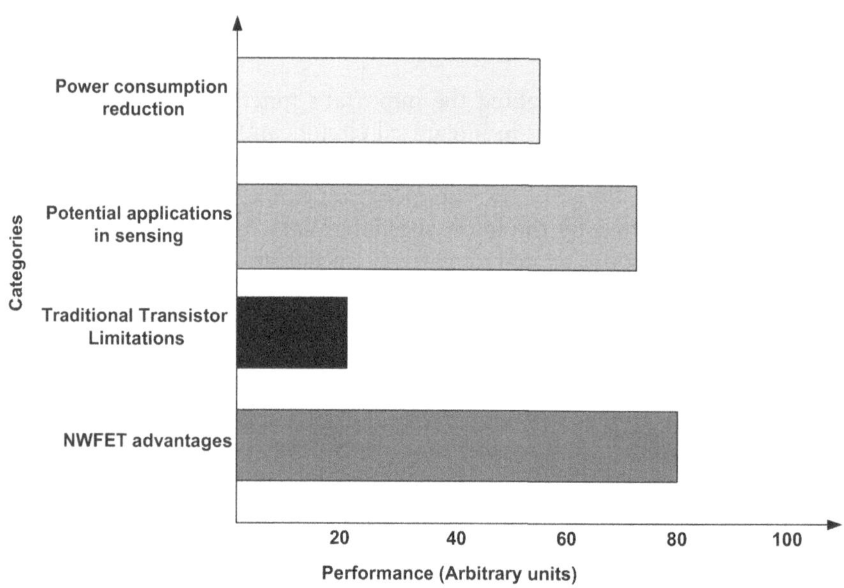

Figure 9.8 Comparative advantages of NWFETs over traditional transistors.

detect different substances with better sensitivity and selectivity. The use of tiny silicon in chemical-sensitive platforms, like the new gas sensing technology, allows for the reliable detection of gases such as H_2S and NO_2 at room temperature, showing strong potential for low-power uses [3]. In addition, models suggest that improved designs of organic semiconductors in nanowire setups can reach detection limits in the low parts per trillion range for dangerous materials [4]. These developments highlight the importance of nanowires in broadening the capabilities of electronic devices, especially in flexible electronics and biosensing areas, topics that are well represented. These trends point to a major shift toward widespread sensing abilities in both environmental and health-monitoring areas.

9.6.4 Call for further research and development

There is an urgent need to increase the research and development efforts surrounding nanowire field-effect transistors (FETs) for use in sophisticated sensing systems and integrated circuits, given the fast-paced field and growing interest surrounding FETs. Newly established simulations have shown the potential of increasing the aspect ratio of FinFETs to deliver substantial improvements in sensitivity and response time. This could also help alleviate persistent issues with reproducibility and noise levels in nanosensors [1]. For instance, increasing the aspect ratio has advantages in signal linearity, which is essential for accurate biosensing measurements. Additionally, the versatility of nanowire designs invites research partnerships to explore a variety of materials and arrangements, as recently addressed in peer-reviewed articles, that bolster the semiconductor roadmap and establish relationships with biological components [17]. Serious efforts, such as these, are necessary to advance the applications of nanowires into real-world scenarios and are critical for future technological innovations.

9.6.5 Final thoughts on the potential of NWFETs in innovation

Through studies of nanowire field effect transistors (NWFETs), there is a continued optimism for future applications, particularly in integrated circuits and sensing applications. Recently, simulations suggested that NWFETs can achieve much higher on-current levels and boast better subthreshold swing than typical planar transistors, ultimately demonstrating enhanced electrical performance. These improved characteristics directly support a new generation of electronics that are more energy efficient and also support enabling smaller electronic components, which allow tighter assembly of wearable technologies and biomedical sensors. Further, the various characteristics of NWFETs provide a broad application potential in areas ranging from fast computing to biological sensing accuracy. As demonstrated, the unique configuration of NWFETs emphasizes their inherent flexibility and efficiency in numerous situations with relevance for the future of electronics. These characteristics position NWFETs at the forefront of nanotechnology developments that will likely change how we engage with electronic devices going forward.[9]

REFERENCES

1. Dahiya, Ravinder, Gupta, Shoubhik, Heidari, Hadi, Vilouras, et al. *Modeling of CMOS devices and circuits on flexible ultrathin chips*. Institute of Electrical and Electronics Engineers (IEEE), 2017
2. Njuguna, Raphael Njoroge Polarization imaging sensors in advanced feature CMOS technologies. 2013.
3. Amani, Matin, Chen, Yu-Ze, Chueh, Yu-Lun, Fahad, et al. *Room temperature multiplexed gas sensing using chemical-sensitive 3.5-nm-thin silicon transistors*. American Association for the Advancement of Science (AAAS), 2017.
4. Bunes, Benjamin R. *Doctor of Philosophy*. University of Utah, 2015.
5. Anikeeva, Polina, Bao, Zhenan, Benfenati, Fabio, Bezanilla, et al. *Roadmap on semiconductor-cell biointerfaces*. eScholarship, University of California, 2018.
6. Clavijo, William, *Nanowire zinc oxide MOSFET pressure sensor*. VCU Scholars Compass, 2014.
7. García, César Pascual, Leturcq, Renaud, Olthuis, Wouter, Rani, et al. *A high aspect ratio Fin-Ion Sensitive Field Effect Transistor: compromises towards better electrochemical bio-sensing*. American Chemical Society (ACS), 2019.
8. Lieber, Charles M., Tian, Bozhi Synthetic nanoelectronic probes for biological cells and tissues. *Annual Reviews*, vol. 6, pp. 31–51, 2013.
9. Pajouhi, Hossein, *Towards single-chip nano-systems*. Purdue University (bepress), 2016.
10. Likharev, K.-K. *CMOL: Second life for silicon?*, Elsevier, 2006.
11. DeMello, John C., Eisenhawer, B., Fagas, Gíorgos, Falk, et al. A miniaturised autonomous sensor based on nanowire materials platform: the SiNAPS mote. *SPIE-Intl Soc Optical Eng*, vol. 8763, pp. 87631Q-1–87631Q-14, 2013.
12. Rahman, Abd, Fatimah, Siti, Hamidon, Mohd Nizar, Sim, Guan Ling, Usman, et al. *Silicon nanowire interface circuit for DNA detection*. Universiti Putra Malaysia Press, 2017.
13. Usman, Kamilu Iman, Hamidon, Mohd Nizar, Rahman, Abd, Fatimah, Siti, Usman, et al. *Silicon nanowire interface circuit for DNA detection*. Universiti Putra Malaysia Press, 2017.
14. Manaf, Abd, Asrulnizam, Abdullah, Makarimi, Ahmad, Alias, Nurain Najihah, Azmi, Mohd, et al. *The effect of nanowire gap for silicon nanowire transistor to the current-voltage (I-Vds)*. Penerbit UTHM, 2023.
15. Bi, Yu Enhanced Hardware Security Using Charge-Based Emerging Device Technology. *Information Bulletin on Variable Stars (IBVS)*, vol. 15, pp. 1–8, 2016.
16. Kim, Saehyun, Li, Pengju, Tian, Bozhi Beyond 25 years of biomedical innovation in nano-bioelectronics. 2024.

17. Choe, Hwan Sung, *Modulated nanowire structures for exploring new nano-processor architectures and approaches to biosensing*. Harvard University Botany Libraries, 2015.
18. Liu, Wen, *Design, characterization and analysis of electrostatic discharge (esd) protection solutions in emerging and modern technologies*. Information Bulletin on Variable Stars (IBVS), 2012.
19. Lindblad, Niklas, *Simulations of III-V TFET double-balanced Gilbert cells with an improved noise model*. Lunds university/Institution för elektro- och informations teknik, 2013.
20. Fagas, Giorgos, Gallagher, John P., Luca, Gammaitoni, Paul et al. *Energy challenges for ICT*. Intech, 2017.
21. Filipovic, L., Grasser, T. *Miniaturized Transistors*. MDPI AG, 2019.
22. Ahnood, A., Amaratunga, G. A. J., Andrew, P., Bonaccorso, et al. *Flexible electronics: The next ubiquitous platform*. Proceedings of the IEEE, 2012.
23. Ayele, G. T. *Developing ultrasensitive and CMOS compatible ISFETs in the BEOL of industrial UTBB FDSOI transistors*. Universite de Sherbrooke, 2019.
24. Convertino, Annalisa, Polese, Davide, Maiolo, Luca, *The rise of flexible electronics in neuroscience, from materials selection to in vitro and in vivo applications*. 2019.
25. Li, Huiping, Li, Dujuan, Yue, Huiyi Chen Xiaojie, Fan, Kai, Dong, Linxi, Wang, Gaofeng Application of silicon nanowire field effect transistor (SiNW-FET) biosensor with high sensitivity. *Journal of MDPI Sensors* 2023, 6808. 10.3390/s23156808
26. Falina, Shaili, Syamsul, Mohd, Rhaffor, Nuha Abd, Hamid, Sofiyah Sal, Zain, Khairu Anuar Mohamed, Manaf, Asrulnizam Abd, Kawarada, Hiroshi Ten years progress of electrical detection of heavy metal ions (HMIs) using various field-effect transistor (FET) nanosensors: A review. *Biosensors* 11(12), 478, 2021. 10.3390/bios11120478
27. Tran, Duy Phu, Pham, Thuy Thi Thanh, Wolfrum, Bernhard, Offenhäusser, Andreas, Thierry, Benjamin CMOS-compatible silicon nanowire field-effect transistor biosensor: Technology development toward commercialization. *Journal of MDPI Materials*, 11050785, 2018. 10.3390/ma11050785
28. Jain, Aakash, Sharma, Sanjeev, Raj, B. Design and analysis of high sensitivity photosensor using cylindrical surrounding gate MOSFET for low power sensor applications. *Engineering Science and Technology, an International Journal*, 19(4), 1864–1870, 2016.
29. Saiphani Kumar, G., Singh, Amandeep, Raj, B. Design and analysis of gate all around CNTFET based SRAM cell design. *Journal of Computational Electronics*, 17(1), 138–145, 2018.
30. Singh, Gurinder Pal, Sohi, B. S., Raj, Balwinder Material properties analysis of graphene base transistor (GBT) for VLSI analog circuits. *Indian Journal of Pure & Applied Physics (IJPAP)*, 55, 896–902, 2017.
31. Kumar, Sunil, Raj, B. Estimation of stability and performance metric for inward access transistor based 6T SRAM cell design using n-type/p-type DMDG-GDOV TFET. *IEEE VLSI Circuits and Systems Letter*, 3(2), 25–39, 2017
32. Sharma, Shashikant, Kumar, Anjan, Pattanaik, Manisha, Raj, B. Forward body biased multimode multi-threshold CMOS technique for ground bounce noise

reduction in static CMOS adders. *International Journal of Information and Electronics Engineering*, 3(3), 567–572, 2013.
33. Singh, Hamendra, Kumar, Pankaj, Raj, Balwinder Performance analysis of majority gate SET based 1-bit full adder. *International Journal of Computer and Communication Engineering (IJCCE)*, 2(4), 1–8, 2013.
34. Bhardwaj, Anil Kumar, Gupta, Sumeet, Raj, B. Investigation of parameters for Schottky barrier (SB) height for Schottky barrier based carbon nanotube field effect transistor device. *Journal of Nanoelectronics and Optoelectronics*, 15, 783–791, 2020.
35. Bansal, Priya, Raj, B. MEMRISTOR: A versatile nonlinear model for dopant drift and boundary issues. *JCTN*, 14 (5), 2319–2325, 2017.
36. Jain, Neeraj, Raj, B. An analog and digital design perspective comprehensive approach on Fin-FET (Fin-field effect transistor) technology - A review. *Reviews in Advanced Sciences and Engineering (RASE)*, 5, 1–14, 2016.
37. Raj, Balwinder, Saxena, A. K., Dasgupta, S. Analytical modeling for the estimation of leakage current and subthreshold swing factor of nanoscale double gate FinFET device. *Microelectronics International*, 26, 53–63, 2009.
38. Soniya, Shailendra Singh, Wadhwa, Girish, Raj, Balwinder An analytical modeling for dual source vertical tunnel field effect transistor. *International Journal of Recent Technology and Engineering (IJRTE)*, 8(2), 603–608, 2019.
39. Singh, Shailendra, Raj, B. Design and analysis of hetrojunction Vertical T-shaped tunnel field effect transistor. *Journal of Electronics Material*, 48(10), 6253–6260, 2019.
40. Goyal, Candy, Ubhi, Jagpal Singh, Raj, B. A low leakage CNTFET based inexact full adder for low power image processing applications. *International Journal of Circuit Theory and Applications*, 47(9), 1446–1458, 2019.
41. Raj, B., Saxena, A. K., Dasgupta, S. A compact drain current and threshold voltage quantum mechanical analytical modeling for FinFETs. *Journal of Nanoelectronics and Optoelectronics (JNO)*, 3(2), 163–170, 2008.
42. Wadhwa, Girish, Raj, B. An analytical modeling of charge plasma based tunnel field effect transistor with impacts of gate underlap region. *Superlattices and Microstructures*, 142, 106512, 2020.
43. Singh, Shailendra, Raj, B. Modeling and simulation analysis of sige hetrojunction double gate vertical t-shaped tunnel FET. *Superlattices and Microstructures*, 142, 106496, 2020.
44. Jain, Neeraj, Raj, Balwinder Device and circuit co-design perspective comprehensive approach on FinFET technology - A review. *Journal of Electron Devices*, 23(1), 1890–1901, 2016.
45. Kumar, Sunil, Raj, B. Analysis of i_{on} and ambipolar current for dual-material gate-drain overlapped DG-TFET. *Journal of Nanoelectronics and Optoelectronics*, 11, 323–333, 2016.
46. Anjum, Naveed, Bali, Tarun, Raj, B. Design and simulation of handwritten multiscript character recognition. *International Journal of Advanced Research in Computer and Communication Engineering*, 2(7), 2544–2549, 2013.
47. Singh, Karmjit, Raj, B. Performance and analysis of temperature dependent multi-walled carbon nanotubes as global interconnects at different technology nodes. *Journal of Computational Electronics*, 14(2), 469–476, 2015.

48. Kumar, Sunil, Raj, B. Compact channel potential analytical modeling of DG-TFET based on evanescent–mode approach. *Journal of Computational Electronics*, 14(2), 820–827, 2015.
49. Singh, Karmjit, Raj, B. Temperature dependent modeling and performance evaluation of multi-walled CNT and single-walled CNT as global interconnects. *Journal of Electronic Materials*, 44(12), 4825–4835, 2015.
50. Sharma, V. K., Pattanaik, M., Raj, B. INDEP approach for leakage reduction in nanoscale CMOS circuits. *International Journal of Electronics*, 102(2), 200–215, 2014.
51. Singh, Karmjit, Raj, B. Influence of temperature on MWCNT bundle, SWCNT bundle and copper interconnects for nanoscaled technology nodes. *Journal of Materials Science: Materials in Electronics*, 26(8), 6134–6142, 2015.
52. Anjum, Naveed, Bali, Tarun, Raj, B. Design and simulation of handwritten Gurumukhi and Devanagri numerical recognition. *International Journal of Computer Applications*, 73(12), 16–21, 2013.
53. Khandelwal, S., Gupta, V., Raj, B., Gupta, R. D. Process variability aware low leakage reliable nano scale DG-FinFET SRAM cell design technique. *Journal of Nanoelectronics and Optoelectronics*, 10(6), 810–817, 2015.
54. Sharma, V. K., Pattanaik, M., Raj, B. ONOFIC approach: Low power high speed nanoscale VLSI circuits design. *International Journal of Electronics*, 101(1), 61–73, 2014.
55. Khandelwal, S., Raj, Balwinder, Gupta, R. D. FinFET based 6T SRAM cell design: Analysis of performance metric, process variation and temperature effect. *Journal of Computational and Theoretical Nanoscience*, 12, 2500–2506, 2015.
56. Singh, Sumit, Yadav, Shekhar, Rahul, Jagdeep, Srivastava, Anurag, Raj, B. Impact of HfO_2 in graded channel dual insulator double gate MOSFET, *Journal of Computational and Theoretical Nanoscience*, 12(6), 950–953, 2015.
57. Sharma, Vijay Kumar, Pattanaik, Manisha, Raj, B. PVT variations aware low leakage INDEP approach for nanoscale CMOS circuits. *Microelectronics Reliability*, 54, 90–99, 2014.
58. Raj, B., Saxena, A. K., Dasgupta, S. Quantum mechanical analytical modeling of nanoscale DG FinFET: Evaluation of potential, threshold voltage and source/drain resistance. *Elsevier's Journal of Material Science in Semiconductor Processing*, 16(4), 1131–1137, 2013.
59. Gopal, Maisagalla, Prasad, Siva Sankar D., Raj, Balwinder 8T SRAM cell design for dynamic and leakage power reduction, *International Journal of Computer Applications*, 71(9), 43–48, 2013.
60. Pattanaik, Manisha, Raj, B., Sharma, Shashikant, Kumar, Anjan Diode based trimode multi-threshold CMOS technique for ground bounce noise reduction in static CMOS adders. *Advanced Materials Research, Trans Tech Publications, Switzerland*, 548, 885–889, 2012.
61. Raj, Balwinder, Saxena, A. K., Dasgupta, S. Nanoscale FinFET based SRAM cell design: Analysis of performance metric, process variation, underlapped FinFET and temperature effect. *IEEE Circuits and System Magazine*, 11(2), 38–50, 2011.
62. Sharma, V. K., Pattanaik, M., Raj, Balwinder Leakage current ONOFIC approach for deep submicron VLSI circuit design. *International Journal of*

Electrical, Computer, Electronics and Communication Engineering, World Academy of Sciences, Engineering and Technology, 7(4), 239–244, 2013.
63. Kaur, Parminder, Gill, Sandeep Singh, Raj, B. Comparative analysis of OFETs materials and devices for sensor applications. *Journal of Silicon*, 14, 4463–4471, 2022.
64. Sharma, Sanjeev Kumar, Kumar, Parveen, Raj, Balwant, Raj, B. $In_{1-x}Ga_xAs$ double metal gate-stacking cylindrical nanowire MOSFET for highly sensitive photo detector. *Journal of Silicon*, 14, 3535–3541, 2022.
65. Raj, B., Saxena, A. K., Dasgupta, S. Analytical modeling of quasi planar nanoscale double gate FinFET with source/drain resistance and field dependent carrier mobility: A quantum mechanical study. *Journal of Computer (JCP)*, 4(9), 1–8, 2009.
66. Bhushan, S., Khandelwal, S., Raj, B. Analyzing different mode finfet based memory cell at different power supply for leakage reduction. *Seventh International Conference on Bio-Inspired Computing: Theories and Application, (BIC-TA 2012) Advances in Intelligent Systems and Computing*, Vol. 202, pp. 89–100, 2013.
67. Singh, Jeetendra, Raj, B. Temperature dependent analytical modeling and simulations of nanoscale memristor. *Journal: Engineering Science and Technology, an International Journal*, 21, 862–868, 2018.
68. Singh, Shradhya, Bala, Shashi, Raj, Balwant, Raj, B. Improved sensitivity of dielectric modulated junctionless transistor for nanoscale biosensor design. *Sensor Letter*, 18, 328–333, 2020.
69. Kumar, Vivek, Vishvakarma, Santosh Kumar, Raj, B. Design and performance analysis of ASIC for IoT applications. *Sensor Letter*, 18, 31–38, 2020.
70. Jaiswal, Akanksha, Sarin, R. K., Raj, B., Sukhija, Shikha A novel circular slotted microstrip-fed patch antenna with three triangle shape defected ground structure for multiband applications. *Advanced Electromagnetic (AEM)*, 7(3), 56–63, 2018.
71. Wadhwa, Girish, Raj, B. Label free detection of biomolecules using charge-plasma-based gate underlap dielectric modulated junctionless TFET. *Journal of Electronic Materials (JEMS)*, 47(8), 4683–4693, 2018.
72. Singh, Gurmohan, Sarin, R. K., Raj, B. Design and performance analysis of a new efficient coplanar quantum-dot cellular automata adder. *Indian Journal of Pure & Applied Physics (IJPAP)*, 55, 97–103, 2017.
73. Kaur, Parminder, Pandey, Vikas, Raj, B. Comparative study of efficient design, control and monitoring of solar power using IoT. *Sensor Letter*, 18, 419–426, 2020.
74. Bhardwaj, Anil Kumar, Gupta, Sumeet, Raj, B. Development & analysis of compact model for double gate Schottky barrier CNTFET. *Journal of Nanoelectronics and Optoelectronics*, 15, 1199–1208, 2020.
75. Wadhwa, Girish, Kamboj, Priyanka, Singh, Jeetendra, Raj, B. Design and investigation of junctionless DGTFET for biological molecule recognition. *Transactions on Electrical and Electronic Materials*, 22, 282–289, 2021.
76. Verma, Sachin Kumar, Singh, Shailendra, Wadhwa, Girish, Raj, B. Detection of biomolecules using charge-plasma based gate underlap dielectric modulated dopingless TFET. *Transactions on Electrical and Electronic Materials (TEEM)*, 21, 528–535, 2020.

77. Jain, Neeraj, Raj, B. Impact of underlap spacer region variation on Electrostatic and analog/RF performance of symmetrical high-k SOI FinFET at 20 nm channel length. *Journal of Semiconductors (JoS)*, 38(12), 122002, 2017.
78. Singh, Shailendra, Raj, B. Analytical modeling and simulation analysis of T-shaped III-V heterojunction vertical T-FET. *Superlattices and Microstructures*, 147, 106717, 2020.
79. Singh, Gurmohan, Sarin, R. K., Raj, B. Design and analysis of area efficient QCA based reversible logic gates. *Journal of Microprocessors and Microsystems*, 52, 59–68, 2017.
80. Singh, Shailendra, Raj, B. Parametric variation analysis on hetero-junction vertical t-shape TFET for supressing ambipolar conduction. *Indian Journal of Pure and Applied Physics*, 58, 478–485 2020.
81. Soniya, Shailendra Singh, Wadhwa, Girish, Raj, Balwinder Design and analysis of dual source vertical tunnel field effect transistor for high performance. *Transactions on Electrical and Electronics Materials*, 21, 74–82, 2019.
82. Kaur, Manjit, Gupta, Neena, Kumar, Sanjeev, Raj, B., Singh, Arun Kumar RF performance analysis of intercalated graphene nanoribbon based global level interconnects. *Journal of Computational Electronics*, 19, 1002–1013, 2020.
83. Wadhwa, Girish, Raj, B. Design and performance analysis of junctionless TFET biosensor for high sensitivity. *IEEE Nanotechnology*, 18, 567–574, 2019.
84. Singh, Jeetendra, Raj, B. Enhanced nonlinear memristor model encapsulating stochastic dopant drift. *JNO*, 14, 958–963, 2019.
85. Kumar, Mohit, Chaudhary, Tarun, Raj, Balwinder Design and comparative analysis of ferroelectric nanowire with dielectric HfO_2 and $Al_2O_3$for low power applications. *Journal of Electronic Materials*, 53, 6785–6791, 2024.
86. Thakur, Yogesh, Raj, Balwant and Raj, Balwinder Design of pentacene thin-film transistor based hydrogen gas sensor for high sensitivity. *ECS Journal of Solid State Science and Technology*, 13, 047005–047012, 2024.
87. Kumar, Manoharan Arun, Ismail, Kamal Batcha Mohamed, Mahalingam, Shanmugam, Raj, Balwinder, Kim, Junghwan, Atchudan, Raji Recent advances in two-dimensional nanomaterials and nanostructures for healthcare monitoring. *ACS Sensors*, 9(4), 1–8, 2024.
88. Batcha, Mohamed Ismail Kamal, Manoharan, Arun Kumar, Mahalingam, Shanmugam, Balwinder, Raj Carbon fiber-reinforced polymers for energy storage applications. *Journal of Energy Storage*, 84(Part B), 110931–110939, 2024.
89. Hussain, Shahraan, Raj, Balwinder, Sharma, Divya Design and parametric optimization of lead free perovskite based tandem solar cell for high efficient. *Engineering Research Express*, 6, 015315–0153122, 2024.
90. Kumar, Parveen, Raj, Balwinder, Wadhwa, Girish, Singh, Balwinder, Kumar, Raj Design and analysis of junctionless based gate all around N+ doped layer nanowire TFET biosensor. *ECS Journal of Solid State Science and Technology*, 13, 017002–017010, 2024.
91. Sharma, Divya, Mehra, Rajesh, Raj, Balwinder Enhancement in efficiency of methyl ammonium tin iodide based perovskite solar cell using SCAPS–1 D. *Nano*, 18(14), 2350109, 2023.
92. Kumar, Parveen, Sharma, Sanjeev Kumar, Raj, Balwant, Raj, Balwinder Analysis of device parameter variations in In1-xGaxAs based gate stacked double metal surrounding gate nanowire MOSFET. *TEEM*, 24, 570–578, 2023.

93. Bhardwaj, Anjana, Kumar, Pradeep, Raj, Balwinder, Kumar, Naveen, Anand, Sunny Design and optimization of vertical nanowire tunnel FET with electrostatic doping. *Engineering Research Express*, 5, 045025–045033, 2023.
94. Thakur, Yogesh, Raj, Balwinder, Gill, Sandeep Singh. Design and performance analysis of pentacene organic field effect transistor with high-K dielectric materials. *Journal of Optoelectronics and Advanced Materials – Rapid Communications*, 17, 7–8, 2023.
95. Kumar, Parveen, Raj, Balwinder Design and simulation of junctionless nanowire tunnel field effect transistor biosensor for high sensitivity. *Microelectronic Journal*, 137, 105826–105834, 2023.
96. Bhardwaj, Anjana, Pradeep Kumar, B. Raj, Sunny Anand, Design and performance optimization of doping-less vertical nanowire TFET using gate stack technique. *Journal of Electronic Materials (JEMS)*, 41(7), 4005–4013, 2022.
97. Singh, Jeetendra, Raj, B. Tunnel current model of asymmetric MIM structure levying various image forces to analyze the characteristics of filamentary memristor. *Applied Physics A*, 125(3), 203.1–203.11, 2019.
98. Goyal, Candy, Ubhi, Jagpal Singh, Raj, B., Low leakage zero ground noise nanoscale full adder using source biasing technique. *Journal of Nanoelectronics and Optoelectronics*, 14, 360–370, 2019.
99. Singh, Gurmohan, Sarin, R. K., Raj, B. A novel robust exclusive-OR function implementation in QCA nanotechnology with energy dissipation analysis. *Journal of Computational Electronics*, 15(2), 455–465, 2016.
100. Wadhwa, Girish, Kamboj, Priyanka, Raj, Balwinder Design optimisation of junctionless TFET biosensor for high sensitivity. *Advances in Natural Sciences: Nanoscience and Nanotechnology*, 10, 045001, 2019.
101. Bansal, Priya, Raj, B. Memristor modeling and analysis for linear dopant drift kinetics. *Journal of Nanoengineering and Nanomanufacturing*, 6, 1–7, 2016.
102. Ashima, D. V., Raj, B. Performance analysis of charge plasma induced graded channel Si nanotube. *Journal of Engineering Research (JER)*, 10, 146–154, 2021.
103. Tomar, Abhishek Singh, Magraiya, Vijay Kumar, Raj, B Scaling of access and data transistor for high performance dram cell design, *Quantum Matter*, 2, 412–416, 2013.
104. Jain, Neeraj, Raj, B. Parasitic capacitance and resistance model development and optimization of raised source/drain SOI FinFET structure for analog circuit applications, *Journal of Nanoelectronics and Optoelectronins*, 13, 531–539, 2018.
105. Singh, Shradhya, Vishvakarma, S. K., Raj, B. Analytical modeling of split-gate junction-less transistor for a biosensor application. *Sensing and Bio-sensing*, 18, 31–36, 2018.
106. Gopal, Maisagalla, Raj, Balwinder, Low power 8T SRAM cell design for high stability video applications. *ITSI Transaction on Electrical and Electronics Engineering*, 1(5), 91–97, 2013.
107. Raj, Balwinder, Mitra, Jatin, Bihani, Deepak Kumar, Rangharajan, V., Saxena, A. K., Dasgupta, S. Analysis of noise margin, power and process variation for 32 nm FinFET based 6T SRAM cell. *Journal of Computer (JCP)*, 5(6), 1–8, 2010.

108. Sharma, Divya, Mehra, Rajesh, Raj, B. Comparative analysis of photovoltaic technologies for high efficiency solar cell design. *Superlattices and Microstructures*, 153, 106861, 2021.
109. Kaur, Pawandeep, Buttar, Avtar Singh, Raj, B. A comprehensive analysis of nanoscale transistor based biosensor: A review. *Indian Journal of Pure and Applied Physics*, 59, 304–318, 2021.
110. Yadav, Divya, Raj, Balwant, Raj, B. Design and simulation of low power microcontroller for IoT applications. *Journal of Sensor Letters*, 18, 401–409, 2020.
111. Singh, Shailendra, Raj, B. A 2-D analytical surface potential and drain current modeling of double-gate vertical t-shaped tunnel FET. *Journal of Computational Electronics*, 19, 1154–1163, 2020.
112. Singh, Jeetendra, Raj, B. An accurate and generic window function for nonlinear memristor model. *Journal of Computational Electronics*, 18 (2), 640–647 2019.
113. Kaur, Manjit, Gupta, Neena, Kumar, Sanjeev, Raj, B, Singh, Arun Kumar Comparative RF and crosstalk analysis of carbon based nano interconnects. *IET Circuits, Devices & Systems*, 15(6), 493–503, 2021.
114. Kandasamy, Nehru, Ahmad, Firdous, Ajitha, D, Raj, B., Telagam, Nagarjuna Quantum dot cellular automata based scan flip flop and boundary scan register. *IETE Journal of Research*, 66, 535–548, 2020.
115. Singh, Jeetendra, Raj, B., Design and investigation of 7T2M NVSARM with enhanced stability and temperature impact on store/restore energy. *IEEE Transactions on Very Large Scale Integration Systems*, 27(6), 1322–1328, 2019.
116. Bhardwaj, Anil Kumar, Gupta, Sumeet, Raj, B., Singh, Amandeep Impact of double gate geometry on the performance of carbon nanotube field effect transistor structures for low power digital design. *Computational and Theoretical Nanoscience*, 16, 1813–1820, 2019.
117. Jain, Neeraj, Raj, B., Thermal stability analysis and performance exploration of asymmetrical dual-k underlap spacer (ADKUS) SOI FinFET for security and privacy applications. *Indian Journal of Pure & Applied Physics (IJPAP)*, 57, 352–360, 2019.
118. Jain, Neeraj, Raj, Balwinder Dual-k spacer region variation at the drain side of asymmetric SOI FinFET structure: Performance analysis towards the analog/RF design applications. *Journal of Nanoelectronics and Optoelectronics*, 14, 349–359, 2019.
119. Singh, Jeetendra, Sharma, Sanjeev, Raj, B., Khosla, Mamta Analysis of barrier layer thickness on performance of In1-xGaxAs based gate stack cylindrical gate nanowire MOSFET, *JNO*, 13, 1473–1477, 2018.
120. Jain, Neeraj, Raj, B. Analysis and performance exploration of high-k SOI FinFETs over the conventional low-k SOI FinFET toward analog/RF design. *Journal of Semiconductors (JoS)*, 39(12), 124002-1-7, 2018.
121. Goyal, Candy, Ubhi, Jagpal Singh, Raj, B. A reliable leakage reduction technique for approximate full adder with reduced ground bounce noise. *Journal of Mathematical Problems in Engineering*, 2018, 3501041, 2018.
122. Anuradha, Jeetendra Singh, Raj, B., Khosla, Mamta Design and performance analysis of nano-scale memristor-based nonvolatile SRAM. *Journal of Sensor Letter*, 16, 798–805, 2018.

123. Wadhwa, Girish, Raj, B. Parametric variation analysis of charge-plasma-based dielectric modulated JLTFET for biosensor application. *IEEE Sensor Journal*, 18(15), 6070–6077, 2018.
124. Singh, Jeetendra, Raj, B. Comparative analysis of memristor models for memories design. *JoS*, 39(7), 074006-1-12, 2018.
125. Yadav, Divya, Chouhan, Shailesh Singh, Vishvakarma, Santosh Kumar, Raj, B. Application specific microcontroller design for IoT based WSN. *Sensor Letter*, 16, 374–385, 2018.
126. Singh, Gurmohan, Sarin, R. K., Raj, B. Fault-tolerant design and analysis of quantum-dot cellular automata based circuits. *IEEE/IET Circuits, Devices & Systems*, 12, 638–664, 2018.
127. Singh, Jeetendra, Raj, B., Modeling of mean barrier height levying various image forces of metal insulator metal structure to enhance the performance of conductive filament based memristor model. *IEEE Nanotechnology*, 17(2), 267–268, 2018.
128. Jain, Aakash, Sharma, Sanjeev, Raj, B. Analysis of triple metal surrounding gate (TM-SG) III-V nanowire MOSFET for photosensing application. *Opto-Electronics Journal*, 26(2), 141–148, 2018.

Chapter 10

IoT-based energy meter for performance evaluation of nanowire-enhanced lithium-ion batteries

Ankit Jain, Anita Shukla, Shailendra Singh, Puspraj Singh Chauhan, and Imran Ullah Khan

10.1 INTRODUCTION

Energy is vital for a country's economic development, the progression of human civilization, and the growth of the global population. The demand for energy is constantly rising due to economic expansion and shifting consumer behavior; fossil fuels have historically been the primary source. However, this rising demand results in the excessive use of fossil fuels, which depletes these resources and poses a significant threat of an energy crisis. The continued reliance on fossil fuels exacerbates the urgency for alternative solutions. Burning fossil fuels has caused the atmospheric concentration of carbon dioxide to rise, which has resulted in extreme weather patterns [1–3]. Global industrialization has disrupted the climate, putting humanity at risk. To address the energy crisis, alternative energy sources have been developed that produce little to no greenhouse gas emissions, such as carbon dioxide (CO_2), as they don't rely on fossil fuels. This helps prevent further exacerbation of the greenhouse effect, a major driver of climate change, by ensuring that energy production from non-conventional sources does not contribute to environmental degradation. Solar energy, geothermal energy, hydroelectric energy, biomass energy, wind energy, and nuclear energy are some examples of alternative energy sources [4]. In many nations around the world, as populations and income levels increase, so does the demand for energy. After decreasing by 4.5% in 2020, the world's energy consumption returned with a 5% growth in 2021 [5, 6]. In the near future, the present energy requirement is expected to rise significantly. While alternative energy sources have been discovered, reliance on fossil fuels remains necessary to meet growing demands, and these resources may be depleted within the next 20 years. Rapid digitization has further accelerated energy demand, presenting significant challenges for sustainable energy solutions. One example of a global attempt to reduce energy consumption in buildings and cities is the EU's 2050 roadmap, which seeks to cut energy and petrol emissions by around 40% [7]. To lessen energy use and the detrimental effects on the

DOI: 10.1201/9781003634645-10

environment, an efficient and effective energy-saving technology must be created.

Lithium-ion batteries (LIBs) with enhanced performance have been developed as a result of rapid technological advancements and the rising need for efficient energy storage. The incorporation of nanowires, which improves the structural and electrical characteristics of LIBs, is one innovative approach that shows promise. Lithium-ion batteries are perfect for Internet of Things (IoT) applications because nanowires greatly increase their energy density, charge-discharge cycles, and overall efficiency. The basics of nanowires, how they enhance LIB performance, and how IoT-based energy meters can evaluate how well they enable intelligent, sustainable energy solutions for a range of applications are all covered in this introduction [8–17].

10.2 NANOWIRES: AN OVERVIEW

Nanowires are minuscule wires at the nanometer scale that exhibit distinct electrical and optical properties not found in the same material in bulk. They are in great demand for designing different kinds of nanosensors because of these qualities. Label-free, direct readouts are perfect for sensing applications thanks to the electronically adjustable characteristics of semiconductor nanowires. This is especially useful because electronic signals from these nanosensors can be easily integrated into small devices and connected to external systems. Nanowire sensors hold great promise for a variety of applications due to their remarkable sensitivity.

Conducting materials like Al, Pt, and Au; semiconductors like Si, Ge, and GaN; and insulating materials like SiO_2, HfO_2, and TiO_2 can all be found in nanowires (NWs) [8]. Both top-down and bottom-up approaches are commonly used to fabricate them; the latter is favored because it allows for precise control of crucial parameters such as length, radius, chemical composition, and growth direction [9–11]. As many studies have shown, nanowires are widely used in the design of different sensors [12]. They are also useful for nano-electromechanical systems (NEMS) due to their high Young's moduli. Nanowires are regarded as versatile nanomaterials because of their exceptional optical, electrical, chemical, and mechanical properties [13].

They are perfect for advanced sensors, random access memory (RAM), and next-generation transistors because their conductivity can be controlled through a field-effect mechanism [14]. Nanowires are one-dimensional nanostructures with nanoscale diameters and micrometer-scale lengths. They are perfect for energy storage systems, photonics, and electronics because of their remarkable mechanical flexibility, high surface area-to-Vol. ratio, and superior electrical conductivity.

Nanowires are mostly used as electrode materials in lithium-ion batteries. Conventional electrode materials frequently have issues like low cycling stability, limited capacity, and gradual degradation. These problems are addressed by nanowires by:

1. **Increasing Surface Area**: Nanowires' large surface area makes it easier for the electrode material and electrolyte to interact, which enhances the kinetics of ion diffusion and charge transfer.
2. **Enhancing Mechanical Stability**: Because nanowires are flexible, they can adapt to Vol.tric changes that take place during cycles of charge and discharge, which lessens electrode degradation.
3. **Increasing Electrical Conductivity**: Nanowires' one-dimensional structure offers effective electron transport channels, reducing resistance and improving battery performance in general [18–24].

10.3 LITHIUM-ION BATTERIES

Lithium-ion batteries (LIBs), commercialized by Sony since 1991, are the most widely used electrochemical energy storage devices [15, 16]. LIBs possess lots of desirable features, including low cost, long life span, high energy density, good reversibility, and pollution-free operation [17]. The LIBs operate following a "rock chair" concept, in which Li-ions are extracted from the anode, diffuse through the electrolyte, and intercalate into the cathode, while the discharging process is the opposite [18]. Consequently, the amount of exchanged ions or electrons and the stability of the material structure during intercalation and de-intercalation determine the LIBs' reversible capacity. Cell phones, laptops, and other electronic devices now frequently use rechargeable LIBs as power sources, and LIBs have begun to enter the market of transportation and large-scale energy storage for the grid [19]. The needs for LIBs, such as greater energy storage density and quicker charge/discharge, are raised by these many applications [25–33].

10.3.1 Nanowire-enhanced lithium-ion batteries

The incorporation of nanowires into lithium-ion batteries has led to significant improvements in key performance metrics, including:

1. **Energy Density**: Nanowires enable higher energy storage capacity by increasing the active surface area available for electrochemical reactions.
2. **Charging Speed**: Improved ion diffusion and electron transport result in faster charging times.
3. **Thermal Regulation**: The high thermal conductivity of nanowires helps in dissipating heat, reducing the risk of overheating and thermal runaway [34–41].

4. **Cycling Stability**: The mechanical resilience of nanowires ensures consistent performance over extended charge-discharge cycles.

These advancements make nanowire-enhanced lithium-ion batteries a promising choice for IoT applications, where reliability, efficiency, and longevity are critical.

10.4 UTILITY OF IOT-BASED ENERGY METERS

IoT-based energy meters are transforming energy consumption monitoring, management, and optimization, benefiting consumers and energy providers. These smart meters leverage Internet of Things (IoT) technology, allowing energy usage to be monitored and evaluated in real time, making energy management easier, better, and smarter [42–54].

1. **Real-time Energy Monitoring**
   The most evident feature of IoT-based energy meters is their capability to provide real-time data on the exercise of energy consumption. Traditional energy meters, at best, offer consumers monthly or bimonthly readings. IoT-based energy meters provide instantaneous feedback on how much electricity is consumed in any window of time. This feedback allows users to continuously observe their energy use and change their consumption to avoid any wastage.
   For consumers, it may provide a much-needed opportunity for deeper contemplation around their energy habits, which may ultimately allow for reduced consumption and reduced energy bills. For business, ongoing nudges and instantaneous feedback might provide better decision-making ability on energy use, in turn allowing them to realize cost savings and sustainable practices [55–64].
2. **Remote Monitoring and Control**
   IoT-based energy meters allow energy suppliers or consumers to access and control energy consumption remotely using an app or online portal. For consumers, this may mean that energy consumption can be viewed and managed remotely, which is particularly helpful for homes with smart home systems.
   For energy suppliers, remote monitoring of energy consumption allows them to detect issues (e.g., failures, overconsumption, or the power going out), capabilities that can lead to faster intervention and timely action. Furthermore, it limits the need for manual meter-reading and saves money and uncertainty due to human error [65–72].
3. **Data-Driven Insights**
   Energy meters based on the Internet of Things collect an immense amount of data about energy consumption patterns; by analyzing this data, users and suppliers receive information on trends regarding

energy usage. For example, users will be able to understand which appliances consume the most energy, and based on this understanding, utilize them in a way that will help them save money. For suppliers, analytics can support expected demand forecasting, improve overall management of their grids, and minimize wasted energy. In addition, IoT meters can provide assurance of service levels through reliable detection of inefficiencies within the system, so they can be rectified ahead of time. As a result, energy is distributed better, and grid congestion will be reduced [73–79].

4. **Improved Billing Accuracy**
   Some traditional energy billing processes can be inaccurate or delayed due to manual errors or the use of estimations. Internet of Things (IoT) energy meters help to eliminate this issue, by providing real-time readings which are accurate, and ensuring that consumers are only billed for the energy they actually consumed. This means fewer chances of a dispute arising between consumers and energy providers, and added transparency into energy usage [80–93].
   In addition, consumers can see their consumption data in detail so that energy usage is understood and that bills are never a surprise!
5. **Load Management and Demand Response**
   Energy meters that are IoT-enabled are an important tool for managing load because they capture consumption data as it occurs. This information allows energy providers to place energy where it is needed when demand exceeds supply; when the demand for energy is less than supply, IoT systems can be used to communicate with smart meters to initiate demand response. Demand response enables utilities to reduce non-essential appliances or shift non-essential activity to off-peak hours.
   Demand response can help utility grids to avoid overload and blackouts; it also allows the consumer to lower rates for off-peak usage as the grid becomes less expensive. Demand response is going to help consumers be more energy efficient while establishing a more reliable and stable energy grid.
6. **Enhanced Energy Efficiency**
   IoT-based energy meters improve efficiency, both for the consumer and energy suppliers, by supplying real-time consumption data. With this information, a consumer may optimize their consumption by making changes to when they use appliances or deciding to invest in energy-efficient devices.
   On an even grander scale, IoT-based energy meters promote energy conservation by providing real-time management of resources. Energy suppliers now have the ability to alter their supply using real-time consumption data, which helps avoid excessive resource consumption, reduces waste, and promotes sustainability in energy consumption [94–103].

7. **Integration with Smart Home Systems**
   IoT energy meters can work in conjunction with more devices, making them a focal point for any smart home of the future. Integration of IoT devices enables users not just to control their lighting, heating, and cooling, but also to have more productive control over energy consumption. Smart thermostats and IoT-enabled appliances can work in tandem with energy meters in ways that deliver energy efficiency, more comfort in the home, and lower costs.

   This integration allows users to treat the IoT energy meter experience more seamlessly while achieving higher levels of intervention in their patterns of energy use, allowing for a broader focus on energy-saving initiatives.
8. **Environmental Impact**
   Energy meters based on Internet of Things devices have the potential to help mitigate the environmental footprint of energy consumption. Merchandise can help promote visibility and controllability of energy use, leading to sustainable behavior (e.g., reductions in energy consumption or moving toward renewable sources of energy where applicable).

   The basic underlying data can help energy providers shift energy generation and distribution to a low-emission capacity and have a cleaner source of energy. Of course, in a future scenario, IoT-based energy meters may become widely adopted and achieve reductions in energy demand, lessening the demand for naturally occurring resources and lowering the production of greenhouse gas emissions [104–113].

10.5 LITERATURE REVIEW

One of the most significant problems that the majority of countries are presently dealing with is the overuse of electricity. The research community has made significant contributions in this area, including the development of low-power technologies that use significantly less energy than conventional systems. Also, they have contributed to developing an efficient methodology that keeps track of this energy consumption, which helps to identify the actual usage of energy by different systems [20, 21].

Industrial automation procedures and the technological landscape have experienced substantial change because of the information and communication technologies (ICT)'s quick development. As a result of the blending of these technical breakthroughs with energy systems, the Internet of Energy (IoE), a cutting-edge idea that enables an energy system to be smart, has advanced [22]. The Internet of Things (IoT) is considered to be a subclass of the Internet of Energy, which encompasses both electrical and thermal sources of energy. Using the internet and advanced communication and technology infrastructures, the IoT's primary objective is to connect distant objects and systems, including devices, people, processes, and data [23]. The use of IoT in power and energy systems points to a bright future for reducing operational and environmental problems with energy systems. However, because of the large investment costs, this has disadvantages [5].

The integration of IoT into smart energy systems (SES) has been the subject of reviews and research findings published in the literature. The system uses Advanced Reduced Instruction Set Computing Machines to handle all of its functions. Using data on humidity and light intensity to wirelessly regulate lighting and fans, Vignesh Mani et al. built an IoT-based smart energy management system in [3]. Instead of just turning the appliances on or off, these sensor data are used to intelligently regulate them. Furthermore, the system continuously determines how much energy the appliances use each day and shows the user that data. In [24], Mishra *et al.* designed an IoT-based smart energy meter using Raspberry Pi devices. They modified the old meter for monitoring it through a web page and a mobile application. Govindarajan *et al.* [25] designed a low-cost energy monitoring system with the aid of Arduino and used a Zigbee communication module to wirelessly monitor the system's health. In a smart city, devices are managed smartly, including the main power grid. With this motivation, in [26] authors have developed a widely used chip-based IoT electrical meter that will record all meter data and wirelessly transmit the necessary data to the main server. The system can also intimate consumers regarding pending dues and can disconnect itself if needed [114–121].

Real-time web-based electrical parameters, such as current, voltage, and power monitoring systems for households, are developed in [27]. For the construction of the sensing peripheral node used for the measurement system, the ESP8266 controller, the PZEM004T current sensor, and relays have been used. A home energy monitoring system for appliances has been developed [27]. To evaluate a household's energy usage and notify the user about power consumption, an Android app that informs about the number of units used and provides an estimate of the bill at the end of the month can be utilized [122–132].

An efficient power monitoring and switching system that reduces the current error of 0.6% from the pre-existing system, which provides a current error of 7.2% is proposed in [29]. Authors in [30, 31] suggested an IoT-based smart energy meter to track energy usage of residential loads, power billing, and smart grid power management automatically. The system is based on an inexpensive ESP32 microcontroller that gets data from sensor nodes via a voltage sensor and non-intrusive Current Transformer (CT) sensors (SCT013 and HLW8012). The Blynk server or a mobile app are the two ways the system transmits the user's consumption data.

10.6 IOT-BASED ENERGY METERS FOR PERFORMANCE EVALUATION

To harness the full potential of nanowire-enhanced lithium-ion batteries, it is essential to monitor and analyze their performance in real time. Energy meters based on the IoT make this possible. These energy meters have advanced sensors and analysis tools that allow for the complete assessment of performance metrics for batteries, such as:

1. **Voltage and Current**: Continuous monitoring of voltage and current helps inform characteristics of the battery's state of charge and discharge.
2. **Energy Density**: IoT-based meters can calculate how much energy density the battery has, as a measure of the ability to store energy.
3. **Charge-Discharge Cycles**: It is useful to track the count and efficiency of charge-discharge cycles to assess battery life and stability.

10.6.1 IoT connectivity and cloud integration

The use of IoT technology empowers energy meters to send real-time data to the cloud. Access through the cloud allows remote monitoring and analysis capabilities that further enable:

1. **Performance Optimization**: Cloud analytics tools process the data to retrieve patterns and optimize the use of the battery.
2. **Predictive Maintenance**: Early detection of potential performance issues allows for reducing downtime and extending the battery's life.
3. **Scalability**: The process can be extended to monitor multiple batteries simultaneously and be replicated across a large-scale IoT ecosystem.

10.6.2 Application based on cloud integration

The integration of nanowire-improved lithium-ion batteries with the IoT-based energy meters will hasten the usage of these technologies in areas of smart applications, such as:

1. **Wearables**: High-performance batteries power wearable devices with extended runtime and compact designs.
2. **Smart Homes**: Better energy storage will provide sustainable energy for smart homes, including sensors, security, and automation devices.
3. **Industrial Automation**: Reliable and durable batteries guarantee the continued operation of industrial IoT devices, leading to greater levels of productivity and safety. A significant breakthrough in energy storage is the incorporation of nanowires into lithium-ion batteries. With the implementation of IoT-based energy meters, the operation of these

enhanced batteries can be tracked and improved in real time, ensuring they are used effectively in smart, sustainable applications. This synergy between nanotechnology and IoT not only validates the potential of nanowire-enhanced batteries but also paves the way for their widespread adoption in the evolving landscape of IoT ecosystems.

10.7 FEATURES OF ENERGY METERS BASED ON LITHIUM-ION BATTERIES

The reliability and functionality of this system have been significantly improved by the addition of numerous features. The Blynk app is used to track data, and we used the ESP32 to create an Internet of Things-based power energy meter. The meter reading room still requires manual measurement recording using the present technology. Consequently, controlling and tracking your electricity use is very labor-intensive. This can be automated with the IoT. It automates remote data collection, saving time and money. Positive comments about the smart energy meter have been coming in lately from people all across the world. Below is a list of these attributes: Choosing the right current and voltage sensors is essential for measuring current and voltage and figuring out how much electricity is being used.

1. The best sensor available right now is the SCT-013. The split-core kind of SCT-013 non-invasive clamp meter sensor is capable of measuring AC current up to 100 A.
2. The best voltage sensor is the ZMPT101B AC voltage sensor module. The ZMPT101B AC voltage sensor is the best choice when utilizing a voltage transformer to precisely detect an AC voltage.
3. The SCT-013 current sensor and ZMPT101B voltage sensor allow us to measure every parameter required for an electricity energy meter. The ESP32 Wi-Fi module will be linked to the ZMPT101B voltage sensor and SCT-013 current sensor, and the Blynk application will receive the data.
4. Data is stored in the controller EEPROM, allowing measurements to continue even when power is not present. The Blynk application dashboard displays the voltage, current, power, and total unit spent in kWh.

10.8 ALGORITHM AND IMPLEMENTATION

In order to address the mentioned problem and incorporate other advantageous features, we propose in this study to develop a model using control engineering. The systematic process used to build the hardware for the system is in phases. The basic objective is to build a comprehensive system that can collect input from external sources (sensors) and modify it to create a specific output. Figure 10.1 displays the suggested AC energy meter's block diagram.

10.8.1 ESP32 pin out

The ESP32 development board shown in Figure 10.2 is based on the ESP WROOM32 WI-FI + BLE Module. The latest ESP-WROOM-32 module powers this small, straightforward system development board, which fits

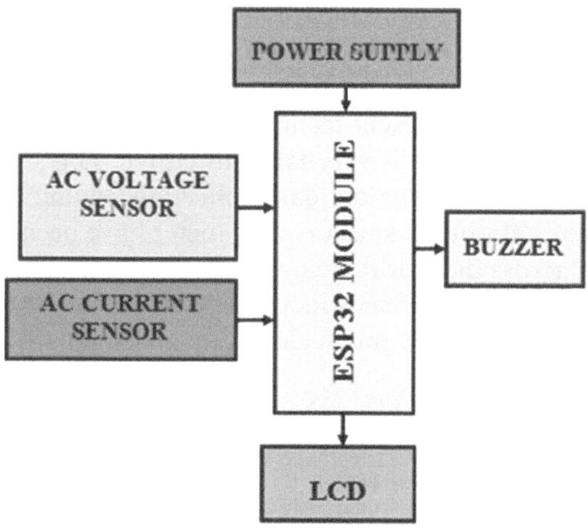

Figure 10.1 Block diagram of the proposed AC energy meter.

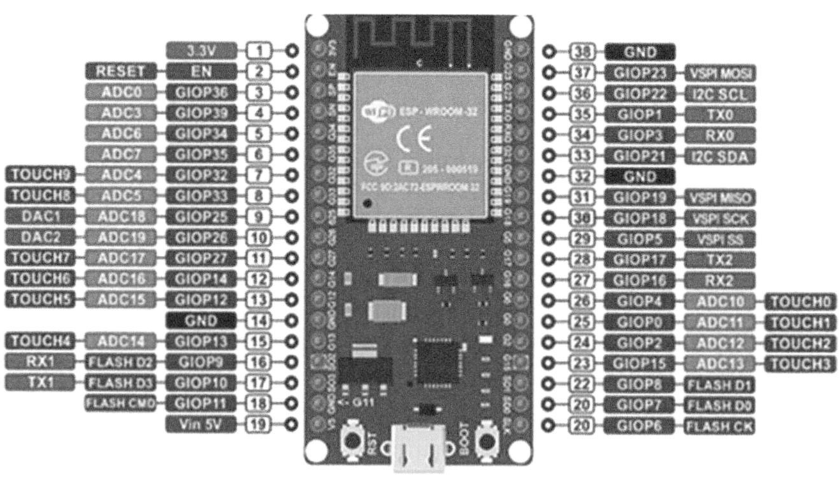

Figure 10.2 Pin mapping of the ESP32 development board.

nicely onto a solderless breadboard. It comes with all of the necessary support parts for the ESP-WROOM-32, including the USB-UART Bridge, reset and boot-mode buttons, an LDO regulator, and a micro-USB connector. Every important GPIO is accessible to the developer.

10.8.2 SCT-013 current sensor

A split-core-type clamp meter sensor, the non-invasive AC current sensor SCT-013 in Figure 10.3, is capable of measuring AC current up to 100 A. Current transformers (CTs) are sensors that monitor alternating current. They are very helpful in determining the total amount of electricity used by a building. High-voltage electrical work is not necessary when the SCT-013 current sensors are clipped directly to the live or neutral wire.

Like any other transformer, a current transformer has a main winding, a secondary winding, and a magnetic core. The secondary winding, consisting of multiple turns of little wire, is located inside the transformer's casing.

10.8.3 ZMPT101B AC single-phase voltage sensor

The high-precision ZMPT101B voltage transformer, which uses a voltage transformer to measure the precise AC voltage, is the foundation of the ZMPT101B AC single-phase voltage sensor module depicted in Figure 10.4. This is the most effective way to use an ESP32 or Arduino to measure the AC voltage. The modules can measure voltage up to 250 V AC and include an analog output that can be adjusted. The analog to digital convertor (ADC) output of the module can be adjusted and calibrated using its multi-turn trim potentiometer.

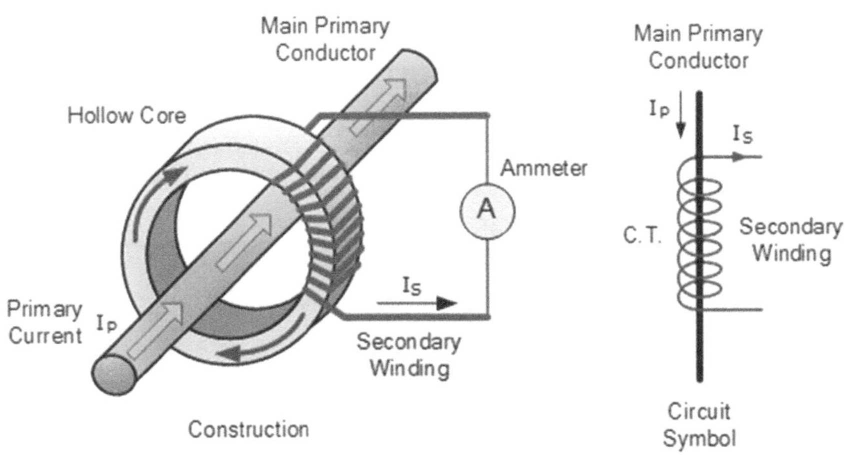

Figure 10.3 SCT-013 AC current sensor. (ZMPT101B).

Figure 10.4 AC single-phase voltage sensor. (ZMPT101B).

10.8.4 16×2 liquid crystal display (LCD)

An example of a multipurpose electronic display module is a liquid crystal display (LCD) screen, such as the one depicted in Figure 10.5. A very simple module that is frequently seen in many different circuits and devices is the 16×2 LCD display. Compared to these modules, multi-segment LEDs with seven segments or more are inferior. Low cost, ease of programming, and the ability to display custom and unique characters, animations, and a variety of other helpful features (unlike in seven segments) are some of the benefits of LCDs. Two lines on a 16×2 LCD can display 16 characters each. This LCD displays each character as a 5×7 pixel matrix. This LCD has data and command registers. The command register contains the command instructions that are transmitted to the LCD. A command is an instruction given to the LCD to carry out a specific task, such as initializing it, cleaning its screen, locating the pointer, controlling the display, etc. The data register contains the information that will be shown on the LCD. The data is the ASCII value of the character that will be shown on the LCD.

10.9 DEVELOPMENT OF CIRCUIT

As seen in the circuit design in Figure 10.6, the system is constructed using an ESP32 module, which was selected due to its small and lightweight size.

Figure 10.5 16×2 LCD.

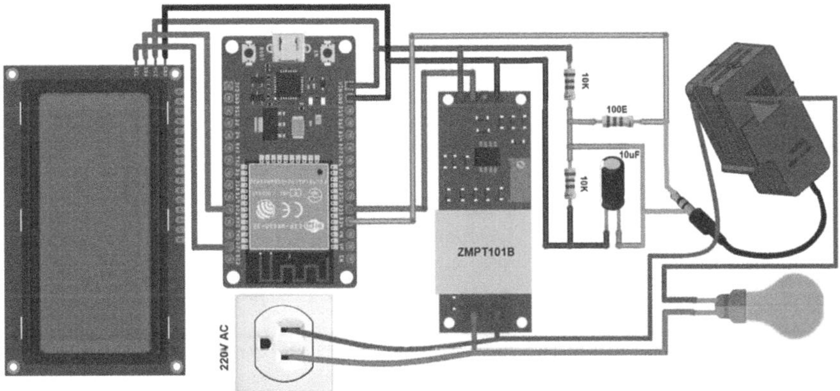

Figure 10.6 Circuit diagram of an AC energy meter.

The design was created using the Fritzing software, and the connection diagram is simple. The VCC and GND pins of the ZMPT101B voltage sensor and SCT-013 current sensor are connected to the Vin and GND pins of the ESP32, which is a 5 V supply. The SCT-013 current sensor's output analogue pin is connected to the ESP32's GPIO34, while the ZMPT101B voltage sensor's output analogue pin is attached to the ESP32's GPIO35.

A 10 uF capacitor, two 10 K resistors, one 100 ohm resistor, and two 10 K resistors are also required for the circuit. To measure current and voltage, the AC wires need to be connected to the voltage sensor's input AC terminal. To utilize the gadget as a current sensor, simply insert one live or neutral wire into the clip section. A 16×2 I2C LCD screen can also be used; however, it needs additional connections. Attach the VCC, GND, SDA, and SCL of the LCD display to the 5 V, GND, GPIO21, and GPIO22 of the ESP32, respectively. The LCD contrast can be adjusted using the 10 K potentiometer on the rear of the I2C module.

Voltage, current, power, and the total number of kWh utilized are all displayed on the Blynk application dashboard.

10.10 SETTING UP THE BLYNK APPLICATION

Simply visit https://blynk.io, click "Start Free," and then sign up using your Gmail address. To sign up for the Blynk IoT cloud server, enter your email address and password. Then, click the "new project" option to provide a project name. Select the Wi-Fi connection type after selecting the ESP32 board. Finally, click the "Create" button as seen in Figure 10.7. The same data can be shown on the Blynk dashboard and LCD as well.

Figure 10.7 Blynk server dashboard.

10.11 RESULTS AND DISCUSSION

In addition to detecting electricity use, the Internet of Things (IoT) can effectively transmit data on wireless energy consumption. The method is time-consuming and ineffective because current technology forces us to manually record values by going to the meter reading room. The tracking and measuring of electricity could be completely transformed by the creation of an Internet of Things (IoT)-based smart electricity energy meter that makes use of the ESP32 and Blynk platforms. This Internet of Things-based system saves time and money by doing away with manual meter readings. The SCT-013 current sensor and ZMPT101B voltage sensor can be used to provide precise measurements of voltage, current, power, and total energy consumption. The Blynk dashboard makes data available anywhere. Furthermore, energy meter data is stored in the ESP32's EEPROM, guaranteeing consistent readings even in the event of a power outage. This project automates and lowers the effort necessary to track electricity usage.

The use of nanowire-based lithium-ion batteries complements the IoT-based smart meter system by providing reliable and efficient power storage. Nanowires will tackle some of the challenges posed by battery technology, such as limited battery capacity and poor cycling stability, due to their high surface area-to-Vol. ratio, high electrical conductivity, and mechanical flexibility. This makes them well-suited for IoT devices, like energy meters, that can require some longevity and faster charging times. Higher energy density and better thermal management in these nanowire-based lithium-ion batteries will give continuous real-time data transmission and storage to enable uninterrupted energy monitoring and use.

10.12 CONCLUSION

To summarize, the application of nanowire technology to lithium-ion batteries has huge promise for energy storage, especially for IoT applications. The use of nanowires to increase the surface area for energy absorption, increase electrical conductivity, and allow flexibility allows for increased energy density and charge-discharge capability, along with longer efficacious battery life, making it ideal for longer-term high-efficiency applications. The advancement of solid nanowire-based electrolytes has an even greater impact on the charging speed and energy efficiency of the batteries, both of which are significant when using IoT devices that require sustained and reliable energy sources.

Nanowire-enhanced batteries pair well with IoT-based energy meters to allow continuous, monitored batteries to be built into smart systems. The continuous collection of battery health data while providing cloud-based operational data is a powerful combination that facilitates both some aspects of predictive maintenance and optimum functionality. The combination of cloud capability, sensors, and analytics enables energy monitoring and, therefore, management and conservation to meet energy efficiency and sustainability targets required to meet the global sustainability agenda.

The implementation of these technologies, through application in smart homes, automated industrial processes, and industry wearables, constitutes a significant step toward the creation of smarter and more sustainable energy. Moreover, with nanotechnology being leveraged in energy resources for energy storage, along with IoT-based monitoring systems, researchers at the University of Cambridge have the precedent for operationalizing more efficient, dependable, and environmentally friendly solutions. This research, ultimately, evaluates the need for innovative energy storage and monitoring to meet the increasing demand for sustainable and intelligent energy.

The use of a nanowire-based lithium-ion battery provides further rationale for the conclusions made above, as it served as an efficient and reliable form of electricity for the smart energy meter and other IoT devices implemented in the library. Nanowires are valuable because of their properties: high surface area-to-Vol. ratio, high conductivity, and mechanical flexibility (Lee et al, 2022). They exhibited higher energy density and thermal control, allowing performance to be efficient and long-lasting. This battery technology supported real-time monitoring and optimization of energy use, which guaranteed that energy-saving measures were effective and contributed to an overall reduction of energy consumption.

REFERENCES

[1] "Annual Energy Outlook, Energy Information Administration, Department of Energy." pp. 1–15, 2020.

[2] T Karthick, Raja S Charles, Drusila Nesamalar, J Jeslin, K Chandrasekaran, "Design of IoT Based Smart Compact Energy Meter for Monitoring and Controlling the Usage of Energy and Power Quality Issues With Demand Side Management for a Commercial Building." *Sustainable Energy, Grids and Networks*, Vol. 26, 2021, 100454.

[3] V Mani, Gunasekhar Abhilasha, Lavanya, S Sankaranarayanan, "IoT Based Smart Energy Management System." *International Journal of Applied Engineering Research*, Vol. 12, no. (6), pp. 5455–5462, 2017.

[4] https://education.nationalgeographic.org/resource/alternative-energy-use/. accessed on s.

[5] Efe Francis Orumwense, Khaled Abo-Al-Ez, "Internet of Things for Smart Energy Systems: A Review on Its Applications, Challenges and Future Trends." *AIMS Electronics and Electrical Engineering*, Vol. 7, no. (1), pp. 50–74, 2022.

[6] Karam M Al-Obaidi, Mohataz Hossain, *et al.*, "A Review of Using IoT for Energy Efficient Buildings and Cities: A Built Environment Perspective." *Energies*, Vol. 15, pp. 1–30, 2022.

[7] https://yearbook.enerdata.net/total-energy/world-consumption-statistics.html, accessed on May 2023.

[8] Y Wu, et al., "Nanowire-Based Nanodevices." *Journal of Nanoscience and Nanotechnology*, Vol. 8, no. 8, pp. 4033–4042, 2008.

[9] W Lu, et al., "Semiconductor Nanowires for Energy Conversion and Storage." *Nature Materials*, Vol. 7, no. 10, pp. 870–879, 2008.

[10] J I Sohn, et al., "High-Performance Single-Nanowire-Based Devices for Sensing and Electronics." *Nature Nanotechnology*, Vol. 4, no. 7, pp. 430–436, 2009.

[11] Z Yu, M Meyyappan, "Carbon Nanotube and Nanowire-Based Field-Effect Transistors." *IEEE Transactions on Nanotechnology*, Vol. 5, no. 3, pp. 284–295, 2006.

[12] T Najam, "Application of Nanowires in Chemical Sensors and Biosensors." *Sensors and Actuators B: Chemical*, Vol. 132, no. 2, pp. 309–317, 2008.

[13] M Muskins, "Optical and Mechanical Properties of Nanowires for Device Applications." *Journal of Applied Physics*, Vol. 103, no. 10, p. 104306, 2008.

[14] A I Hochbaum, et al., "Enhanced Thermoelectric Performance of Nanostructured Materials." *Nature*, Vol. 451, no. 7175, pp. 164–167, 2008.

[15] J B Goodenough, "Electrochemical Energy Storage in a Sustainable Modern Society." *Energy & Environmental Science*, Vol. 7, pp. 14–18, 2014. doi:10.1039/c3ee42613k

[16] J B Goodenough, K-S Park, "The Li-Ion Rechargeable Battery: A Perspective." *Journal of the American Chemical Society*, Vol. 135, pp. 1167–1176, 2013. doi:10.1021/ja3091438

[17] B Dunn, H Kamath, J-M Tarascon, "Electrical Energy Storage for the Grid: A Battery of Choices." *Science*, Vol. 334, pp. 928–935, 2011. doi:10.1126/science.1212741

[18] J-M Tarascon, M Armand, "Issues and Challenges Facing Rechargeable Lithium Batteries." *Nature*, Vol. 414, pp. 359–367, 2001. doi:10.1038/35104644

[19] S Chu, A Majumdar, "Opportunities and Challenges for a Sustainable Energy Future." *Nature*, Vol. 488, pp. 294–303, 2012. doi:10.1038/nature11475
[20] R R Mohassel, A Fung, F Mohammadi, K Raahemifar, "A survey of Advanced Metering Infrastructure." *International Journal of Electrical Power & Energy Systems*, Vol. 63, pp. 473–484, 2014.
[21] S M Amin, B F Wollenberg, "Toward a Smart Grid: Power Delivery for the 21st Century." *IEEE Power and Energy Magazine*, Vol.3, no. (5), pp. 34–41, 2005.
[22] Y Saleem, N Crespi, M H Rehmani, *et al.*, "Internet of Things-Aided Smart Grid: Technologies, Architectures, Applications, Prototypes, Future Research Directions." *IEEE Access*, Vol. 7, pp. 62962–63003, 2017.
[23] S A Chelloug, M A El-Zawawy, "Middleware for Internet of Things: Survey and Challenges." *Intelligent Automation and Soft Computing*, Vol. 3, pp. 70–95, 2017.
[24] J K Mishra, S Goyal, V A Tikkiwal, A Kumar, "An IoT based smart energy management system." *2018 4th International Conference on Computing Communication and Automation (ICCCA)*, Greater Noida, India, 2018, pp. 1–3.
[25] R Govindarajan, S Meikandasivam, D Vijayakumar, "Energy Monitoring System Using Zigbee and Arduino." *International Journal of Engineering & Technology*, Vol. 7, pp. 608–611, 2018.
[26] M M Haque, Z H Choudhury, F M Alamgir, "IoT Based Smart Energy Metering System for Power Consumers." *2019 2nd International Conference on Innovation in Engineering and Technology (ICIET)*, Dhaka, Bangladesh, 2019, pp. 1–6.
[27] M Y Muliadi I S Fahrezi E Palantel Areni, A Achmad, "A Smart Home Energy Consumption Monitoring System Integrated With Internet Connection." *2020 IEEE International Conference on Communication, Networks and Satellite (Comnetsat)*, Batam, Indonesia, 2020, pp. 75–80.
[28] R S Hariharan, Reema Agarwal, Madhurya Kandamuru, H Abdul Gaffa, "Energy Consumption Monitoring in Smart Home System." *IOP Conference Series: Materials Science and Engineering*, Vol. 1085, 2021.
[29] M K Hasan, M M Ahmed, B Pandey, H Gohel, S Islam, I F Khalid, "Internet of Things-Based Smart Electricity Monitoring and Control System Using Usage Data." *Wireless Communications and Mobile Computing*, 2021.
[30] P S Macheso, D Thotho, "ESP32 Based Electric Energy Consumption Meter." *International Journal of Computer Communication and Informatics*, Vol. 4, pp. 23–35, 2022.
[31] M Pimplea, S Thopateb, A Nikamc, S Gadekard, "IOT Based Smart Energy Meter Using ESP 32." *3rd International Conference on Communication and Information Processing (ICCIP-2021)*, 2021.
[32] Anjana Bhardwaj, B Pradeep Kumar Sunny Anand Raj, "Design and Performance Optimization of Doping-less Vertical Nanowire TFET Using Gate Stack Technique." *Journal of Electronic Materials (JEMS)*, Vol. 41, no. 7, pp. 4005–4013, 2022.
[33] Jeetendra Singh, B Raj, "Tunnel Current Model of Asymmetric MIM Structure Levying Various Image Forces to Analyze the Characteristics of Filamentary Memristor." *Applied Physics A*, Vol. 125, No. 3, pp. 203.1–203.11, February 2019.

[34] Candy Goyal, Jagpal Singh Ubhi, B Raj, "Low Leakage Zero Ground Noise Nanoscale Full Adder Using Source Biasing Technique." *Journal of Nanoelectronics and Optoelectronics*, Vol. 14, pp. 360–370, March 2019.

[35] Gurmohan Singh, R K Sarin, B Raj, "A Novel Robust Exclusive-OR Function Implementation in QCA Nanotechnology With Energy Dissipation Analysis." *Journal of Computational Electronics*, Vol. 15, No. 2, pp. 455–465, June 2016.

[36] Girish Wadhwa, Priyanka Kamboj, Balwinder Raj "Design Optimisation of Junctionless TFET Biosensor for High Sensitivity." *Advances in Natural Sciences: Nanoscience and Nanotechnology*, Vol. 10, p. 045001, 2019.

[37] Priya Bansal, B Raj, "Memristor Modeling and Analysis for Linear Dopant Drift Kinetics." *Journal of Nanoengineering and Nanomanufacturing*, Vol. 6, pp. 1–7, 2016.

[38] D Vaithiyanathan Ashima, B Raj, "Performance Analysis of Charge Plasma Induced Graded Channel Si Nanotube." *Journal of Engineering Research (JER)*, pp. 146–154, August 2021.

[39] Abhishek Singh Tomar, Vijay Kumar Magraiya, B Raj, "Scaling of Access and Data Transistor for High Performance DRAM Cell Design." *Quantum Matter*, Vol. 2, pp. 412–416, Oct. 2013.

[40] Neeraj Jain, B Raj, "Parasitic Capacitance and Resistance Model Development and Optimization of Raised Source/Drain SOI FinFET Structure for Analog Circuit Applications." *Journal of Nanoelectronics and Optoelectronins*, Vol. 13, pp. 531–539, April 2018.

[41] Shradhya Singh, S K Vishvakarma, B Raj, "Analytical Modeling of Split-Gate Junction-Less Transistor for a Biosensor Application." *Sensing and Bio-Sensing*, Vol. 18, pp. 31–36, April 2018.

[42] Maisagalla Gopal, Balwinder Raj, "Low Power 8T SRAM Cell Design for High Stability Video Applications." *ITSI Transaction on Electrical and Electronics Engineering*, Vol. 1, no. 5, pp. 91–97, 2013.

[43] Balwinder Raj, Jatin Mitra, Deepak Kumar Bihani, V Rangharajan, A K Saxena, S Dasgupta, "Analysis of Noise Margin, Power and Process Variation for 32 nm FinFET Based 6T SRAM Cell." *Journal of Computer (JCP)*, Vol. 5, No. 6, pp. 1–8, 2010.

[44] Divya Sharma, Rajesh Mehra, B Raj, "Comparative Analysis of Photovoltaic Technologies for High Efficiency Solar Cell Design." *Superlattices and Microstructures*, Vol. 153, pp. 106861, May 2021.

[45] Pawandeep Kaur, Avtar Singh Buttar, B Raj, "A comprehensive Analysis of Nanoscale Transistor Based Biosensor: A Review." *Indian Journal of Pure and Applied Physics*, Vol. 59, pp. 304–318, April 2021.

[46] Divya Yadav, Balwant Raj, B Raj, "Design and Simulation of Low Power Microcontroller for IoT Applications." *Journal of Sensor Letters*, Vol. 18, pp. 401–409, May 2020.

[47] Shailendra Singh, B Raj, "A 2-D Analytical Surface Potential and Drain Current Modeling of Double-Gate Vertical t-Shaped Tunnel FET." *Journal of Computational Electronics*, Vol. 19, pp. 1154–1163, April 2020.

[48] Jeetendra Singh, B Raj, "An Accurate and Generic Window Function for Non-linear Memristor Model" *Journal of Computational Electronics*, Vol. 18, No. 2, pp. 640–647 June 2019.

[49] Manjit Kaur, Neena Gupta, Sanjeev Kumar, B Raj, Arun Kumar Singh, "Comparative RF and Crosstalk Analysis of Carbon Based Nano Interconnects." *IET Circuits, Devices and Systems*, Vol.15, No. 6, pp. 493–503, February 2021.
[50] Nehru Kandasamy, Firdous Ahmad, D Ajitha, B Raj, Nagarjuna Telagam "Quantum Dot Cellular Automata Based Scan Flip Flop and Boundary Scan Register." *IETE Journal of Research*, Vol. 66, pp. 535–548, 2020.
[51] Jeetendra Singh, B Raj, "Design and Investigation of 7T2M NVSARM with Enhanced Stability and Temperature Impact on Store/Restore Energy." *IEEE Transactions on Very Large Scale Integration Systems*, Vol. 27, No. 6, pp. 1322–1328, June 2019.
[52] Anil Kumar Bhardwaj, Sumeet Gupta, B Raj, Amandeep Singh, "Impact of Double Gate Geometry on the Performance of Carbon Nanotube Field Effect Transistor Structures for Low Power Digital Design." *Computational and Theoretical Nanoscience*, Vol. 16, pp. 1813–1820, 2019.
[53] Neeraj Jain, B Raj, "Thermal Stability Analysis and Performance Exploration of Asymmetrical Dual-k underlap Spacer (ADKUS) SOI FinFET for Security and Privacy Applications." *Indian Journal of Pure & Applied Physics (IJPAP)*, Vol. 57, pp. 352–360, May 2019.
[54] Neeraj Jain, Balwinder Raj, "Dual-k Spacer Region Variation at the Drain Side of Asymmetric SOI FinFET Structure: Performance Analysis Towards the Analog/RF Design Applications." *Journal of Nanoelectronics and Optoelectronics*, Vol. 14, pp. 349–359, March 2019.
[55] Jeetendra Singh, Sanjeev Sharma, B Raj, Mamta Khosla, "Analysis of Barrier Layer Thickness on Performance of In1-xGaxAs Based Gate Stack Cylindrical Gate Nanowire MOSFET", *JNO*, Vol. 13, pp 1473–1477, October 2018.
[56] Neeraj Jain, B Raj, "Analysis and Performance Exploration of High-k SOI FinFETs Over the Conventional Low-k SOI FinFET toward Analog/RF Design." *Journal of Semiconductors (JoS)*, Vol. 39, No. 12, p. 124002-1-7, December 2018.
[57] Candy Goyal, Jagpal Singh Ubhi, B Raj, "A Reliable Leakage Reduction Technique for Approximate Full Adder With Reduced Ground Bounce Noise." *Journal of Mathematical Problems in Engineering*, Vol. 2018, p. 3501041, October 7, 2018.
[58] Jeetendra Singh Anuradha, B Raj, Mamta Khosla, "Design and Performance Analysis of Nano-scale Memristor-based Nonvolatile SRAM." *Journal of Sensor Letter*, Vol. 16, pp. 798–805, October 2018.
[59] Girish Wadhwa, B Raj, "Parametric Variation Analysis of Charge-Plasma-based Dielectric Modulated JLTFET for Biosensor Application." *IEEE Sensors Journal*, Vol. 18, No. 15, pp. 6070–6077, August 1, 2018.
[60] Jeetendra Singh, B Raj, "Comparative Analysis of Memristor Models for Memories Design." *JoS*, Vol. 39, No. 7, p. 074006-1-12, July 2018.
[61] Divya Yadav, Shailesh Singh Chouhan, Santosh Kumar Vishvakarma, B Raj, "Application Specific Microcontroller Design for IoT Based WSN." *Sensor Letters*, Vol. 16, pp. 374–385, May 2018.
[62] Gurmohan Singh, R K Sarin, B Raj, "Fault-Tolerant Design and Analysis of Quantum-Dot Cellular Automata Based Circuits." *IEEE/IET Circuits, Devices & Systems*, Vol. 12, pp. 638–664, 2018.

[63] Jeetendra Singh, B Raj, "Modeling of Mean Barrier Height Levying Various Image Forces of Metal Insulator Metal Structure to Enhance the Performance of Conductive Filament Based Memristor Model." *IEEE Nanotechnology*, Vol. 17, No. 2, pp. 267–268, March 2018.
[64] Aakash Jain, Sanjeev Sharma, B Raj, "Analysis of Triple Metal Surrounding Gate (TM-SG) III-V Nanowire MOSFET for Photosensing Application." *Opto-electronics Journal*, Vol. 26, No. 2, pp. 141–148, May 2018.
[65] Aakash Jain, Sanjeev Sharma, B Raj, "Design and Analysis of High Sensitivity Photosensor Using Cylindrical Surrounding Gate MOSFET For Low Power Sensor Applications." *Engineering Science and Technology, an International Journal*, Vol. 19, No. 4, pp. 1864–1870, December 2016.
[66] G Saiphani Kumar, Amandeep Singh, B Raj, "Design and Analysis of Gate All Around CNTFET based SRAM Cell Design." *Journal of Computational Electronics*, Vol. 17, No. 1, pp 138–145, March 2018.
[67] Gurinder Pal Singh, B S Sohi, Balwinder Raj, "Material Properties Analysis of Graphene Base Transistor (GBT) for VLSI Analog Circuits." *Indian Journal of Pure & Applied Physics (IJPAP)*, Vol. 55, pp. 896–902, December 2017.
[68] Sunil Kumar, B Raj, "Estimation of Stability and Performance Metric for Inward Access Transistor based 6T SRAM Cell Design Using n-Type/p-Type DMDG-GDOV TFET." *IEEE VLSI Circuits and Systems Letter*, Vol. 3, No. 2, pp. 25–39, June 2017
[69] Shashikant Sharma, Anjan Kumar, Manisha Pattanaik, B Raj, "Forward Body Biased Multimode Multi-Threshold CMOS Technique for Ground Bounce Noise Reduction in Static CMOS Adders." *International Journal of Information and Electronics Engineering*, Vol. 3, No. 3, pp. 567–572, 2013.
[70] Hamendra Singh, Pankaj Kumar, Balwinder Raj, "Performance Analysis of Majority Gate SET Based 1-bit Full Adder." *International Journal of Computer and Communication Engineering (IJCCE)*, Vol. 2, No. 4, 2013.
[71] Anil Kumar Bhardwaj, Sumeet Gupta, B Raj, "Investigation of Parameters for Schottky Barrier (SB) Height for Schottky Barrier Based Carbon Nanotube Field Effect Transistor Device." *Journal of Nanoelectronics and Optoelectronics*, Vol. 15, pp. 783–791, July 2020.
[72] Priya Bansal, B Raj, "Memristor: A Versatile Nonlinear Model for Dopant Drift and Boundary Issues." *JCTN*, Vol. 14, No. 5, pp. 2319–2325, May 2017.
[73] Neeraj Jain, B Raj, "An Analog and Digital Design Perspective Comprehensive Approach on Fin-FET (Fin-Field Effect Transistor) Technology – A Review." *Reviews in Advanced Sciences and Engineering (RASE)*, Vol. 5, pp. 1–14, 2016.
[74] Balwinder Raj, A K Saxena, S Dasgupta, "Analytical Modelling for the Estimation of Leakage Current and Subthreshold Swing Factor of Nanoscale Double Gate FinFET Device." *MicroElectronics International*, Vol. 26, pp. 53–63, 2009.
[75] Shailendra Singh Soniya, Girish Wadhwa, Balwinder Raj, "An Analytical Modeling for Dual Source Vertical Tunnel Field Effect Transistor." *International Journal of Recent Technology and Engineering (IJRTE)*, Vol. 8, No. 2, pp. 603–608, July 2019.
[76] Shailendra Singh, B Raj, "Design and Analysis of Hetrojunction Vertical T-shaped Tunnel Field Effect Transistor." *Journal of Electronics Material*, Vol. 48, No. 10, pp 6253–6260, October 2019.

[77] Candy Goyal, Jagpal Singh Ubhi, B Raj, "A Low Leakage CNTFET Based Inexact Full Adder for Low Power Image Processing Applications." *International Journal of Circuit Theory and Applications*, Vol. 47, No. 9, pp. 1446–1458, September 2019.
[78] B Raj, A K Saxena, S Dasgupta, "A Compact Drain Current and Threshold Voltage Quantum Mechanical Analytical Modeling for FinFETs." *Journal of Nanoelectronics and Optoelectronics (JNO)*, Vol. 3, No. 2, pp. 163–170, 2008.
[79] Girish Wadhwa, B Raj, "An Analytical Modeling of Charge Plasma Based Tunnel Field Effect Transistor With Impacts of Gate Underlap Region." *Superlattices and Microstructures*, Vol. 142, pp. 106512, June 2020.
[80] Shailendra Singh, B Raj, "Modeling and Simulation Analysis of SiGe hetrojunction Double GateVertical t-Shaped Tunnel FET." *Superlattices and Microstructures*, Vol. 142, pp. 106496, June 2020.
[81] Neeraj Jain, Balwinder Raj, "Device and Circuit Co-Design Perspective Comprehensive Approach on FinFET Technology – A Review." *Journal of Electron Devices*, Vol. 23, No. 1, pp 1890–1901, 2016.
[82] Sunil Kumar, B Raj, "Analysis of I_{ON} and Ambipolar Current for Dual-Material Gate-Drain Overlapped DG-TFET." *Journal of Nanoelectronics and Optoelectronics*, Vol. 11, pp. 323–333, June 2016.
[83] Naveed Anjum, Tarun Bali, B Raj, "Design and Simulation of Handwritten Multiscript Character Recognition." *International Journal of Advanced Research in Computer and Communication Engineering*, Vol. 2, No. 7, pp. 2544–2549, July 2013.
[84] Karmjit Singh, B Raj, "Performance and Analysis of Temperature Dependent Multi-walled Carbon Nanotubes as Global Interconnects at Different Technology Nodes." *Journal of Computational Electronics*, Vol. 14 No. 2, pp. 469–476, June 2015.
[85] Sunil Kumar, B Raj, "Compact Channel Potential Analytical Modeling of DG-TFET Based on Evanescent-mode Approach." *Journal of Computational Electronics*, Vol. 14, No. 2, pp.820–827, July 2015.
[86] Karmjit Singh, B Raj, "Temperature Dependent Modeling and Performance Evaluation of Multi-Walled CNT and Single-Walled CNT as Global Interconnects." *Journal of Electronic Materials*, Vol. 44, No. 12, pp 4825–4835, Dec 2015.
[87] V K Sharma, M Pattanaik, B Raj, "INDEP Approach for Leakage Reduction in Nanoscale CMOS Circuits." *International Journal of Electronics*, Vol. 102, No.2, pages 200–215, 2014.
[88] Karmjit Singh, B Raj, "Influence of Temperature on MWCNT Bundle, SWCNT Bundle and Copper Interconnects for Nanoscaled Technology Nodes." *Journal of Materials Science: Materials in Electronics*, Vol. 26, No. 8, pp. 6134–6142, 2015.
[89] Naveed Anjum, Tarun Bali, B Raj, "Design and Simulation of Handwritten Gurumukhi and Devanagri Numerical Recognition." *International Journal of Computer Applications*, Vol. 73, No. 12, pp. 16–21, 2013.
[90] Balwinder Raj, A K Saxena, S Dasgupta, "Nanoscale FinFET Based SRAM Cell Design: Analysis of Performance Metric, Process Variation, Underlapped FinFET and Temperature effect." *IEEE Circuits and Systems Magazine*, Vol. 11, No. 2, pp. 38–50, 2011.

[91] V K Sharma, M Pattanaik, Balwinder Raj, "Leakage Current ONOFIC Approach for Deep Submicron VLSI Circuit Design." *International Journal of Electrical, Computer, Electronics and Communication Engineering*, Vol. 7, No. 4, pp. 239–244, 2013.

[92] Parminder Kaur, Sandeep Singh Gill, B Raj, "Comparative Analysis of OFETs Materials and Devices for Sensor Applications." *Journal of Silicon*, Vol 14, pp. 4463–4471, 2022.

[93] Sanjeev Kumar Sharma, Parveen Kumar, Balwant Raj, B Raj, "$In_{1-x}Ga_xAs$ Double Metal Gate-Stacking Cylindrical Nanowire MOSFET for Highly Sensitive Photo Detector." *Journal of Silicon*, Vol.14, pp. 3535–3541, 2022.

[94] B Raj, A K Saxena, S Dasgupta, "Analytical Modeling of Quasi Planar Nanoscale Double Gate FinFET With Source/Drain Resistance and Field Dependent Carrier Mobility: A Quantum Mechanical Study." *Journal of Computer (JCP)*, Vol. 4, No. 9, pp. 1–8, 2009.

[95] S Bhushan, S Khandelwal, B Raj, "Analyzing Different Mode FinFET Based Memory Cell at Different Power Supply for Leakage Reduction." *Seventh International Conference on Bio-Inspired Computing: Theories and Application, (BIC-TA 2012) Advances in Intelligent Systems and Computing*, Vol. 202, pp. 89–100 (2013).

[96] Jeetendra Singh, B Raj, "Temperature Dependent Analytical Modeling and Simulations of Nanoscale Memristor." *Journal: Engineering Science and Technology, an International Journal*, Vol. 21, pp. 862–868, October 2018.

[97] Shradhya Singh, Shashi Bala, Balwant Raj, B Raj, "Improved Sensitivity of Dielectric Modulated Junctionless Transistor for Nanoscale Biosensor Design." *Sensor Letters*, Vol. 18, pp. 328–333, April 2020.

[98] Vivek Kumar, Santosh Kumar Vishvakarma, B Raj, "Design and Performance Analysis of ASIC for IoT Applications." *Sensor Letters*, Vol. 18, pp. 31–38, January 2020.

[99] Akanksha Jaiswal, R K Sarin, B Raj, Shikha Sukhija, "A Novel Circular Slotted Microstrip-fed Patch Antenna with Three Triangle Shape Defected Ground Structure for Multiband Applications." *Advanced Electromagnetic*, Vol. 7, No. 3, pp. 56–63, August 2018.

[100] Girish Wadhwa, B Raj, "Label Free Detection of Biomolecules Using Charge-Plasma-Based Gate Underlap Dielectric Modulated Junctionless TFET." *Journal of Electronic Materials (JEMS)*, Vol. 47, No. 8, pp. 4683–4693, August 2018.

[101] Gurmohan Singh, R K Sarin, B Raj, "Design and Performance Analysis of a New Efficient Coplanar Quantum-Dot Cellular Automata Adder." *Indian Journal of Pure & Applied Physics (IJPAP)*, Vol. 55, pp. 97–103, February 2017.

[102] Parminder Kaur, Vikas Pandey, B Raj, "Comparative Study of Efficient Design, Control and Monitoring of Solar Power Using IoT." *Sensor Letters*, Vol. 18, pp. 419–426, May 2020.

[103] Anil Kumar Bhardwaj, Sumeet Gupta, B Raj, "Development & Analysis of Compact Model for Double Gate Schottky Barrier CNTFET." *Journal of Nanoelectronics and Optoelectronics*, Vol. 15, pp. 1199–1208, August 2020.

[104] Girish Wadhwa, Priyanka Kamboj, Jeetendra Singh, B Raj, "Design and Investigation of Junctionless DGTFET for Biological Molecule Recognition."

Transactions on Electrical and Electronic Materials, Vol. 22, pp. 282–289, 2021.

[105] Sachin Kumar Verma, Shailendra Singh, Girish Wadhwa, B Raj, "Detection of Biomolecules Using Charge-Plasma Based Gate Underlap Dielectric Modulated Dopingless TFET." *Transactions on Electrical and Electronic Materials (TEEM)*, Vol. 21, pp. 528–535, June 2020.

[106] Neeraj Jain, B Raj, "Impact of Underlap Spacer Region Variation on Electrostatic and Analog/RF Performance of Symmetrical High-k SOI FinFET at 20 nm Channel Length." *Journal of Semiconductors (JoS)*, Vol. 38, No. 12, pp. 122002, December 2017.

[107] Shailendra Singh, B Raj, "Analytical Modeling and Simulation Analysis of T-shaped III-V Heterojunction Vertical T-FET." *Superlattices and Microstructures*, Vol. 147, pp. 106717, November 2020.

[108] Gurmohan Singh, R K Sarin, B Raj, "Design and Analysis of Area Efficient QCA Based Reversible Logic Gates." *Journal of Microprocessors and Microsystems*, Vol. 52, pp. 59–68, May 2017.

[109] Shailendra Singh, B Raj, "Parametric Variation Analysis on Hetero-Junction Vertical t-Shape TFET for Supressing Ambipolar Conduction." *Indian Journal of Pure and Applied Physics*, Vol. 58, pp. 478–485 June 2020.

[110] Shailendra Singh Soniya, Girish Wadhwa, Balwinder Raj, "Design and Analysis of Dual Source Vertical Tunnel Field Effect Transistor for High Performance." *Transactions on Electrical and Electronic Materials*, Vol. 21, pp. 74–82, October 2019.

[111] Manjit Kaur, Neena Gupta, Sanjeev Kumar, B Raj and Arun Kumar Singh, "RF Performance Analysis of Intercalated Graphene Nanoribbon Based Global Level Interconnects." *Journal of Computational Electronics*, Vol. 19, pp. 1002–1013, June 2020

[112] Girish Wadhwa, B Raj, "Design and Performance Analysis of Junctionless TFET Biosensor for High Sensitivity." *IEEE Nanotechnology*, Vol. 18, pp. 567–574, 2019.

[113] Jeetendra Singh, B Raj, "Enhanced Nonlinear Memristor Model Encapsulating Stochastic Dopant Drift." *JNO*, Vol. 14, pp. 958–963, 2019.

[114] Mohit Kumar, Tarun Chaudhary, Balwinder Raj, "Design and Comparative Analysis of Ferroelectric Nanowire With Dielectric HfO_2 and Al_2O_3 for Low Power Applications." *Journal of Electronic Materials*, Vol. 53, pp. 6785–6791, 2024.

[115] Yogesh Thakur, Balwant Raj, Balwinder Raj, "Design of Pentacene Thin-Film Transistor Based Hydrogen Gas Sensor for High Sensitivity." *ECS Journal of Solid State Science and Technology*, Vol.13, pp. 047005–047012, April 24, 2024.

[116] Manoharan Arun Kumar, Kamal Batcha Mohamed Ismail, Shanmugam Mahalingam, Balwinder Raj, Junghwan Kim, Raji Atchudan, "Recent Advances in Two-Dimensional Nanomaterials and Nanostructures for Healthcare Monitoring." *ACS Sensors*, Vol. 9, No. 4, pp. 1–8, April 2, 2024.

[117] Mohamed Ismail Kamal Batcha, Arun Kumar Manoharan, Shanmugam Mahalingam, Raj Balwinder, "Carbon Fiber-Reinforced Polymers for Energy Storage Applications." *Journal of Energy Storage*, Vol. 84, Part B, pp. 110931–110939, April 20, 2024.

[118] Shahraan Hussain, Balwinder Raj, Divya Sharma, "Design and Parametric Optimization of Lead Free Perovskite Based Tandem Solar Cell for High Efficient", *Engineering Research Express*, Vol. 6, pp. 015315–0153122, March 13, 2024.
[119] Parveen Kumar, Balwinder Raj, Girish Wadhwa, Balwinder Singh, Raj Kumar "Design and Analysis of Junctionless Based Gate All Around N+ Doped Layer Nanowire TFET Biosensor." *ECS Journal of Solid State Science and Technology*, Vol. 13, pp. 017002–017010, 2024.
[120] Divya Sharma, Rajesh Mehra, Balwinder Raj, "Enhancement in Efficiency of Methyl Ammonium Tin Iodide Based Perovskite Solar Cell Using SCAPS-1 D." *Nano*, Vol. 18, No. 14, pp. 2350109, December 14, 2023.
[121] Parveen Kumar, Sanjeev Kumar Sharma, Balwant Raj, Balwinder Raj, "Analysis of Device Parameter Variations in In1-xGaxAs based Gate Stacked Double Metal Surrounding Gate Nanowire MOSFET." *TEEM*, Vol. 24, pp. 570–578, October 2023.
[122] Anjana Bhardwaj, Pradeep Kumar, Balwinder Raj, Naveen Kumar, Sunny Anand, "Design and Optimization of Vertical Nanowire Tunnel FET With Electrostatic Doping." *Engineering Research Express*, Vol. 5, pp. 045025–045033, 2023.
[123] Yogesh Thakur, Balwinder Raj, Sandeep Singh Gill, "Design and Performance Analysis of Pentacene Organic Field Effect Transistor With High-K dielectric Materials." *Journal of Optoelectronics and Advanced Materials – Rapid Communications*, Vol. 17, No. 7–8, July-August 2023.
[124] Parveen Kumar, Balwinder Raj, "Design and Simulation of Junctionless Nanowire Tunnel Field Effect Transistor Biosensor for High Sensitivity." *Microelectronics Journal*, Vol. 137, pp. 105826–105834, July 2023.
[125] S Khandelwal, V Gupta, B Raj, R D Gupta, "Process Variability Aware Low Leakage Reliable Nano Scale DG-FinFET SRAM Cell Design Technique." *Journal of Nanoelectronics and Optoelectronics*, Vol. 10, No. 6, pp. 810–817, December 2015.
[126] V K Sharma, M Pattanaik, B Raj, "ONOFIC Approach: Low Power High Speed Nanoscale VLSI Circuits Design." *International Journal of Electronics*, Vol. 101, No. 1, pp. 61–73, 2014.
[127] S Khandelwal, Balwinder Raj, R D Gupta, "FinFET Based 6T SRAM Cell Design: Analysis of Performance Metric, Process Variation and Temperature Effect". *Journal of Computational and Theoretical Nanoscience*, Vol. 12, pp. 2500–2506, 2015.
[128] Sumit Singh, Shekhar Yadav, Jagdeep Rahul, Anurag Srivastava, B Raj, "Impact of HfO_2 in Graded Channel Dual Insulator Double Gate MOSFET." *Journal of Computational and Theoretical Nanoscience*, Vol. 12, No. 6, pp. 950–953, April 2015.
[129] Vijay Kumar Sharma, Manisha Pattanaik, B Raj, "PVT Variations Aware Low Leakage INDEP Approach for Nanoscale CMOS Circuits." *Microelectronics Reliability*, Vol. 54, pp. 90–99, 2014.
[130] B Raj, A K Saxena, S Dasgupta, "Quantum Mechanical Analytical Modeling of Nanoscale DG FinFET: Evaluation of Potential, Threshold Voltage and Source/Drain Resistance." *Elsevier's Journal of Material Science in Semiconductor Processing*, Vol. 16, No. 4, pp. 1131–1137, 2013.

[131] Maisagalla Gopal, Siva Sankar D Prasad, Balwinder Raj, "8T SRAM Cell Design for Dynamic and Leakage Power Reduction." *International Journal of Computer Applications*, Vol. 71, No. 9, pp. 43–48, June 2013.

[132] Manisha Pattanaik, B Raj, Shashikant Sharma, Anjan Kumar, "Diode Based Trimode Multi-Threshold CMOS Technique for Ground Bounce Noise Reduction in Static CMOS Adders." *Advanced Materials Research*, Vol. 548, pp 885–889, 2012.

Chapter 11

A comprehensive review on nanowires

Synthesis, utility, and technical hurdles

Ankit Jain, Anita Shukla, Vivek Kumar, Puspraj Singh Chauhan, Shailendra Singh, and Imran Ullah Khan

11.1 INTRODUCTION

Nanowires are cylindrical nanostructures, and their diameter ranges from 10 to 100 nm. Their lengths can be hundreds to thousands of times longer (1-10 μm) [1]. This high length-to-diameter ratio is responsible for their unique properties, including electrical conductivity and energy structure [2]. Nanowires come in various types depending on their constituents. Conducting nanowires comprise metals such as nickel, platinum, gold, copper, and silver, while non-conducting nanowires are made from insulators like SiO_2 and polymers. Moreover, fabrication of molecular nanowires can be done from organic or inorganic deoxyribonucleic acids. A variety of nanowires have been created, for example, insulating nanowires (like SiO_2 and TiO_2), metallic nanowires (like Ni, Pt, and Au), superconducting nanowires (like YBCO), and semiconducting nanowires [like silicon nanowires, SiNWs, Indium Phosphide (InP), and Gallium Nitride (GaN)]. Elements from groups III-V are the most researched for nanowire systems. These include nanowires from single semiconductors (Si and Ge), composite semiconductors (InP, CdS, and GaAs), nitrides (GaN and Si_3N_4), carbides (SiC), functional oxides [zinc oxide (ZnO) and SnO_2], ceramics (SiO_2 and Al_2O_3), and composites (Si-Ge, SiC-SiO_2) [1, 2]. Core-shell structure nanowires are created from two distinct materials. For example, silicon oxide nanowires comprise an amorphous oxide shell that prevents damage and contamination of the silicon core. Such hybrid nanostructures offer improved functionalities and novel properties [2]. The nanowire form, the diameter-dependent band gap, the state's carrier density, and other factors all affect their physical characteristics. Nanowires are considered one-dimensional (1-D) materials since they exhibit aspect ratios (length-to-width ratios) of 1000 or greater. Neither the bulk materials nor their larger, three-dimensional structures exhibit the conductive qualities that nanowires do. In contrast to the typical continuum of bands or energy levels observed in bulk materials, electrons in nanowires occupy energy levels that are different due to their quantum confinement laterally. Field Effect Transistors (FETs), logic gates, junction diodes, and transistors are examples of basic

 DOI: 10.1201/9781003634645-11

electronic devices made by using semiconductor and superlattice nanowires. Nanowires can be employed for both bioelectrical and mechanical and biomedical sensing applications [3–9].

Metallic nanowires can be used to create thermoelectric cooling systems. There are several opto-electronic uses for semiconductor nanowire junctions, and structured arrays of magnetic nanowires hold great promise for recording media applications. In biotechnological applications, molecular nanowires can be made from both organic (peptides, RNA, DNA, sugars, etc.) and inorganic (Mo6S9–xIx) repeating molecular units. Their small size makes it easier for electrons to flow through them, making them good conductors. This characteristic has made significant strides in computer science possible [10–16]. For instance, computer speed could be significantly increased by creating an optical photonic switch with specific cadmium sulfide nanowires that let photons flow through them and function as binary signals (0 and 1). In genetics, scientists have synthesized synthetic protein-coding DNA using nanowires. The building blocks of proteins and DNA are amino acids, which are used to create these nanowires. Nanowires are helpful in many different industries, such as electronics, optoelectronics, energy storage, sensing, and medical devices, due to their special qualities.

However, controlled synthesis, assembly, and electronics integration problems need to be addressed [3–5]. Because of their high aspect ratio, nanowires possess unique mechanical, optical, and electrical properties [6]. Metals, semiconductors, oxides, and polymers can all be used to create nanowires by techniques including electrospinning and vapor deposition [7]. Nanowires are appropriate for use as circuit elements and interconnects in nanoscale electronic devices due to their minuscule diameter [8]. They have a variety of applications, including data storage, photonics, chemical and biological sensing, drug delivery, and energy storage [9–11]. Innovations in a variety of industries, including biomedicine and flexible electronics, could arise from developments in the production, characterization, and integration of nanowires [11]. Because of their small size and adaptability, nanowires can be used in a variety of potentially revolutionary technological applications [17–25].

11.2 TYPES OF NANOWIRES

There are presently many varieties of nanostructures, such as carbon nanotubes, nanoparticles, and nanocomposites. One type of nanostructure that has been the object of considerable attention and study is nanowires. They expand in one direction (longitudinally) and are limited along the two directions that are transverse to the direction of expansion/not limited by other factors. With two outputs, electrical conductivity is an important part of those designs. Within these structures, electrical charge transfer occurs when an electrical potential drop is applied at each end of the structure, as well as along its length [26–37]. The formation of nanometer-sized wires is interesting from both a technological perspective and with regard to scientific interest because they need certain special characteristics. The ratio of length to diameter for nanowires is extreme (L >> D). When creating biological markers, chemical and biological sensors, magnetic devices, or internal connections in nanoelectronics, such as joining two pieces of aluminum superconductor with silver nanowires, nanowires provide useful properties. Bulk and three-dimensional materials seem inadequate compared to the properties of nanowires. Types of nanowires can include fabric that is made from a range of various types of nanowires that include metal nanowires (nickel, platinum, gold, or a range of various metal-based alloys); semiconductors (GaN, InP, Si, etc.); non-conductors (TiO_2 and SiO_2); polymers; and nanowires. Molecular nanowires (such as organic or inorganic DNA) can be categorized using the following categories [38–44].

11.3 TYPES OF SEMICONDUCTORS AND METALS

It contains various materials such as single-element semiconductors (silicon and germanium), compound semiconductors (InP, CdS, and GaAs, as well as heterogeneous), nitride (GaN and Si_3N_4), carbide (SiC), and noble metals (gold, silver, and copper) [12, 13].

11.4 FUNCTIONAL TYPES

Nanowires are structurally controlled for growth directions and lateral surfaces in several materials, including functional oxides like ZnO and SnO_2, which yield programmable electrical and optical properties with many applications in nanotechnology. [14, 15]

11.4.1 Elemental (metal) nanowires

These nanostructures may provide better performance in electronic components because of their size. Future advancements in electronics and the ongoing miniaturization of electronic components will be critical to the development of power and electronic technologies.

However, as researchers continue to seek to miniaturize their conventional technology, they will be limited by quantum physics and the limits of production methods, and thus, the cost of production will continue to escalate. In recent years, there has been considerable interest in nanoscopic materials driven by significant interest in alternative technologies. The most important types of nanowires include silicon, germanium, boron, tin, lead, indium, bismuth, selenium, tellurium, and gold structural ceramics (MgO, SiO_2, and Al_2O_3), and composite materials (Si-Ge, SiC-SiO_2), which integrate polymers. Figure 11.1 [16, 17] illustrates the schematics of some types

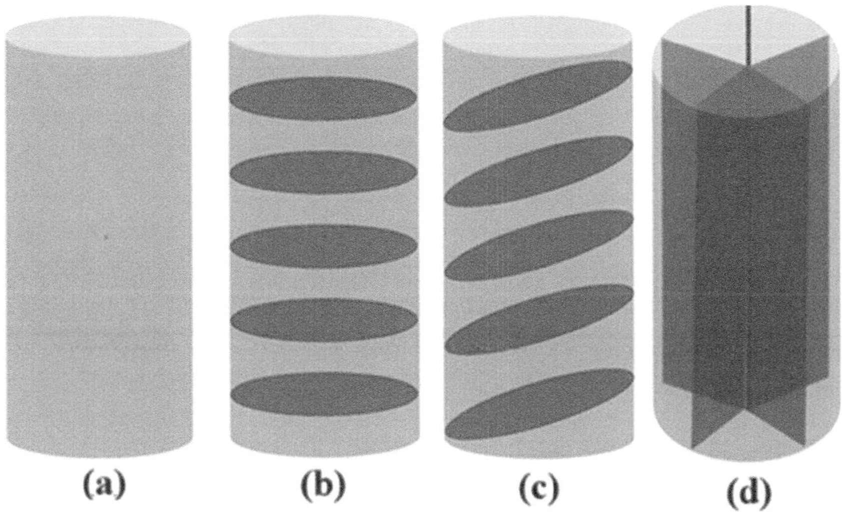

Figure 11.1 Schematics of metal nanowires with various ordered TB configurations, highlighted in brown. (a) One TB-free crystal nanowire. (b) Horizontal TB nanowire. (c) An inclined TB nanowire. (d) Five-fold vertical TBs on a nanowire.

of metal nanowires with different types of ordered twin boundary (TB) arrangements, shown in brown [45–52].

11.4.2 Nanowires made of polymers

The progress that nanotechnology provides in the fields of photonics has unique scientific and technological potential. A number of advances have been made to discover new materials, develop technologies that are easy to manufacture, reduce the size of photonic devices, increase device integration density, and fabricate low-cost nanoscale devices. In this context, we will discuss the development of polymer nanowires. Nanometer-scale photonic wires are highly desired for applications such as high-density and compact photon integrated circuits, whereby silica fibers, or bulk glasses, are drawn and heated to form wires no thicker than a wavelength. The most widely used polymer for nanophotonic applications is polyethylene terephthalate (PTT) nanowires [18, 19]. Figure 11.2 shows microscope images of PTT nanowires [53–60].

11.4.3 Nanowires made of two distinct materials (shell and core)

The x value in SiOx can be anywhere from 1 to 2 based on the atomic ratios of silicon and oxygen present in silicon oxide nanowires. The chemical reactivity of silicon oxide nanowires is reduced by the amorphous shell

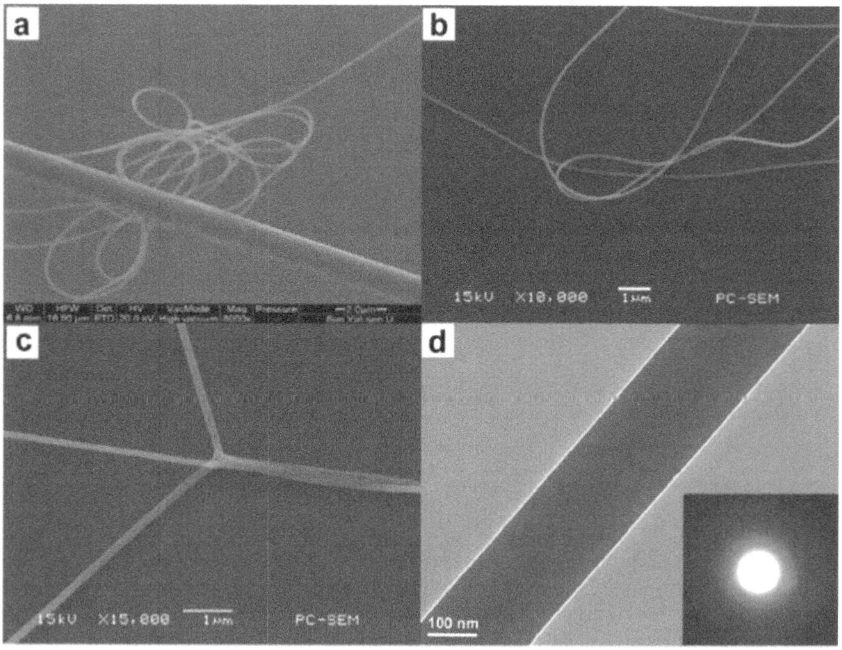

Figure 11.2 Scanning electron microscope images (a), (b), and (c), as well as electron microscopy images (d) that pass through a 190 nm-diameter nanowire.

of this particular nanowire, which may help alleviate the effects of radiation or mechanical damage that can lead to oxidation and contamination of silicon nanowires. SiOx nanowires can be used like Si/SiOx core-shell nanowires [61–69]. The fact that SiOx nanowires emit blue-red light means they could be integrated into photo-luminescence materials, such as integrated optical nano-instruments, high-resolution optical tip near-field scanning light microscopes, and effective and powerful blue light transmitters. Due to the improved functionality and unique properties of alloy or hybrid nanostructures derived from two or more chemically distinct materials, they are classed as desirable.

11.4.4 Gold nanowires

Gold nanowires are thermally unstable; hence, when heated, they may break up into nanoparticles. In order to stabilize gold nanowires, a coating of organic molecules, like oleylamine (OA), is applied. Pure gold nanowires lacking protective coatings may be important for sensing and catalytic applications due to their exceptional conductivity and potential as a catalyst. A summary of gold nanowire features and applications is provided in Table 11.1 [20]. Gold nanowires provide excellent waveguide applications

Table 11.1 List of properties of gold nanowires and their potential applications

Characteristics of gold nanowires	*Possible use*
Vast surface area (Glas, 2006)	(Glas, 2006) Catalytic nanoelectrode sensor
Catalytic nanoelectrode sensor	Electron transport using an organic OA layer
Electron transport using an organic OA layer	High resistivity, temperature-dependent nonlinear behavior, decreased catalytic activity, and 106 Ω
Transport of electrons without an OA layer	Strong catalytic ability, electrochemical, low resistivity, 103 Ω, strong electron transport, and a nanoelectrode sensor
Mechanical characteristics	Strong engineering, interconnects
Chiral gold clusters	Asymmetric sensors, catalysts, and drugs
Asymmetric sensors, catalysts, and drugs	Self-assembling active substrates for SERS
Self-building	

because they exhibit minimal scattering losses, hold coherence well, and can achieve deep sub-wavelength confinement. Also, chemical and biological sensors would benefit a lot from the surface-enhanced optical diffraction and fluorescence offered by gold nanowires.

11.5 NANOWIRE CHARACTERIZATION

The geometrical and structural properties of nanowires have a pronounced influence on their resultant electrical and optical performance. A variety of characterization techniques have been developed to study these properties in the micrometer and nanometer realm. However, due to the nanowire dimensions being similar to or smaller than the wavelength of visible light, traditional optical microscopy has limitations in tracing the details of surface and morphological features [70–83]. Therefore, it remains common in the literature to use electron microscopy techniques, principally scanning and transmission electron microscopy [SEM and Transmission Electron Microscopy (TEM)]. SEM primarily provides surface morphology and general overall structure, while TEM provides more detailed information about internal structure in terms of length, diameter, and crystallinity. Energy-Dispersive X-ray Spectroscopy (EDX) can be used in tandem with TEM for chemical compositional analysis. An example of full characterization includes simply identifying the growth direction, cross-sectional geometry, and morphology of the surface, as well as identifying defects such as dislocations and twinning. It is also crucial to know the growth direction and the geometry in cross-section obtained, especially in the case of catalyst-assisted synthesis, as it provides information about the dynamics of energy on the surface and the formation of the crystal. Such information is critical to understanding how the structure of nanowires relates to their functional properties, such as electrical conductivity or optical response. This becomes important when we wish to design nanowire-based devices and optimize their performance – specifically for applications in the areas of electronics, photonics, and sensing [84–90].

11.6 PROPERTIES OF NANOWIRES

11.6.1 Quantum mechanical effects and conductivity

Ohm's law describes the conductance of a regular piece of copper wire as inversely proportional to length and directly proportional to cross-section. However, for nanowires, conductance does not scale with cross-section because the width of the nanowire is comparable to the wavelength of an electron. Conductance is limited by the quantum of conductance, equal to $2e^2/h$.

A few examples of quantum transport phenomena that can be studied because of the nanowire geometry and various materials that naturally confine electrons in a 1D conductance channel are ballistic flow of electrons, Coulomb blockade, quantum Hall effect, induced superconductivity, and spin transport. However, nanowires may have lower conductivity than macroscale fractions due to scattering from collisions with impurities, phonons, charge carriers, and crystal defects. Surface scattering is reduced even for ideal nanowires due to their high surface-to-ratio [91–104].

11.6.2 Transport properties

Understanding the electrical transport properties of nanowires is important for characterizing the nanowires, using them in electronic devices, and studying the unexpected transport phenomena caused by the one-dimensional effect of quantum mechanics. The transport properties of nanowires depend on many critical parameters: the wire diameter (this is very important for both classical and quantum sizes), materials, surface, crystal quality, and crystallographic orientation of the wire axis (very important for anisotropic materials with parameters such as carrier mobility, Fermi surface, or effective mass tensor). The predominant types of electrical transport phenomena for low-dimensional systems are ballistic and diffusive transport [105–117].

Ballistic transport is when electrons flow through the nanowire without scattering. In this phenomenon, conduction is determined primarily by the contacts between the nanowire and its external circuit. The conductance is quantized in terms of an integer number of universal conductance units, $G0 = 2e^2/h$. The phenomenon of ballistic transport is generally seen in very short quantum wires, such as the mechanically controlled break junctions (MCBJ) that represent conduction as a true quantum process where the electron mean free path is much longer than the length of the nanowire. However, when electrons (or holes) move down nanowires that are significantly longer than the carrier mean free path, they experience a lot of scattering events. The conduction in this instance is dominated by carrier scattering within the wires as a result of phonon boundary scattering, lattice and other structural defects, and impurity atoms. The diffusive regime is where the transport occurs [118–121].

11.6.3 Optical properties

Optical measurements provide a simple and sensitive technique for determining the electrical structure of nanowires because they are sensitive to quantum behaviors and they do not require any sample preparation (contacts, etc.). Carbon nanotubes and other 1-D systems' optical spectra almost always have significant features at certain energies close to singularities of the combined density of states that are created under high quantum confinement situations. Nanowires are generally not equivalent to their bulk material analogues, as verified by an assembly of optical experiments.

Optical properties are shown to be an important methodology for characterizing nanowires, although not always easy to interpret. Often, the wavelength of light used to characterize the sample is less than the wire length but greater than the wire diameter. This means that the probe light used in an optical measurement probe both the wire and the substrate that it rests on simultaneously, and not only the wire. If the substrate does not absorb or luminescence within the photonic range we are experimenting with in the experiments, then the substrate can be ignored in experiments like photo-luminescence (PL). PL directly detects the photo-luminescence of the nanowires. For transmission and reflection experiments, even a non-absorbing substrate can subsequently change the spectra of a nanowire.

11.7 TECHNIQUES FOR CREATING NANOWIRES

One-dimensional nanostructured materials can be generated and manipulated in many different ways. Many of these ways can be classified as either:

1. **Spontaneous growth**
    - Evaporation (or dissolution) – condensate;
    - Recrystallization.
    - Vapor growth (or solution) – liquid – solid [vapor-liquid-solid (VLS) or SLS];
2. **Template-based methods**
    - Electroplating.
    - Conversion with chemical reaction.
    - Filling with melt or colloidal nanoparticles or solution.

The predominant methods of nanowire fabrication involve top-down and bottom-up approaches. The top-down approach uses techniques, such as lithography, etching, and electrophoresis, to break down bulk materials to the nanoscale. These methods have shortcomings that involve lithography constraints and complicated vertical properties that often limit modification of orientation and crystal structure. However, nanowires originating from top-down processes produce arrays with precise control over ordering and density.

Bottom-up approaches can facilitate the growth of nanowires in a stepwise manner from simple building blocks. Though many bottom-up methods exist, the most common is the vapor-liquid-solid (VLS) method due to its versatility. VLS uses metal nanoparticles, for instance, gold as the catalyst, to promote development from a material delivered to the nanoscale in the vapor phase.

The vapor-solid-solid (VSS) concept is a version of the VLS process, where the catalyst remains solid during the growth cycle to produce vertical nanowires. In the selective-area epitaxy (SAE) method, nanoholes act as nucleation sites for the growth of nanowires using non-catalytic methods at masked substrates.

Nanowires can also form spontaneously through evaporation and recrystallization methods or be encouraged through surfactant molecules that help promote growth in particular directions. Template methods for nanowire synthesis are conducted using porous materials that are filled with colloidal nanoparticles through electrodeposition.

11.8 APPLICATIONS OF NANOWIRES

Nanowires are suitable for many applications due to their well-defined electrical and optical properties. Carbon-based nanowires are widely employed in electronics, such as porous nanowires as high-performance anodes in storage batteries and graphite-diamond composite nanowires in electromechanical nano-switches. Semiconducting nanowires are widely employed in many optical processes in photonics, including sub-wavelength waveguiding, room-temperature ultraviolet (UV) lasers, and nonlinear optical mixing. Nanowires are also used in sophisticated spectroscopy and microscopy systems.

Research activities are focused on reducing the size of optical components and enhancing device integration density, where polymer nanowires have incredible potential for compact photonic integrated circuits. Nanowires can downsize existing transistors, addressing the limiting features of silicon electronics. For example, nanowires do not produce heat in conduction and address the critical issue of increasing transistor density on integrated circuits. Due to their advanced properties, Ag and Cu metal

Table 11.2 Applications of different types of nanowires

S. No	*Types of nanowires*	*Applications of nanowires*
1.	Silicon	Electronics, biosensors, Electroluminescence
2.	Germanium	Electronics, IR detectors
3.	GaAs	High speed electronics, Photoconductive roles
4.	InAs	Photodetector, Solar cell, NWFET, Tunnel diode, TFET, SET
5.	InP	NWFET, Single-photodetectors, Single-photon detector, Tunnel diode
6.	GaP	Single-photonsource, Photodetector, Solar cell
7.	Ag NWs	Wearable devices
8.	CuO NWs	Electronic-conductor
9.	Zn NWs	Optoelectronics, light emitters and lasers
10.	DNWs	Radiation particle detectors, UV-light detectors and emitters, high-speed and high-power field effect transistors, field emission sources, position-sensitive biochemical substrates, and room temperature-stabilized high-efficiency single-photon emitters.
11.	Cadmium telluride	IR detectors for various wavelengths

nanowires have potential as complex stretchable and transparent electrodes in the fields of optoelectronics and microelectromechanics. Flexibly transparent conductors such as these have high optical transmittance and low sheet resistance, making them ideal for use in optoelectronic devices, including photovoltaic cells and touch panels. Additionally, nanowires can also be high-tech medical devices. As they transmit ultraviolet photons with no loss mechanism, they can be found within the human body serving as sensitive and effective sensors that identify poisons and help remove them. Additionally, nanowires can also be implemented within blood arteries to aid in stimulating the nerves of the brain, which speeds up the process of identifying the causes of neurological diseases such as Parkinson's. Different forms of nanowires can be used in the applications demonstrated in Table 11.2.

11.9 CONCLUSION

As a significant class of nanomaterials, nanowires have tremendous opportunities to help improve technology in various areas, including health care, electronics, optics, and magnetism. Nanowires differ from the bulk materials in chemical and physical properties due to their very high surface-to-ratio. The high surface-to ratio gives rise to different features that allow many new nanoscale applications and capabilities.

Both bottom-up and top-down methods are widely used in the manufacturing and processing of nanowires. Top-down methods, such as lithography, etching, or mechanical milling, reduce the size of bulk materials to a nanoscale. Alternatively, bottom-up methods manipulate chemical reactions or physical interactions, such as chemical vapor deposition, solution-phase synthesis, or self-assembly mechanisms, to create nanowires by the controlled assembly of atoms or molecules. Nanowires can be produced using metals, oxides, carbides, nitrides, polymers, and composite materials. Each material type has unique advantages based on its electrical, optical, thermal, and mechanical properties and is suitable for particular applications.

Nanowires have diverse applications that are rapidly expanding to include ultra-sensitive chemical and biological sensors, mechanical and optoelectronic systems, nanoelectromechanical systems (NEMS), and nanoelectronic devices. Nanowires' small size and tunable properties are ideal for next-generation technologies that require high performance within small form factors.

One of the most overlooked aspects of developing these technologies is understanding and mastering the surface chemistry and interfacial interactions of nanowires that determine how they will interact with the surrounding materials and environments, as these interactions can drastically impact the overall function and reliability of the device based on the nanowires. Future work must be dedicated to developing and maximizing the mechanical robustness and reproducibility of nanowires, especially when incorporated in complex systems, as each research group will have subtly different materials and different environmental sensitivities that may inhibit economic success and large-scale applications. For nanowires to reach their potential, there needs to be reliable nanowire diagrams and controlled and scalable production processes. There have been numerous advancements in the synthesis, characterization, and functionalization of nanowires to date, and it is imperative for technological innovations and multidisciplinary collaborations to develop real pathways to tackle the challenges we face and bring the tantalizing potential caused by the extraordinary characteristics of these nanostructures into the reality of commercial use.

REFERENCES

1. C Y Hsu Nanowires Properties and Applications: A Review Study. *South African Journal of Chemical Engineering*, 2023. 10.1016/j.sajce.2023.08.006
2. G E Badawy Electronic Transport and Quantum Phenomena in Nanowires. *Chemical Reviews*, 2024. 10.1021/acs.chemrev.3c00656
3. J Wang, M S Hassan, J Liu, S Yu Nanowire Assemblies for Flexible Electronic Devices: Recent Advances and Perspectives. *Advanced Materials*, 30(48), 2018. 10.1002/adma.201803430

4. Y Sun, T Dong, L Yu, J Xu, K Chen Planar Growth, Integration, and Applications of Semiconducting Nanowires. *Advanced Materials*, 32 (27), 2019. 10.1002/adma.201903945
5. T Arjmand, M Legallais, T T H Nguyen, P Serre, M Vallejo-Perez, F Morisot, B Salem, C Ternon Functional Devices From Bottom-Up Silicon Nanowires: A Review. *Nanomaterials*, 12(7), 1043, 2022. 10.3390/nano12071043
6. Y Zhu, S Kim, X Ma, P Byrley, N Yu, Q Liu, X Sun, D Xu, S Peng, M C Hartel, S Zhang, V Jucaud, M R Dokmeci, A Khademhosseini, R Yan Ultrathinshell Epitaxial Ag@Au Core-Shell Nanowires for High-Performance and Chemically-Stable Electronic, Optical, and Mechanical Devices. *Nano Research*, 14(11), 4294–4303, 2021. 10.1007/s12274-021-3718-z
7. D George, A Garcia, Q N Pham, M R Perez, J Deng, M T Nguyen, T Zhou, S O Martínez-Chapa, Y Won, C Liu, R C Lo, R Ragan, M Madou Fabrication of Patterned Graphitized Carbon Wires Using Low Voltage Near-Field Electrospinning, Pyrolysis, Electrodeposition, and Chemical Vapor Deposition. *Microsystems & Nanoengineering*, 6 (1), 2020. 10.1038/s41378-019-0117-7
8. T Adam, T S Dhahi, S C Gopinath, U Hashim Novel Approaches in Fabrication and Integration of Nanowire for Micro/Nano Systems. *Critical Reviews in Analytical Chemistry*, 52(8), 1913–1929, 2021. 10.1080/10408347.2021.1925523
9. S Raman, A R Sankar, M Sindhuja Advances in Silicon Nanowire Applications in Energy Generation, Storage, Sensing, and Electronics: A Review. *Nanotechnology*, 34(18), 182001, 2023. 10.1088/1361-6528/acb320
10. Z Azam, A Singh *Various Applications of Nanowires*. Advances in Computer and Electrical Engineering Book Series, (pp. 17–53), 2021. 10.4018/978-1-7998-6467-7.ch002
11. A Mukhtar, K Wu, X Cao, L Gu Magnetic Nanowires in Biomedical Applications. *Nanotechnology*, 31(43), 433001, 2020. 10.1088/1361-6528/aba1ba
12. B Liu, K P Chen, N P Kherani, S Zukotynski Power-Scaling Performance of a Three-Dimensional Tritium Betavoltaic Diode. *Applied Physics Letters*, 95, 233112, 2009. 10.1063/1.3272105
13. A K Hyder, S Sabripour, D J Flood, G Halpert, R L Wiley, A K Hyder *Sspacecraft Power Technologies*. Imperial College Press (2000).
14. Y Yang, K Wang, H W Liang, G Q Liu, M Feng, L Xu, ..., S H Yu A New Generation of Alloyed/Multimetal Chalcogenide Nanowires by Chemical Transformation Science Advances, *Nation Library of Medicine*, 1(10), e1500714, 2015.
15. X Zhang Top-k Feature Selection Framework Using Robust 0–1 Integer Programming. *IEEE Transactions on Neural Networks and Learning Systems*, 2020. 10.1109/TNNLS.2020.3009209
16. Y Y Liu, L Liu, L Xue, L B Mao, H L Gao, L Xu, S H Yu Charged Inorganic Nanowire-Directed Mineralization of Amorphous Calcium Carbonate. *ChemNanoMat*, 2(4), 259–263, 2016.
17. C H Thai, A J M Ferreira, H Nguyen-Xuan, P Phung-Van A Size Dependent Meshfree Model for Functionally Graded Plates Based on the Nonlocal Strain Gradient Theory, *Composite Structures*, 272, 114169, 2021.

18. B Andrei Bernevig, Taylor L Hughes, Shou-Cheng Zhang Quantum Spin Hall Effect and Topological Phase Transition in HgTe Quantum Wells. *Science*, 314, 5806, 2006.
19. R X Yan, D Gargas, P D Yang Nanowire Photonics. *Nature Photonics*, 3, 569–576, 2009.
20. A Miraoui, M A Didi Neodymium (III) Sorption Onto Sodium Bentonite and Magnetic Activated Carbon: A Comparative Study. *European Chemical Bulletin*, 9(6), 138–147, 2020.
21. Anjana Bhardwaj, Pradeep Kumar, B Raj, Sunny Anand Design and Performance Optimization of Doping-Less Vertical Nanowire TFET Using Gate Stack Technique. *Journal of Electronic Materials (JEMS)*, 41(7), 4005–4013, 2022.
22. Jeetendra Singh, B Raj Tunnel Current Model of Asymmetric MIM Structure Levying Various Image Forces to Analyze the Characteristics of Filamentary Memristor. *Applied Physics A*, 125(3), 203.1–203.11, 2019.
23. Candy Goyal, Jagpal Singh Ubhi, B Raj Low Leakage Zero Ground Noise Nanoscale Full Adder Using Source Biasing Technique. *Journal of Nanoelectronics and Optoelectronics*, 14, 360–370, 2019.
24. Gurmohan Singh, R K Sarin, B Raj A Novel Robust Exclusive-OR Function Implementation in QCA Nanotechnology With Energy Dissipation Analysis. *Journal of Computational Electronics*, 15(2), 455–465, 2016.
25. Girish Wadhwa, Priyanka Kamboj, Balwinder Raj Design Optimisation of Junctionless TFET Biosensor for High Sensitivity. *Advances in Natural Sciences: Nanoscience and Nanotechnology*, 10(7), 045001, 2019.
26. Priya Bansal, B Raj Memristor Modeling and Analysis for Linear Dopant Drift Kinetics. *Journal of Nanoengineering and Nanomanufacturing*, 6, 1–7, 2016.
27. D V Ashima, B Raj Performance Analysis of Charge Plasma Induced Graded Channel Si Nanotube. *Journal of Engineering Research (JER)*, EMSME Special 146–154, 2021.
28. Abhishek Singh Tomar, Vijay Kumar Magraiya, B Raj Scaling of Access and Data Transistor for High Performance DRAM Cell Design. *Quantum Matter*, 2, 412–416, 2013.
29. Neeraj Jain, B Raj Parasitic Capacitance and Resistance Model Development and Optimization of Raised Source/Drain SOI FinFET Structure for Analog Circuit Applications. *Journal of Nanoelectronics and Optoelectronins*, 13, 531–539, 2018.
30. Shradhya Singh, S K Vishvakarma, B Raj Analytical Modeling of Split-Gate Junction-Less Transistor for a Biosensor Application. *Sensing and Bio-Sensing*, 18, 31–36, 2018.
31. Maisagalla Gopal, Balwinder Raj Low Power 8T SRAM Cell Design for High Stability Video Applications. *ITSI Transaction on Electrical and Electronics Engineering*, 1(5)), 91–97, 2013.
32. Balwinder Raj, Jatin Mitra, Deepak Kumar Bihani, V Rangharajan, A K Saxena, S Dasgupta Analysis of Noise Margin, Power and Process Variation for 32 nm FinFET Based 6T. *SRAM Cell Journal of Computer (JCP)*, 5(6), 1–8, 2010.
33. Divya Sharma, Rajesh Mehra, B Raj Comparative Analysis of Photovoltaic Technologies for High Efficiency Solar Cell Design. *Superlattices and Microstructures*, 153, 106861, 2021.

34. Pawandeep Kaur, Avtar Singh Buttar, B Raj A Comprehensive Analysis of Nanoscale Transistor Based Biosensor: A Review. *Indian Journal of Pure and Applied Physics*, 59), 304–318, 2021.
35. Divya Yadav, Balwant Raj, B Raj Design and Simulation of Low Power Microcontroller for IoT Applications. *Journal of Sensor Letters*, 18, 401–409, 2020.
36. Shailendra Singh, B Raj A 2-D Analytical Surface Potential and Drain Current Modeling of Double-Gate Vertical t-Shaped Tunnel FET. *Journal of Computational Electronics*, 19, 1154–1163, 2020.
37. Jeetendra Singh, B Raj An Accurate and Generic Window Function for Nonlinear Memristor Model. *Journal of Computational Electronics*, 18(2), 640–647, 2019.
38. Manjit Kaur, Neena Gupta, Sanjeev Kumar, B Raj, Arun Kumar Singh Comparative RF and Crosstalk Analysis of Carbon Based Nano Interconnects IET Circuits. *Devices & Systems*, 15(6), 493–503, 2021.
39. Nehru Kandasamy, Firdous Ahmad, D Ajitha, B Raj, Nagarjuna Telagam Quantum Dot Cellular Automata based Scan Flip Flop and Boundary Scan Register. *IETE Journal of Research*, 66, 535–548, 2020.
40. Jeetendra Singh, B Raj Design and Investigation of 7T2M NVSARM With Enhanced Stability and Temperature Impact on Store/Restore Energy. *IEEE Transactions on Very Large Scale Integration Systems*, 27(6), 1322–1328, 2019.
41. Anil Kumar Bhardwaj, Sumeet Gupta, B Raj, Amandeep Singh Impact of Double Gate Geometry on the Performance of Carbon Nanotube Field Effect Transistor Structures for Low Power Digital Design. *Computational and Theoretical Nanoscience*, 16, 1813–1820, 2019.
42. Neeraj Jain, B Raj Thermal Stability Analysis and Performance Exploration of Asymmetrical Dual-k Underlap Spacer (ADKUS) SOI FinFET for Security and Privacy Applications. *Indian Journal of Pure & Applied Physics (IJPAP)*, 57, 352–360, 2019.
43. Neeraj Jain, Balwinder Raj, Dual-k Spacer Region Variation at the Drain Side of Asymmetric SOI FinFET Structure: Performance Analysis towards the Analog/RF Design Applications. *Journal of Nanoelectronics and Optoelectronics*, 14, 349–359, 2019.
44. Jeetendra Singh, Sanjeev Sharma, B Raj, Mamta Khosla Analysis of Barrier Layer Thickness on Performance of In1-xGaxAs Based Gate Stack Cylindrical Gate Nanowire MOSFET. *JNO*, 13, 1473–1477, 2018.
45. Neeraj Jain, B Raj Analysis and Performance Exploration of High-k SOI FinFETs Over the Conventional Low-k SOI FinFET Toward Analog/RF Design. *Journal of Semiconductors (JoS)*, 39(12), 124002-1-7, 2018.
46. Candy Goyal, Jagpal Singh Ubhi, B Raj A Reliable Leakage Reduction Technique for Approximate Full Adder With Reduced Ground Bounce Noise. *Journal of Mathematical Problems in Engineering*, 2018, 3501041, 2018.
47. Anuradha, Jeetendra Singh, B Raj, Mamta Khosla Design and Performance Analysis of Nano-Scale Memristor-based Nonvolatile SRAM, *Journal of Sensor Letter*, 16, 798–805, 2018.
48. Girish Wadhwa, B Raj Parametric Variation Analysis of Charge-Plasma-based Dielectric Modulated JLTFET for Biosensor Application. *IEEE Sensors Journal*, 18, 6070–6077, 2018.

49. Jeetendra Singh, B Raj Comparative Analysis of Memristor Models for Memories Design. *JoS*, 39(7), 074006-1-12, 2018.
50. Divya Yadav, Shailesh Singh Chouhan, Santosh Kumar Vishvakarma, B Raj Application Specific Microcontroller Design for IoT Based WSN. *Sensor Letters*, 16, 374–385, 2018.
51. Gurmohan Singh, R K Sarin, B Raj Fault-Tolerant Design and Analysis of Quantum-Dot Cellular Automata Based Circuits. *IEEE/IET Circuits, Devices & Systems*, 12, 638–664, 2018.
52. Jeetendra Singh, B Raj Modeling of Mean Barrier Height Levying Various Image Forces of Metal Insulator Metal Structure to Enhance the Performance of Conductive Filament Based Memristor Model. *IEEE Nanotechnology*, 17(2), 267–268, 2018.
53. Aakash Jain, Sanjeev Sharma, B Raj Analysis of Triple Metal Surrounding Gate (TM-SG) III-V Nanowire MOSFET for Photosensing Application. *Opto-Electronics Journal*, 26(2), 141 148, 2018.
54. Aakash Jain, Sanjeev Sharma, B Raj Design and Analysis of High Sensitivity Photosensor Using Cylindrical Surrounding Gate MOSFET For Low Power Sensor Applications. *Engineering Science and Technology, an International Journal*, 19(4), 1864–1870, 2016.
55. G Saiphani Kumar, Amandeep Singh, B Raj Design and Analysis of Gate All Around CNTFET based SRAM Cell Design. *Journal of Computational Electronics*, 17(1), 138–145, 2018.
56. Gurinder Pal Singh, B S Sohi, Balwinder Raj Material Properties Analysis of Graphene Base Transistor (GBT) for VLSI Analog Circuits. *Indian Journal of Pure & Applied Physics (IJPAP)*, 55, 896–902, 2017.
57. Sunil Kumar, B Raj Estimation of Stability and Performance Metric for Inward Access Transistor Based 6T SRAM Cell Design Using n-type/p-type DMDG-GDOV TFET. *IEEE VLSI Circuits and Systems Letter*, 3(2), 25–39, 2017.
58. Shashikant Sharma, Anjan Kumar, Manisha Pattanaik, B Raj Forward Body Biased Multimode Multi-Threshold CMOS Technique for Ground Bounce Noise Reduction in Static CMOS Adders. *International Journal of Information and Electronics Engineering*, 3(3), 567–572, 2013.
59. Hamendra Singh, Pankaj Kumar, Balwinder Raj Performance Analysis of Majority Gate SET Based 1-bit Full Adder. *International Journal of Computer and Communication Engineering (IJCCE)*, 2(4), 2013.
60. Anil Kumar Bhardwaj, Sumeet Gupta, B Raj Investigation of Parameters for Schottky Barrier (SB) Height for Schottky Barrier Based Carbon Nanotube Field Effect Transistor Device. *Journal of Nanoelectronics and Optoelectronics*, 15, 783–791, 2020.
61. Priya Bansal, B Raj Memristor: A Versatile Nonlinear Model for Dopant Drift and Boundary Issues. *JCTN*, 14(5), 2319–2325, 2017.
62. Neeraj Jain, B Raj An Analog and Digital Design Perspective Comprehensive Approach on Fin-FET (Fin-Field Effect Transistor) Technology – A Review. *Reviews in Advanced Sciences and Engineering (RASE)*, 5, 1–14, 2016.
63. Balwinder Raj, A K Saxena, S Dasgupta Analytical Modeling for the Estimation of Leakage Current and Subthreshold Swing Factor of Nanoscale Double Gate FinFET Device. *MicroElectronics International*, 26), 53–63, 2009.

64. Soniya, Shailendra Singh, Girish Wadhwa, Balwinder Raj An Analytical Modeling for Dual Source Vertical Tunnel Field Effect Transistor. *International Journal of Recent Technology and Engineering (IJRTE)*, 8(2), 603–608, 2019.
65. Shailendra Singh, B Raj Design and analysis of hetrojunction Vertical T-Shaped Tunnel Field Effect Transistor. *Journal of Electronics Material*, 48(10), 6253–6260, 2019.
66. Candy Goyal, Jagpal Singh Ubhi, B Raj A Low Leakage CNTFET Based Inexact Full Adder for Low Power Image Processing Applications. *International Journal of Circuit Theory and Applications*, 47(9), 1446–1458, 2019.
67. B Raj, A K Saxena, S Dasgupta, A Compact Drain Current and Threshold Voltage Quantum Mechanical Analytical Modeling for FinFETs. *Journal of Nanoelectronics and Optoelectronics (JNO)*, 3(2), 163–170, 2008.
68. Girish Wadhwa, B Raj An Analytical Modeling of Charge Plasma Based Tunnel Field Effect Transistor With Impacts of Gate Underlap Region. *Superlattices and Microstructures*, 142, 106512, 2020.
69. Shailendra Singh, B Raj Modeling and Simulation Analysis of SiGe Hetrojunction Double GateVertical t-Shaped Tunnel FET. *Superlattices and Microstructures*, 142, 106496, 2020.
70. Neeraj Jain, Balwinder Raj Device and Circuit Co-Design Perspective Comprehensive Approach on FinFET Technology – A Review. *Journal of Electron Devices*, 23(1), 1890–1901, 2016.
71. Sunil Kumar, B Raj Analysis of I_{ON} and Ambipolar Current for Dual-Material Gate-drain Overlapped DG-TFET. *Journal of Nanoelectronics and Optoelectronics*, 11, 323–333, 2016.
72. Naveed Anjum, Tarun Bali, B Raj Design and Simulation of Handwritten Multiscript Character Recognition. *International Journal of Advanced Research in Computer and Communication Engineering*, 2 (7), 2544–2549, 2013.
73. Karmjit Singh, B Raj Performance and Analysis of Temperature Dependent Multi-Walled Carbon Nanotubes as Global Interconnects at Different Technology Nodes. *Journal of Computational Electronics*, 14 (2), 469–476, 2015.
74. Sunil Kumar, B Raj Compact Channel Potential Analytical Modeling of DG-TFET Based on Evanescent–Mode Approach. *Journal of Computational Electronics*, 14(2), 820–827, 2015.
75. Karmjit Singh, B Raj Temperature Dependent Modeling and Performance Evaluation of Multi-Walled CNT and Single-Walled CNT as Global Interconnects. *Journal of Electronic Materials*, 44(12), 4825–4835, 2015.
76. V K Sharma, M Pattanaik, B Raj INDEP Approach for Leakage Reduction in Nanoscale CMOS Circuits. *International Journal of Electronics*, 102(2), 200–215, 2014.
77. Karmjit Singh, B Raj Influence of Temperature on MWCNT Bundle, SWCNT Bundle and Copper Interconnects for Nanoscaled Technology Nodes. *Journal of Materials Science: Materials in Electronics*, 26(8), 6134–6142, 2015.
78. Naveed Anjum, Tarun Bali, B Raj Design and Simulation of Handwritten Gurumukhi and Devanagri Numerical Recognition. *International Journal of Computer Applications*, 73(12), 16–21, 2013.

79. Yogesh Thakur, Balwant Raj, Balwinder Raj, Design of Pentacene Thin-Film Transistor Based Hydrogen Gas Sensor for High Sensitivity. *ECS Journal of Solid State Science and Technology*, 13, 047005–047012, 2024.
80. Manoharan Arun Kumar, Kamal Batcha Mohamed Ismail, Shanmugam Mahalingam, Balwinder Raj, Junghwan Kim, Raji Atchudan Recent Advances in Two-Dimensional Nanomaterials and Nanostructures for Healthcare Monitoring. *ACS Sensors*, 9(4), 1–8, 2024.
81. Mohamed Ismail Kamal Batcha, Arun Kumar Manoharan, Shanmugam Mahalingam, Raj Balwinder Carbon Fiber-Reinforced Polymers for Energy Storage Applications. *Journal of Energy Storage*, 84(Part B), 110931–110939, 2024.
82. Shahraan Hussain, Balwinder Raj, Divya Sharma Design and Parametric Optimization of Lead Free Perovskite Based Tandem Solar Cell for High Efficient. *Engineering Research Express*, 6, 015315–0153122, 2024.
83. Parveen Kumar, Balwinder Raj, Girish Wadhwa, Balwinder Singh, Raj Kumar Design and Analysis of Junctionless Based Gate All Around N+ Doped Layer Nanowire TFET Biosensor. *ECS Journal of Solid State Science and Technology*, 13, 017002–017010, 2024.
84. Divya Sharma, Rajesh Mehra, Balwinder Raj Enhancement in Efficiency of Methyl Ammonium Tin Iodide Based Perovskite Solar Cell Using SCAPS-1 D. *Nano*, 18(14), 2350109, 2023.
85. Parveen Kumar, Sanjeev Kumar Sharma, Balwant Raj, Balwinder Raj Analysis of Device Parameter Variations in In1-xGaxAs Based Gate Stacked Double Metal Surrounding Gate Nanowire MOSFET. *TEEM Springer*, 24, 570–578, 2023.
86. Anjana Bhardwaj, Pradeep Kumar, Balwinder Raj, Naveen Kumar, Sunny Anand Design and Optimization of Vertical Nanowire Tunnel FET With Electrostatic Doping. *Engineering Research Express*, 5, 045025–045033, 2023.
87. Yogesh Thakur, Balwinder Raj, Sandeep Singh Gill Design and Performance Analysis of Pentacene Organic Field Effect Transistor With High-K Dielectric Materials. *Journal of Optoelectronics and Advanced Materials – Rapid Communications*, 17, 7–8, 2023.
88. Parveen Kumar, Balwinder Raj Design and Simulation of Junctionless Nanowire Tunnel Field Effect Transistor Biosensor for High Sensitivity. *Microelectronics Journal*, 137, 105826–105834, 2023.
89. S Khandelwal, V Gupta, B Raj, R D Gupta Process Variability Aware Low Leakage Reliable Nano Scale DG-FinFET SRAM Cell Design Technique. *Journal of Nanoelectronics and Optoelectronics*, 10(6), 810–817, 2015.
90. V K Sharma, M Pattanaik, B Raj ONOFIC Approach: Low Power High Speed Nanoscale VLSI Circuits Design, *International Journal of Electronics*, 101(1), 61–73, 2014.
91. S Khandelwal, Balwinder Raj, R D Gupta FinFET based 6T SRAM Cell Design: Analysis of Performance Metric, Process Variation and Temperature Effect. *Journal of Computational and Theoretical Nanoscience*, 12, 2500–2506, 2015.
92. Sumit Singh, Shekhar Yadav, Jagdeep Rahul, Anurag Srivastava, B Raj Impact of HfO_2 in Graded Channel Dual Insulator Double Gate MOSFET. *Journal of Computational and Theoretical Nanoscience*, 12(6), 950–953, 2015.

93. Vijay Kumar Sharma, Manisha Pattanaik, B Raj PVT Variations Aware Low Leakage INDEP Approach for Nanoscale CMOS Circuits. *Microelectronics Reliability*, 54, 90–99, 2014.
94. B Raj, A K Saxena, S Dasgupta Quantum Mechanical Analytical Modeling of Nanoscale DG FinFET: Evaluation of Potential, Threshold Voltage and Source/Drain Resistance, *Elsevier's Journal of Material Science in Semiconductor Processing*, 16(4), 1131–1137, 2013.
95. Maisagalla Gopal, Siva Sankar D Prasad, Balwinder Raj 8T SRAM Cell Design for Dynamic and Leakage Power Reduction. *International Journal of Computer Applications*, 71(9), 43–48, 2013.
96. Manisha Pattanaik, B Raj, Shashikant Sharma, Anjan Kumar Diode Based Trimode Multi-Threshold CMOS Technique for Ground Bounce Noise Reduction in Static CMOS Adders. *Advanced Materials Research*, 548, 885–889, 2012.
97. Balwinder Raj, A K Saxena, S Dasgupta, Nanoscale FinFET Based SRAM Cell Design: Analysis of Performance metric, Process variation, Underlapped FinFET and Temperature Effect. *IEEE Circuits and Systems Magazine*, 11(2), 38–50, 2011.
98. V K Sharma, M Pattanaik, Balwinder Raj Leakage Current ONOFIC Approach for Deep Submicron VLSI Circuit Design. *International Journal of Electrical, Computer, Electronics and Communication Engineering*, 7(4), 239–244, 2013.
99. Parminder Kaur, Sandeep Singh Gill, B Raj Comparative Analysis of OFETs Materials and Devices for Sensor Applications. *Journal of Silicon*, 14, 4463–4471, 2022.
100. Sanjeev Kumar Sharma, Parveen Kumar, Balwant Raj, B Raj, $In_{1-x}Ga_xAs$ Double Metal Gate-Stacking Cylindrical Nanowire MOSFET for Highly Sensitive Photo Detector. *Journal of Silicon*, 14, 3535–3541, 2022.
101. B Raj, A K Saxena, S Dasgupta Analytical Modeling of Quasi Planar Nanoscale Double Gate FinFET With Source/Drain Resistance and Field Dependent Carrier Mobility: A Quantum Mechanical Study. *Journal of Computer (JCP)*,4(9), 1–8, 2009.
102. S Bhushan, S Khandelwal, B Raj Analyzing Different Mode FinFET Based Memory Cell at Different Power Supply for Leakage Reduction, Seventh International Conference on Bio-Inspired Computing: Theories and Application, (BIC-TA 2012). *Adv. Intell. Syst. Comput.*, 202, 89–100, 2013.
103. Jeetendra Singh, B Raj Temperature Dependent Analytical Modeling and Simulations of Nanoscale Memristor. *Journal: Engineering Science and Technology, an International Journal*, 21, 862–868, 2018.
104. Shradhya Singh, Shashi Bala, Balwant Raj, B Raj Improved Sensitivity of Dielectric Modulated Junctionless Transistor for Nanoscale Biosensor Design. *Sensor Letters*, 18, 328–333, 2020.
105. Vivek Kumar, Santosh Kumar Vishvakarma, B Raj Design and Performance Analysis of ASIC for IoT Applications. *Sensor Letters*, 18, 31–38.
106. Akanksha Jaiswal, R K Sarin, B Raj, Shikha Sukhija A Novel Circular Slotted Microstrip-Fed Patch Antenna With Three Triangle Shape Defected Ground Structure for Multiband Applications. *Advanced Electromagnetic (AEM)*, 7(3), 56–63, 2018.
107. Girish Wadhwa, B Raj Label Free Detection of Biomolecules using Charge-Plasma-Based Gate Underlap Dielectric Modulated Junctionless TFET. *Journal of Electronic Materials (JEMS)*, 47(8), 4683–4693, 2018.

108. Gurmohan Singh, R K Sarin, B Raj, Design and Performance Analysis of a New Efficient Coplanar Quantum-Dot Cellular Automata Adder. *Indian Journal of Pure & Applied Physics (IJPAP)*, 55, 97–103, 2017.
109. Parminder Kaur, Vikas Pandey, B Raj Comparative Study of Efficient Design, Control and Monitoring of Solar Power Using IoT. *Sensor Letters*, 18, 419–426, 2020.
110. Anil Kumar Bhardwaj, Sumeet Gupta, B Raj Development & Analysis of Compact Model for Double Gate Schottky Barrier CNTFET. *Journal of Nanoelectronics and Optoelectronics*, 15, 1199–1208, 2020.
111. Girish Wadhwa, Priyanka Kamboj, Jeetendra Singh, B Raj Design and Investigation of Junctionless DGTFET for Biological Molecule Recognition. *Transactions on Electrical and Electronic Materials*, 22, 282–289, 2021.
112. Sachin Kumar Verma, Shailendra Singh, Girish Wadhwa, B Raj Detection of Biomolecules Using Charge-Plasma Based Gate Underlap Dielectric Modulated Dopingless TFET. *Transactions on Electrical and Electronic Materials (TEEM)*, 21, 528–535, 2020.
113. Neeraj Jain, B Raj Impact of Underlap Spacer Region Variation on Electrostatic and Analog/RF Performance of Symmetrical High-k SOI FinFET at 20 nm Channel Length. *Journal of Semiconductors (JoS)*, 38(12), 122002, 2017.
114. Shailendra Singh, B Raj Analytical Modeling and Simulation Analysis of T-Shaped III-V Heterojunction Vertical T-FET. *Superlattices and Microstructures*, 147, 106717 2020.
115. Gurmohan Singh, R K Sarin, B Raj Design and Analysis of Area Efficient QCA Based Reversible Logic Gates. *Journal of Microprocessors and Microsystems*, 52, 59–68, 2017.
116. Shailendra Singh, B Raj Parametric Variation Analysis on Hetero-Junction Vertical t-Shape TFET for Supressing Ambipolar Conduction. *Indian Journal of Pure and Applied Physics*, 58, 478–485, 2020.
117. Soniya, Shailendra Singh, Girish Wadhwa, Balwinder Raj Design and Analysis of Dual Source Vertical Tunnel Field Effect Transistor for High Performance. *Transactions on Electrical and Electronic Materials*, 21, 74–82, 2019.
118. Manjit Kaur, Neena Gupta, Sanjeev Kumar, B Raj, Arun Kumar Singh RF Performance Analysis of Intercalated Graphene Nanoribbon Based Global Level Interconnects. *Journal of Computational Electronics*, 19, 1002–1013, 2020.
119. Girish Wadhwa, B Raj Design and Performance Analysis of Junctionless TFET Biosensor for High Sensitivity. *IEEE Nanotechnology*, 18, 567–574, 2019.
120. Jeetendra Singh, B Raj Enhanced Nonlinear Memristor Model Encapsulating Stochastic Dopant Drift. *JNO*, 14, 958–963, 2019.
121. Mohit Kumar, Tarun Chaudhary, Balwinder Raj Design and Comparative Analysis of Ferroelectric Nanowire With Dielectric HfO_2 and Al_2O_3 for Low Power Applications. *Journal of Electronic Materials*, 53, 6785–6791, 2024.

Chapter 12

Nano-enhanced biosensors

Anita Shukla, Ankit Jain, Puspraj Singh Chauhan, Shailendra Singh, and Imran Ullah Khan

12.1 INTRODUCTION

Nanowires are nanostructures with an unrestricted length but a diameter or thickness limited to a few tens of nanometers or less. We call them "quantum wires" due to the overwhelming dominance of quantum mechanical effects at these scales. Nanowires come in a variety of forms, such as metallic (such as gold, silver, nickel, and platinum) [1], superconducting, semiconducting, and insulating (such as TiO_2 and SiO_2) [2]. The production of nanoscale probes for a range of small-scale biomedical applications, including biosensing, electrophysiology, intracellular delivery, physical transduction, and immunomodulation, has relied heavily on nanowires in recent years [3]. Nanowires' tunable physical and chemical characteristics are the main characteristic that sets them apart from other materials. For instance, in terms of their surface, chemical composition, optical characteristics, and electrical conductivity. This tunability greatly enhances the application of nanowires to detect a range of biochemical targets, including glucose, proteins, nucleic acids, and small molecules, such as hydrogen peroxide [4]. Furthermore, nanowire sensors demonstrate high throughput and high spatial and temporal resolution, enabling detection at the cellular level. They are especially valuable for detecting numerous extracellular biophysical signals, like contraction forces and cellular mechanical responses, as well as providing measurements of biomolecular interactions [5]. Importantly, these nanowire-based platforms can not only be used for sensing but also for stimulating biologically active cells, thus enabling both real-time assessments and interactions with biological systems [6]. The nanoscale dimensions of the probes have the additional advantage of being minimally invasive during a cell interaction, making for a highly biocompatible option for long-lived studies of cells [7]. Patterned nanowires are uniquely suited for monitoring for intracellular signals and for longer-duration, high-throughput measurements. Their high surface-area-to-volume ratio allows for enough receptor molecules or cell sites for enhanced sensitivity [8]. In three-dimensional sensor arrays, this geometry allows for multichannel detection of weak and spatially localized biosignals [9]. Additionally, their complex shapes and hierarchical structure

DOI: 10.1201/9781003634645-12

can allow specific binding to complex cell structures, like neurites or synaptic junctions, which adds a further degree of selectivity. Along with customizable bio-recognition layers, which add specificity toward the interactions of target biomolecules with the surfaces of the nanowires [10], affordability remains a critical characteristic of the measurement systems that require detection of biological and/or chemical species, for example, in healthcare applications to detect diseases or in the screening of new therapeutic agents [11]. Therefore, developing sensing tools for such compounds that can be developed to analyze rapidly, directly, and sensitively is highly vital [12]. Nanowire-based devices are coming to the forefront as incredibly promising electrical sensors due to their potential for ultra-sensitive real-time detection of target biomolecules [13]. These biosensors can have a tremendous impact on advancing diagnostic technologies related to test kits, vaccines, and antiviral therapies. The rapid analysis, continuous monitoring, low number of reagents, and sensitivity and specificity are incredibly positive attributes that put these sensors in line to be used as immunosensors for biomolecule detection in the bodily fluids [14]. However, the current biosensor technology may still have unreliable detection limits in low-concentration matrices in clinically relevant samples, especially in biomarker concentrations less than 1 ng/mL. Without a means to improve positivity at these concentrations, high specificity for true positives may not support reliable diagnostics. Thus, there is an urgent need for biosensor systems with improved detection limits and dynamic range for clinically relevant targets [15]. This review examines recent advances in nanowire-based biosensors. It discusses reliable sensing mechanisms for identifying targets that consist of biological analytes, including RNA, DNA, proteins, neurotransmitters, and viruses. The review also provides extensive coverage of the nanowire biosensor: sensor design strategies and fabrication platforms to facilitate the advancement of nanowire applications across numerous biomedical fields. The distinctions made in this review between applications utilizing specific materials or diagnostics highlight the potential implementations of nanowires in various areas of biomedicine.

Furthermore, we outline certain limitations that will need to be addressed, as well as recent and relevant advancements in nanowire technologies to improve their roles as device biosensors for biological analytes, with improvements discussed in long-term performance, high ionic strength solvents, reusability, and self-powered nanowire platforms. The improvements in the fields of biosensing are an essential step toward the movement of nanowire biosensors out of the lab space and into real-world biomedical practice. The review also discusses the significance of making nanowire sensor arrays as a platform for the future of diagnostic tools, particularly for applications in precision medicine, early detection of diseases, and telemedicine monitoring. To sum up, nanowire biosensors provide a unique combination of extremely sensitive instruments and adaptable designs that are increasingly being used in biological and medical research. Their growing applications will support new advancements in therapeutic and diagnostic monitoring and enable precision medicine and point-of-care diagnostics [16–21].

12.2 BIOSENSORS

A physiochemical analytical tool used to measure the interactions between an analyte and a biological component is called a biosensor. It consists of a biological recognition element capable of detecting either a chemical or physical change in the environment, or both, and a transducer to transform those changes into measurable signals. The characteristic feature of a biosensor is the distance between the transducer and the bioreceptor, which allows real-time assessment of the surrounding changes [22–29].

A biosensor is composed of two major components, namely, the recognition part and the transducer (Figure 12.1). Through particular biochemical recognition reactions, a bio-recognition element interacts with a target analyte. The recognition is transformed into an electrical, optical, thermal, or mass-based signal by the transducer. Enzymes, antibodies, lectins, microbes, and nucleic acids are the most prevalent types of bioreceptors. Biosensors have several advantages over conventional diagnostic tools, including ease of use, high specificity, fast response time, affordability, portability, and the possibility of continuous and real-time monitoring. Biosensors are useful for biomedical diagnostic applications, food safety testing, and environmental monitoring. A relatively new but fascinating area of biosensor technology is low-density lipoprotein (LDL) sensors.

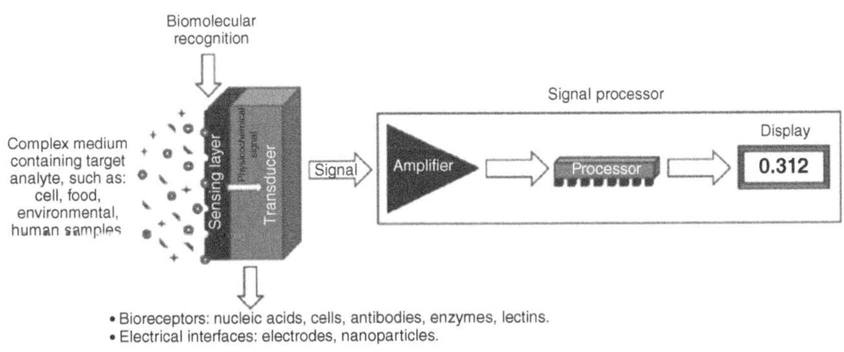

Figure 12.1 The schematic diagram for the basic components of a biosensor. The bioreceptor, the transducer, and the amplifier are the relevant components of a biosensor. The schematic diagram provides a well-defined understanding of the general components that work together to provide accurate and sensitive biosensing capabilities.

LDL sensors are drawing interest with their high selectivity and stability, factors critical to assessing plasma LDL levels, a primary contributor to atherosclerosis. Given that cardiovascular diseases continue to rank among the world's leading causes of death, the development of biosensors for precise LDL detection is crucial. LDL biosensor platforms frequently use contemporary analytical methods like surface plasmon resonance (SPR), cyclic voltammetry (CV), electrochemical impedance spectroscopy (EIS), and quartz crystal microbalance (QCM). The specificity and consistency of LDL-based biomedical diagnostics can be enhanced by these methods, which offer both qualitative and quantitative measurements (Figure 12.1).

12.3 CHARACTERISTICS OF BIOSENSORS

The biosensor prototype must also exhibit the following features and characteristics in the ideal scenario so that the outcome of achieving better and improved overall health of the society can be fulfilled.

- **Selectivity**: One primary consideration of the biosensor designer before designing any biosensor is to consider how selective the biosensor is in detecting the desired analyte from a sample containing different or closely related analytes/contaminants as well [16]. Thus, selectivity is the essential feature for the biosensor.
- **Reproducibility**: The reproducibility of the biosensor results in terms of repeated experiments is a major issue [17]. A biosensor that has high reproducibility and reliability for those using it, and it is a material that is definitely in high demand. In addition to reproducibility, the results must also be accurate and precise, so in the end, they are even more reliable as a biosensor [30–38].
- **Stability**: The stability of the biosensor is a very important factor in contributing to the commercial success of biosensors. It is evident that biosensors normally lose their signal strength with aging, which makes stability an important factor that is worthy of further consideration and attention [16]. In addition, the stability of the biosensor is typically accelerated with an increase in temperature, or, in other words, stability is typically related to a temperature-dependent interaction mechanism.
- **Sensitivity and Linearity**: The biosensors are rated highly only if they have high sensitivity. In this modern society, for air, water, and soil pollutant detection, you are required to operate at the ppm level; however, in the medical field, it can be from nanograms per milliliter to femtograms per milliliter [17]. Following on, the linearity of the device and the accuracy of response to a given set of measurements at different concentrations of analyte.

12.4 WORKING PRINCIPLE AND FABRICATION PROCESS

12.4.1 Fundamentals of nanowire field-effect transistors

Nanowire-based biosensors usually work by detecting changes in the nanowire's electrical, optical, or electrochemical characteristics as a result of interactions with a target analyte. In a SiNW-FET (Field-Effect Transistor) biosensor, for instance, when a target molecule docks onto the nanowire's surface, the conductivity of the nanowire changes and is once more measured by the FET [39–45].

Nanowire field-effect transistors (FETs) are nanoscale advanced sensors that are adapted from traditional planar FETs. Nanowire FETs retain the essential features of FETs, which include the source, drain, gate, and body. However, instead of the traditional doped channels and doped gates associated with planar FETs, nanowire FETs use semiconductor nanowires and biological or chemical receptors (Figure 12.2a and b). These receptors will bind specific charged species, which modifies the conductivity of the nanowire, allowing for increased sensitivity of detection. Traditional planar FETs are limited by lower sensitivity, larger sample volumes, and poor practicality, whereas nanowire-based designs improve detection challenges by taking advantage of the increased surface-to-volume ratio.

Despite differing in terms of structure from conventional FETs, the fundamental mechanism is similar: changes in conductivity in the nanowire channel take place in relation to charged species in the surrounding environment. Silicon and silica nanowires, among all materials, are the most used because they are compatible with CMOS (complementary metal-oxide-semiconductor) technology. Additional benefits include that silicon oxide forms naturally as an insulation layer on silicon and that there is more flexibility for surface modification to engineer sensing functionalities. Even though the biomolecules are charged, the fact that the biomolecules accumulate on the isolation layer does not change the conductivity of the nanowires, as neither the original silicon nor the silica surfaces are sensitive to biomolecules. To detect specific charged species (e.g., DNA, RNA, viruses), the nanowire FET surfaces are functionalized with specific receptors. These receptors bind the target molecules, which create an electric field that changes the conductivity of the nanowires – just like a gate changes current. A typical approach is to use (3-Aminopropyl)triethoxysilane (APTES), which forms a silane layer (Si–O-Si-X) on the surface of the nanowire, immobilizing receptors such as DNA, PNA, or antibodies. When the nanowire's isolation layer undergoes a change in surface charge due to the pH of the surrounding gas or liquid, conductivity will also be affected. Low pH converts $-NH_2$ groups to $-NH_3^+$ groups to lower conductivity. High pH converts –SiOH groups to $-SiO^-$ groups to increase conductivity. If these pH-induced conformational changes also make relationship-based measurements for sensor readings as a

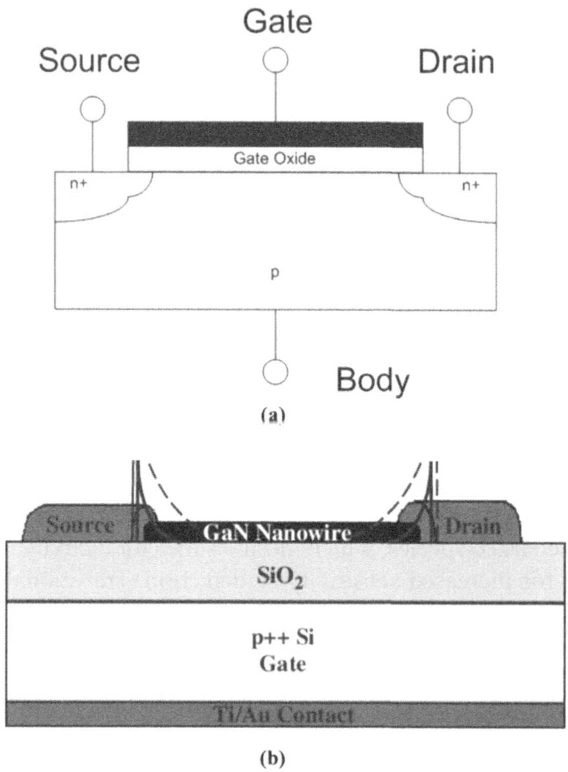

Figure 12.2 (a) Schematics of standard field-effect transistors (b). Schematics of nanowire field-effect transistors.

variable. Therefore, maintaining controlled pH levels during measurements is necessary for obtaining accurate readings [46–53].

The ion condensation effect on the counterions bound to the nanowire will also change the conductivity. When biomolecules that are negatively charged bind to the nanowire in a solvent, they will attract a positively charged counterion to neutralize their charge over an effective distance (called Debye length, λD). Anything beyond this Debye length does not impact the conductivity of the nanowire because electrostatic effects have been screened and will no longer contribute to a detected signal. As such, a longer Debye length will lead to improved sensor sensitivity, while a shorter Debye length will prevent detection. Debye length can be estimated based on the ionic strength (I) of the solution, where $\lambda D=0.304/\sqrt{I}$. Diluting commonly used conductive solutions like phosphate-buffered saline (PBS) with deionized water, increasing the sanitizing effect, will also increase the Debye length and therefore the sensitivity of the sensor. Designing nanowire sensors requires careful control of the Debye length to ensure effective signal detection.

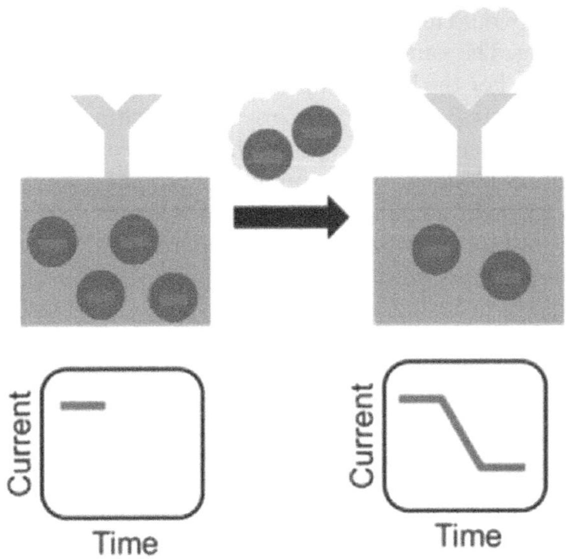

Figure 12.3 The working principle of the nanowire sensor.

Figure 12.3 shows the principle of operation of a nanowire sensor. The negative ions of the aqueous solution bind to the receptor on the Si nanowire FET and induce positive charges on the nanowire surface. Alternatively, we can think of this as a loss of electrons, or an increase in holes, resulting in a difference in conductivity between the source and drain. For n-type Si nanowires, where electrons are the major carriers, the conductivity is decreased. For p-type Si nanowires, where hole carriers are being utilized, the current increases [54–61].

12.5 FABRICATION PROCESS

Nanowire sensors can be made through either bottom-up or top-down methods of synthesis. A bottom-up approach would likely yield high-quality nanowires grown on silicon wafers (Figure 12.4a) [18], as shown in Figure 12.3a Most nanowires made this way would have a cylindrical cross-section. However, bottom-up methods can produce other cross-sectional shapes, including round, square, and triangular shapes. Nanowire fabrication would begin by growing silicon nanowires using chemical vapor deposition (CVD) in a flame or from other materials, using the vapor–liquid–solid (VLS) mechanism. The silicon nanowires would then be suspended in an ethanol suspension and deposited onto a silicon substrate. A photoresist would be spin-coated on the silicon nanowires and, using lift-off, patterned into the metal electrode [62–69]. The last part of the nanowire sensor synthesis would involve passivation and surface modification through receptor binding. The nanowire surface would be passivated by exposing it to air or oxygen. For the bottom-up way of making a sensor, the main issue is that the nanowires are randomly oriented, resulting in poor uniformity and low yield as a sensor. To improve the alignment of nanowires, methods of Langmuir–Blodgett, blown-bubble, microfluidic flow, contact printing, and electric-field alignment are utilized. However, these methods for alignment are also not compatible with standard CMOS fabrication, which would complicate mass production of aligned nanowire sensors.

Unlike bottom-up techniques, top-down methods utilize micro-fabrication on either silicon-on-insulator (SOI) or single-crystalline silicon (SCS) wafers to produce strongly aligned nanowires, while Chipron manufactures Wafer SCS nanowires, and all are compatible with current CMOS technology (Figure 12.4b). First, the top silicon layer of an SOI wafer would be lightly doped with boron or phosphorus to give the top silicon layer the

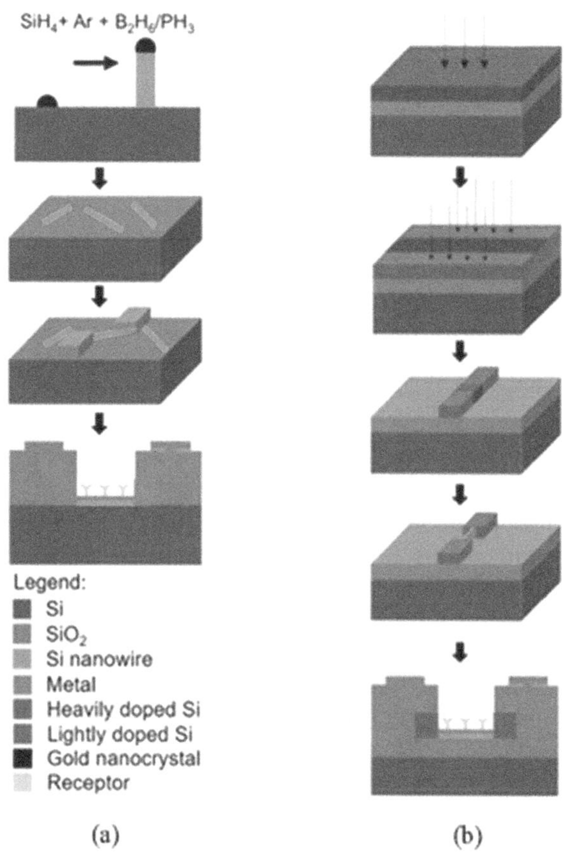

Figure 12.4 (a) The bottom-up approach and (b) the top-down approach.

potential to have a p-type or n-type conductivity. Second, using photolithography, the source and drain would be heavily doped in the patterned areas to form those regions. Micron-sized electrodes are formed through reactive ion etching (RIE) of the doped source and drain regions. Silicon nanowires with nanometer-sized diameters are produced through electron-beam (E-beam) lithography, and a thermal evaporator is used to pattern metal contacts on the silicon wafer after the silicon nanowires are created. The process finishes with passivation and surface alteration, generally the same as what occurs in a bottom-up method [49, 50]. Top-down methods will produce strongly oriented, CMOS-compatible nanowires, while the diameters of the nanowires in all methods are generally bigger than they would be for a bottom-up nanowire. When comparing bottom-up and top-down processes, the orientation may vary and the source and drain materials may vary, but generally the devices all have the same architecture and look similar (drain/nanowire/source) [70–82].

12.6 RECENT ADVANCES IN NANOWIRE BIOSENSORS

This section discusses some of the latest major developments in overcoming the main issues of traditional nanowire-based biosensors, including in vivo sensing, compatibility with inexpensive portable devices, and better signal processing approaches. These developments have certainly increased the usability, sensitivity, and flexibility of nanowire biosensors and their practical use in a much wider array of real-world biomedical applications.

12.6.1 In vivo sensing

Traditionally, nanowires have been utilized exclusively for in vitro uses. However, new developments have broadened interest in utilizing nanowires for continuous in vivo biosignal monitoring. The Lieber group has published pilot studies suggesting that the injection of nanowire sensors directly into the human body is viable for long-term recording of bio-electric signals, for example, recording of neural activity in the brain. The in vivo environment poses serious challenges. High ionic concentrations may degrade the nanowire's natural silicon oxide passivation layer, significantly hindering stability. From the early studies, preliminary research validates that nanowires will degrade rapidly upon use with cells.

To mitigate this, [19] a 10 nm-thick Al_2O_3 shell coating was applied, which dramatically improved the durability of the nanowires. The nanowires with this coating were able to hold their form for over 100 days in 1xPBS at 37°C, while the uncoated nanowires dissolved. Not only did this method work for nanowires of gold, but it was also successful for other nanowire materials, including Si–Ge and InAs complexes. Although the electrochemical sensors had lower sensitivities when the shell thickness increased, a cohort of wire thicknesses allowed for a suitable balance between electrochemical performance and long-term stability. Despite these constraints, protective coatings are a promising first step toward stable and injectable nanowire electronics for in vivo applications [83–89].

Beyond chemical stability, the physical profile of nanowires across biological environments raises questions around cytotoxicity and interactions with cells. Some studies show little to no impact on cell viability or proliferation, while others suggest that 1D nanostructures can elicit cellular responses, including elongation or differentiation. For instance, Chen et al. showed that silicon carbide (SiC) nanowires caused toxicity to human mesenchymal stem cells (hMSCs), reducing adhesion, proliferation, and differentiation, but had little to no effect on the breast cancer cell line, MCF-7. The toxicity was attributed to morphological changes, such as stress, which were confirmed through changes in cytokine gene expressions. In another example, [20] showed that nickel (Ni) nanowires caused DNA damage and oxidative stress, although a direct link was not made to mutagenesis.

In summary, while nanowires have much potential for in vivo biomedical applications, the complexity of their interactions with cells and potential toxicity highlight the need for more systematic studies. It is paramount to demonstrate both long-term stability and biocompatibility in order to drive the development of nanowires as the basis for sensing technologies in clinical applications.

12.6.2 Integration

Besides using traditional microfluidic channel-based devices, nanowires have recently been coupled with paper-based analytical devices (PADs) to allow for inexpensive and portable biomarker detection. PADs are simple, low-cost, materials and fabrication, recyclable/biodegradable, and easy to handle which make them well suited for point-of-care and/or rural applications. One of the more intriguing approaches is to grow zinc oxide (ZnO) nanowires directly onto the paper substrate with conductive inks and low-temperature hydrothermal growth. This method allows for large-scale production and provides good control over nanowire length [90–97].

These nanowire-enhanced PADs have demonstrated promising results in detecting biomarkers like glucose, showing decent sensitivity and a practical detection range. However, their detection limits are still higher compared to more advanced sensors, which limits their use in applications requiring ultra-low concentration detection. Key challenges include maintaining consistent surface temperature and addressing the high surface roughness of paper, which hampers low concentration detection. Nanofibrillated cellulose (NFC) paper has emerged as a potential solution, offering improved surface uniformity at the nanoscale.

To advance remote and portable sensing, future development needs a strong emphasis on wireless and self-powered technology, achieved by converging PADs and energy harvesting technologies to enable efficient biomarker monitoring on-site without external power.

12.6.3 Self-powering

There are two major classifications of self-powering approaches to nanowire devices: those using triboelectric or piezoelectric nanogenerators and those utilizing biofuels. Piezoelectric nanogenerators based on ZnO nanowires have been extensively studied since the fortuitous discovery of the piezoelectric effect in ZnO nanowires in 2006, and they are utilized to generate energy. There are physical constraints limiting in vivo mechanical energy sources, such as the lung, muscle, and cardiac movements. For example, one can apply mechanical stress to a well-aligned array of ZnO nanowires and harvest up to 1.26 volts, which is enough to power various devices, including nanowire-based pH sensors and wireless data transmission systems. However, there are limits to how you can use this mechanical energy from the many in vivo sources [98–105].

Consequently, biofuels, very much a niche fuel source, may be a more favorable in vivo source of energy. The glucose/oxygen and glucose/air research/reaction systems have been researched, and many glucose-based systems have been developed to generate power internally. Hansen et al. proposed a hybrid self-powered system that combined the piezoelectric and biofuel approaches. In this system, nanofibers of poly(vinylidene fluoride) could harvest the mechanical energy of movements associated with breathing and heartbeats, while a flexible enzymatic biofuel cell could harvest the biochemical energy from its surrounding body fluids. This dual approach was believed to increase the power outputs and extend overall operation time, conferring extremely important/desirable properties for in vivo (in-body) applications and processes.

12.6.4 Signal processing and data analysis

Disease diagnostics often require analysis of multiple biomarkers, and there have been a lot of different proposals for multiplexing biosensors, but most have stayed at the proof-of-concept stage, usually validated by spiked samples and not clinical samples. To unlock the maximum potential of multivariable and multiplexed biosensors for practical diagnostics, more advanced signal processing for diagnostics is needed to help extract meaningful diagnostic information priory from sensor datasets.

The concept of using signal processing to look at data from nanowire sensor arrays was first investigated in gas-sensing-type sensors. For example, a group of researchers used sensor arrays made from platinum, copper, indium, and nickel to detect vapors from explosive precursors. Once the signals from the sensor arrays were collected, machine learning models, such as decision trees, were used to classify signals to determine the concentrations of chemicals and to assess the likelihood of detonating or releasing a potential explosive. In the end, these sensors could predict with a high degree of detection accuracy [106–112].

A modified application for medical diagnostics was developed from this. In one application, an array of silicon nanowires was functionalized to detect a breath biomarker for specific diseases, from gastric cancer to lung cancer, asthma, and chronic obstructive pulmonary disease (COPD). After characterizing the sensors with simulated breath samples, an artificial neural network (ANN) was trained with the sensor data to classify the disease states. The system was validated with real patient breath samples, obtaining over 80% correct results when differentiating various disease states [113–121].

As these examples show, the combination of nanowire sensor arrays with machine learning provides many opportunities for enhanced diagnostic performance with more dependable multi-biomarker detection methods and applications in a clinical setting.

12.7 CONCLUSION

Overall, biosensors based on nanomaterials are a game-changing technology at the intersection of nanotechnology, biotechnology, and analytical science. Their capacity to detect a wide variety of analytes with enhanced sensitivity, selectivity, and stability makes biosensors indispensable for contemporary diagnostics. This review highlighted key examples of nanostructures, including metal oxide nanoparticles, carbon nanotubes, quantum dots, and nanowires, which have been very effective at improving the performance of biosensors. Their unique physicochemical properties allow for enhanced charge (electron or ion) transfer, enabling a strong surface interaction and signal amplification. As biosensor technology continues to evolve toward customization with respect to miniaturization, automation, and integration with Internet of Things (IoT) and AI, and deep learning systems become widely employed, the landscape of commercialization of biosensor technology becomes clearer with greater access to Point-of-Care Testing (POCT) in real-time. The limitations, including interference due to ionic strength concentrations provided by physiological samples, encourage researchers to think outside the box for the development of innovation, including sample preprocessing methods. These challenges do not detract from the potential capability of nanobiosensors in healthcare or environmental monitoring, food safety, or biosecurity. Nano biosensors are transforming diagnostics and establishing intelligent, connected sensing platforms utilizing nanomaterials. The future of biosensing will be together and in this convergence of technology, for rapid, accessible, and accurate analytical platform providing accurate and rapid analysis in the future.

REFERENCES

1. Shao, Q., Lu, K., Huang, X. Platinum Group Nanowires for Efficient Electrocatalysis. *Small Methods*, 3, 1800545, 2019.
2. Li, X., Mo, J., Fang, J., Xu, D., Yang, C., Zhang, M., Li, H., Xie, X., Hu, N., Liu, F. Vertical Nanowire Array-Based Biosensors: Device Design Strategies and Biomedical Applications. *Journal of Materials Chemistry B*, 8, 7609–7632, 2020.
3. Tran, V.A., Tran, N.T., Doan, V.D., Nguyen, T.-Q., Thi, H.H.P., Vo, G.N.L. Application Prospects of MXenes Materials Modifications for Sensors. *Micromachines*, 14, 247, 2023.
4. Adam, T., Dhahi, T.S., Gopinath, S.C.B., Hashim, U. Novel Approaches in Fabrication and Integration of Nanowire for Micro/Nano Systems. *Critical Reviews in Analytical Chemistry*, 52, 1913–1929, 2022.
5. Shi, J., Sun, C., Liang, E., Tian, B. Semiconductor Nanowire-Based Cellular and Subcellular Interfaces. *Advanced Functional Materials* 2022, 32, 2107997.
6. Smith, A.F., Liu, X., Woodard, T.L., Fu, T., Emrick, T., Jiménez, J.M., Lovley, D.R., Yao, J. Bioelectronic Protein Nanowire Sensors for Ammonia Detection. *Nano Research*, 13, 1479–1484, 2020.
7. Fanton, L., Loria, F., Amores, M., Pazos, M.R., Adán, C., García-Muñoz, R.A., Marugán, J. Proliferation of Osteoblast Precursor Cells on the Surface of TiO2 Nanowires Anodically Grown on a-Type Biomedical Titanium Alloy. *Scientific Reports*, 12, 7895, 2022.
8. Liu, D., Liu, Y., Liu, X., Xu, C., Zhu, J., Chen, H. Growth of Uniform CuCo2O4 Porous Nanosheets and Nanowires for High Performance Hybrid Supercapacitors. *Journal of Energy Storage*, 52, 105048, 2022.
9. Prameswati, A., Han, J.W., Kim, J.H., Wibowo, A.F., Entifar, S.A.N., Park, J., Lee, J., Kim, S., Lim, D.C., Moon, M.-W., et al. Highly stretchable and mechanically robust silver nanowires on surface-functionalized wavy elastomers for wearable healthcare electronics. *Organic Electronics*, 108, 106584, 2022.
10. Chang, T.-Y., Kim, H., Zutter, B.T., Lee, W.-J., Regan, B.C., Huffaker, D.L. Orientation-Controlled Selective-Area Epitaxy of III–V Nanowires on (001) Silicon for Silicon Photonics. *Advanced Functional Materials*, 30, 2002220, 2020.
11. Tran, V.A., Lee, S.-W. pH-Triggered Degradation and Release of Doxorubicin From Zeolitic Imidazolate Framework-8 (ZIF8) Decorated With Polyacrylic Acid. *RSC Advances*, 11, 9222–9234, 2021.
12. Tran, V.A., Lee, S.-W. A Prominent Anchoring Effect on the Kinetic Control of Drug Release From Mesoporous Silica Nanoparticles (MSNs). *Journal of Colloid and Interface Science*, 510, 345–356, 2018.
13. Kim, K., Park, C., Kwon, D., Kim, D., Meyyappan, M., Jeon, S., Lee, J.-S. Silicon Nanowire Biosensors for Detection of Cardiac Troponin I (cTnI) With High Sensitivity. *Biosensors & Bioelectronics*, 77, 695–701, 2016.
14. Zhang, H., Kikuchi, N., Ohshima, N., Kajisa, T., Sakata, T., Izumi, T., Sone, H. Design and Fabrication of Silicon Nanowire-Based Biosensors with Integration of Critical Factors: Toward Ultrasensitive Specific Detection of Biomolecules. *ACS Applied Materials & Interfaces*, 12, 51808–51819, 2020.
15. Takahashi, H., Baba, Y., Yasui, T. Oxide Nanowire Microfluidics Addressing Previously-Unattainable Analytical Methods for Biomolecules Towards Liquid Biopsy. *Chemical Communications*, 57, 13234–13245, 2021.

16. Naresh, V., Lee, N., A Review on Biosensors and Recent Development of Nanostructured Materials-Enabled Biosensors. *Sensors*, 21(4), 1109, 2021. doi:10.3390/s21041109
17. Karunakaran, C., Rajkumar, R., Bhargava, K. Introduction to Biosensors. In *Biosensors and Bioelectronics* (pp. 1–68), Elsevier, 2015.
18. Baraban, L., Zörgiebel, F., Pahlke, C., Baek, E. *Nanowire Field Effect Transistors: Principles and Applications*. Springer International Publishing, 2014.
19. Zhou, W., Dai, X., Lieber, C.M. Advances in Nanowire Bioelectronics. *Reports on Progress in Physics*, 80, 016701, 2017.
20. Alaraby, M., Hernández, A., Marcos, R. Systematic In Vivo Study of NiO Nanowires and Nanospheres: Biodegradation, Uptake and Biological Impacts. *Nanotoxicology*, 12, 1–18, 2018.
21. Bhardwaj, A., Kumar, P., Raj, B., Anand, S. Design and Performance Optimization of Doping-Less Vertical Nanowire TFET Using Gate Stack Technique. *Journal of Electronic Materials (JEMS)*, 41(7), 4005–4013, 2022.
22. Singh, J., Raj, B. Tunnel Current Model of Asymmetric MIM Structure Levying Various Image Forces to Analyze the Characteristics of Filamentary Memristor *Applied Physics A*, 125(3), 203.1–203.11, 2019.
23. Goyal, C., Ubhi, J. S., Raj, B. Low Leakage Zero Ground Noise Nanoscale Full Adder using Source Biasing Technique. *Journal of Nanoelectronics and Optoelectronics*, 14, 360–370, 2019.
24. Singh, G., Sarin, R. K., Raj, B. A Novel Robust Exclusive-OR Function Implementation in QCA Nanotechnology With Energy Dissipation Analysis. *Journal of Computational Electronics*, 15(2), 455–465, 2016.
25. Wadhwa, G., Kamboj, P., Raj, B. Design optimisation of junctionless TFET biosensor for high sensitivity *Advances in Natural Sciences: Nanoscience and Nanotechnology*, 10(7), 045001, 2019.
26. Bansal, P., Raj, B. Memristor modeling and Analysis for Linear Dopant Drift Kinetics. *Journal of Nanoengineering and Nanomanufacturing*, 6, 1–7, 2016.
27. Ashima, D. V., Raj, B. Performance Analysis of Charge Plasma Induced Graded Channel Si Nanotube. *Journal of Engineering Research (JER)*, EMSME Special Issue 146–154, 2021.
28. Tomar, A. S., Magraiya, V. K., Raj, B. Scaling of Access and Data Transistor for High Performance DRAM Cell Design. *Quantum Matter*, 2, 412–416, 2013.
29. Jain, N., Raj, B. Parasitic Capacitance and Resistance Model Development and Optimization of Raised Source/Drain SOI FinFET Structure for Analog Circuit Applications. *Journal of Nanoelectronics and Optoelectronins*, 13, 531–539, 2018.
30. Singh, S., Vishvakarma, S. K., Raj, B. Analytical Modeling of Split-Gate Junction-Less Transistor for a Biosensor Application. *Sensing and Bio-Sensing*, 18, 31–36, 2018.
31. Gopal, M., Raj, B., Low Power 8T SRAM Cell Design for High Stability Video Applications. *ITSI Transaction on Electrical and Electronics Engineering*, 1(5), 91–97, 2013.
32. Raj, Balwinder, Mitra, Jatin, Bihani, Deepak Kumar, Rangharajan, V., Saxena, A.K., Dasgupta, S. Analysis of Noise Margin, Power and Process Variation for 32 nm FinFET Based 6T. *SRAM Cell Journal of Computer (JCP)*, 5(6), 1–8, 2010.

33. Sharma, D., Mehra, R., Raj, B. Comparative Analysis of Photovoltaic Technologies for High Efficiency Solar Cell Design. *Superlattices and Microstructures*, 153, 106861, 2021.
34. Kaur, P., Buttar, A. S., Raj, B., A comprehensive Analysis of Nanoscale Transistor based Biosensor: A Review. *Indian Journal of Pure and Applied Physics*, 59, 304–318, 2021.
35. Yadav, D., Raj, Balwant, Raj, B. Design and Simulation of Low Power Microcontroller for IoT Applications. *Journal of Sensor Letters*, 18, 401–409, May 2020.
36. Singh, Shailendra, Raj, B. A 2-D Analytical Surface Potential and Drain current Modeling of Double-Gate Vertical t-Shaped Tunnel FET. *Journal of Computational Electronics*, 19, 1154–1163, 2020.
37. Singh, J., Raj, B., An Accurate and Generic Window Function for Non-linear Memristor Model. *Journal of Computational Electronics*, 18(2), 640–647, 2019.
38. Kaur, Manjit, Gupta, Neena, Kumar, Sanjeev, Raj, B., Singh, Arun Kumar Comparative RF and Crosstalk Analysis of Carbon Based Nano Interconnects. *IET Circuits, Devices and Systems*, 15(6), 493–503, 2021.
39. Kandasamy, Nehru, Ahmad, Firdous, Ajitha, D., Raj, B., Telagam, N. Quantum Dot Cellular Automata based Scan Flip Flop and Boundary Scan Register. *IETE Journal of Research*, 66, 535–548, 2020.
40. Singh, J., Raj, B., Design and Investigation of 7T2M NVSARM With Enhanced Stability and Temperature Impact on Store/Restore Energy. *IEEE Transactions on Very Large Scale Integration Systems*, 27(6), 1322–1328, 2019.
41. Bhardwaj, Anil Kumar, Gupta, Sumeet, Raj, B., Singh, A. Impact of Double Gate Geometry on the Performance of Carbon Nanotube Field Effect Transistor Structures for Low Power Digital Design. *Computational and Theoretical Nanoscience*, 16, 1813–1820, 2019.
42. Jain, Neeraj, Raj, B. Thermal Stability Analysis and Performance Exploration of Asymmetrical Dual-k Underlap Spacer (ADKUS) SOI FinFET for Security and Privacy Applications. *Indian Journal of Pure & Applied Physics (IJPAP)*, 57, 352–360, 2019.
43. Jain, N., Raj, Balwinder Dual-k Spacer Region Variation at the Drain Side of Asymmetric SOI FinFET Structure: Performance Analysis Towards the Analog/RF Design Applications. *Journal of Nanoelectronics and Optoelectronics*, 14, 349–359, 2019.
44. Singh, Jeetendra, Sharma, Sanjeev, Raj, B., Khosla, M. Analysis of Barrier Layer Thickness on Performance of In1-xGaxAs Based Gate Stack Cylindrical Gate Nanowire MOSFET. *JNO*, 13, 1473–1477, 2018.
45. Jain, Neeraj, Raj, B. Analysis and Performance Exploration of High-k SOI FinFETs Over the Conventional Low-k SOI FinFET Toward Analog/RF *Design Journal of Semiconductors (JoS)*, 39(12), 124002-1-7, 2018.
46. Goyal, Candy, Ubhi, Jagpal Singh, Raj, B. A Reliable Leakage Reduction Technique for Approximate Full Adder With Reduced Ground Bounce Noise. *Journal of Mathematical Problems in Engineering*, 3501041. https://doi.org/10.1155/2018/3501041
47. Anuradha, Jeetendra Singh, Raj, B., Khosla, M., Design and Performance Analysis of Nano-Scale Memristor-Based Nonvolatile SRAM. *Journal of Sensor Letter*, 16, 798–805, 2018.

48. Wadhwa, Girish, Raj, B., Parametric Variation Analysis of Charge-Plasma-based Dielectric Modulated JLTFET for Biosensor Application. *IEEE Sensors Journal*, 18(15), 6070–6077, 2018.
49. Singh, Jeetendra, Raj, B. Comparative Analysis of Memristor Models for Memories Design, *JoS*, 39(7), 074006-1-12, 2018.
50. Yadav, Divya, Chouhan, Shailesh Singh, Vishvakarma, Santosh Kumar, Raj, B. Application Specific Microcontroller Design for IoT Based WSN. *Sensor Letters*, 16, 374–385, 2018.
51. Singh, Gurmohan, Sarin, R. K., Raj, B. Fault-Tolerant Design and Analysis of Quantum-Dot Cellular Automata Based Circuits. *IEEE/IET Circuits, Devices & Systems*, 12, 638–664, 2018.
52. Singh, Jeetendra, Raj, B. Modeling of Mean Barrier Height Levying Various Image Forces of Metal Insulator Metal Structure to Enhance the Performance of Conductive Filament Based Memristor Model. *IEEE Nanotechnology*, 17(2), 267–268, 2018.
53. Jain, Aakash, Sharma, Sanjeev, Raj, B. Analysis of Triple Metal Surrounding Gate (TM-SG) III-V Nanowire MOSFET for Photosensing Application. *Opto-Electronics Journal*, 26(2), 141–148, 2018.
54. Jain, Aakash, Sharma, Sanjeev, Raj, B. Design and Analysis of High Sensitivity Photosensor Using Cylindrical Surrounding Gate MOSFET For Low Power Sensor Applications. *Engineering Science and Technology, an International Journal*, 19(4), 1864–1870, 2016.
55. Saiphani Kumar, G., Singh, Amandeep, Raj, B. Design and Analysis of Gate All Around CNTFET based SRAM cell Design. *Journal of Computational Electronics*, 17(1), 138–145, 2018.
56. Singh, Gurinder Pal, Sohi, B. S., Raj, Balwinder Material Properties Analysis of Graphene Base Transistor (GBT) for VLSI Analog Circuits. *Indian Journal of Pure & Applied Physics (IJPAP)*, 55, 896–902, 2017.
57. Kumar, Sunil, Raj, B. Estimation of Stability and Performance Metric for Inward Access Transistor Based 6T SRAM Cell Design Using n-Type/p-Type DMDG-GDOV TFET *IEEE VLSI Circuits and Systems Letter*, 3(2), 25–39, 2017.
58. Sharma, Shashikant, Kumar, Anjan, Pattanaik, Manisha, Raj, B. Forward Body Biased Multimode Multi-Threshold CMOS Technique for Ground Bounce Noise Reduction in Static CMOS Adders. *International Journal of Information and Electronics Engineering*, 3(3), 567–572, 2013.
59. Singh, Hamendra, Kumar, Pankaj, Raj, Balwinder Performance Analysis of Majority Gate SET Based 1-bit Full Adder. *International Journal of Computer and Communication Engineering (IJCCE)*, 2(4), 2010–3743, 2013.
60. Bhardwaj, A. K., Gupta, S., Raj, B. Investigation of Parameters for Schottky Barrier (SB) Height for Schottky Barrier Based Carbon Nanotube Field Effect Transistor Device. *Journal of Nanoelectronics and Optoelectronics*, 15, 783–791, 2020.
61. Bansal, P., Raj, B. Memristor: A Versatile Nonlinear Model for Dopant Drift and Boundary Issues. *JCTN*, 14(5), 2319–2325, 2017.
62. Jain, N., Raj, B. An Analog and Digital Design Perspective Comprehensive Approach on Fin-FET (Fin-Field Effect Transistor) Technology – A Review. *Advanced Sciences and Engineering (RASE)*, 5, 1–14, 2016.

63. Raj, Balwinder, Saxena, A. K., Dasgupta, S. Analytical Modeling for the Estimation of Leakage Current and Subthreshold Swing Factor of Nanoscale Double Gate FinFET. *Device Microelectronics International*, 26, 53–63, 2009.
64. Soniya, S. S., Wadhwa, G., Raj, Balwinder An Analytical Modeling for Dual Source Vertical Tunnel Field Effect Transistor. *International Journal of Recent Technology and Engineering (IJRTE)*, 8(2), 603–608, 2019.
65. Singh, S., Raj, B. Design and Analysis of Hetrojunction Vertical T-Shaped Tunnel Field Effect Transistor. *Journal of Electronics Material*, 48(10), 6253–6260, 2019.
66. Goyal, C., Ubhi, J. S., Raj, B. A Low Leakage CNTFET Based Inexact Full Adder for Low Power Image Processing Applications. *International Journal of Circuit Theory and Applications*, 47(9), 1446–1458, 2019.
67. Raj, B., Saxena, A. K., Dasgupta, S. A Compact Drain Current and Threshold Voltage Quantum Mechanical Analytical Modeling for FinFETs. *Journal of Nanoelectronics and Optoelectronics (JNO)*, 3(2), 163–170, 2008.
68. Wadhwa, G., Raj, B. An Analytical Modeling of Charge Plasma based Tunnel Field Effect Transistor With Impacts of Gate underlap Region. *Superlattices and Microstructures*, 142, 106512, 2020.
69. Singh, S., Raj, B. Modeling and Simulation analysis of SiGe hetrojunction Double GateVertical t-shaped Tunnel FET. *Superlattices and Microstructures* 142, 106496, 2020.
70. Jain, N., Raj, B. Device and Circuit Co-Design Perspective Comprehensive Approach on FinFET Technology – A Review. *Journal of Electron Devices*, 23(1), 1890–1901, 2016.
71. Kumar, S., Raj, B. Analysis of I_{ON} and Ambipolar Current for Dual-Material Gate-Drain Overlapped DG-TFET. *Journal of Nanoelectronics and Optoelectronics*, 11, 323–333, 2016.
72. Anjum, N., Bali, T., Raj, B. Design and Simulation of Handwritten Multiscript Character Recognition. *International Journal of Advanced Research in Computer and Communication Engineering*, 2 (7), 2544–2549, 2013.
73. Singh, Karmjit, Raj, B. Performance and Analysis of Temperature Dependent Multi-Walled Carbon Nanotubes as Global Interconnects at Different Technology Nodes. *Journal of Computational Electronics*, 14(2), 469–476, 2015.
74. Kumar, Sunil, Raj, B. Compact Channel Potential Analytical Modeling of DG-TFET Based on Evanescent–Mode Approach *Journal of Computational Electronics*, 14(2), 820–827, 2015.
75. Singh, K., Raj, B. Temperature Dependent Modeling and Performance Evaluation of Multi-Walled CNT and Single-Walled CNT as Global Interconnects. *Journal of Electronic Materials*, 44(12), 4825–4835, 2015.
76. Sharma, V. K., Pattanaik, M., Raj, B. INDEP Approach for Leakage Reduction in Nanoscale CMOS Circuits. *International Journal of Electronics*, 102(2), 200–215, 2014.
77. Singh, K., Raj, B. Influence of Temperature on MWCNT Bundle, SWCNT Bundle and Copper Interconnects for Nanoscaled Technology Nodes *Journal of Materials Science: Materials in Electronics*, 26(8), 6134–6142, 2015.
78. Anjum, N., Bali, T., Raj, B. Design and Simulation of Handwritten Gurumukhi and Devanagri Numerical Recognition. *International Journal of Computer Applications*, 73(12), 16–21, 2013.

79. Khandelwal, S., Gupta, V., Raj, B., Gupta, R. D. Process Variability Aware Low Leakage Reliable Nano Scale DG-FinFET SRAM Cell Design Technique. *Journal of Nanoelectronics and Optoelectronics*, 10(6), 810–817, 2015.
80. Sharma, V. K., Pattanaik, M., Raj, B. ONOFIC Approach: Low Power High Speed Nanoscale VLSI Circuits Design. *International Journal of Electronics*, 101(1), 61–73, 2014.
81. Khandelwal, S., Raj, Balwinder, Gupta, R. D. FinFET Based 6T SRAM Cell Design: Analysis of Performance Metric, Process Variation and Temperature Effect. *Journal of Computational and Theoretical Nanoscience*, 12, 2500–2506, 2015.
82. Singh, S., Yadav, S., Rahul, J., Srivastava, A., Raj, B. Impact of HfO_2 in Graded Channel Dual Insulator Double Gate MOSFET. *Journal of Computational and Theoretical Nanoscience*, 12(6), 950–953, 2015.
83. Sharma, V. K., Pattanaik, M., Raj, B. PVT Variations Aware Low Leakage INDEP Approach for Nanoscale CMOS Circuits. *Microelectronics Reliability*, 54, 90–99, 2014.
84. Raj, B., Saxena, A. K., Dasgupta, S. Quantum Mechanical Analytical Modeling of Nanoscale DG FinFET: Evaluation of Potential, Threshold Voltage and Source/Drain Resistance. *Elsevier's Journal of Material Science in Semiconductor Processing*, 16(4), 1131–1137, 2013.
85. Gopal, Maisagalla, Prasad, S. S. D., Raj, B. 8T SRAM Cell Design for Dynamic and Leakage Power Reduction. *International Journal of Computer Applications*, 71(9), 43–48, 2013.
86. Pattanaik, M., Raj, B., Sharma, S., Kumar, A. Diode Based Trimode Multi-Threshold CMOS Technique for Ground Bounce Noise Reduction in Static CMOS Adders. *Advanced Materials Research*, 548, 885–889, 2012.
87. Raj, B., Saxena, A. K., Dasgupta, S. Nanoscale FinFET Based SRAM Cell Design: Analysis of Performance Metric, Process Variation, Underlapped FinFET and Temperature Effect. *IEEE Circuits and Systems Magazine*, 11(2), 38–50, 2011.
88. Wadhwa, G., Raj, B. Label Free Detection of Biomolecules using Charge-Plasma-Based Gate Underlap Dielectric Modulated Junctionless TFET. *Journal of Electronic Materials (JEMS)*, 47(8), 4683–4693, 2018.
89. Singh, Gurmohan, Sarin, R. K., Raj, B. Design and Performance Analysis of a New Efficient Coplanar Quantum-Dot Cellular Automata Adder. *Indian Journal of Pure & Applied Physics (IJPAP)*, 55, 97–103, 2017.
90. Kaur, P., Pandey, Vikas, Raj, B. Comparative Study of Efficient Design, Control and Monitoring of Solar Power Using IoT. *Sensor Letters* 18, 419–426, 2020.
91. Bhardwaj, Anil Kumar, Gupta, Sumeet, Raj, B. Development & Analysis of compact model for Double Gate Schottky Barrier CNTFET, *Journal of Nanoelectronics and Optoelectronics*, 15, 1199–1208, 2020.
92. Wadhwa, Girish, Kamboj, Priyanka, Singh, Jeetendra, Raj, B. Design and Investigation of Junctionless DGTFET for Biological Molecule Recognition. *Transactions on Electrical and Electronic Materials*, 22, 282–289, 2021.
93. Verma, S. K., Singh, S., Wadhwa, G., Raj, B. Detection of Biomolecules using Charge-Plasma based Gate Underlap Dielectric Modulated Dopingless TFET. *Transactions on Electrical and Electronic Materials (TEEM)*, 21, 528–535, 2020.

94. Jain, Neeraj, Raj, B. Impact of Underlap Spacer Region Variation on Electrostatic and Analog/RF Performance of Symmetrical High-k SOI FinFET at 20 nm Channel Length. *Journal of Semiconductors (JoS)*, 38(12), 122002, 2017.
95. Singh, Shailendra, Raj, B. Analytical Modeling and Simulation Analysis of T-Shaped III-V heterojunction Vertical T-FET. *Superlattices and Microstructures*, 147, 106717, 2020.
96. Singh, Gurmohan, Sarin, R. K., Raj, B. Design and Analysis of Area Efficient QCA Based Reversible Logic Gates. *Journal of Microprocessors and Microsystems*, 52, 59–68, 2017.
97. Singh, Shailendra, Raj, B. Parametric Variation Analysis on Hetero-Junction Vertical t-Shape TFET for Supressing Ambipolar Conduction. *Indian Journal of Pure and Applied Physics*, 58, 478–485, 2020.
98. Soniya, Singh, Shailendra, Wadhwa, G., Raj, B. Design and Analysis of Dual Source Vertical Tunnel Field Effect Transistor for High Performance. *Transactions on Electrical and Electronic Materials*, 21, 74–82, 2019.
99. Kaur, M., Gupta, N., Kumar, S., Raj, B., Singh, A. K., RF Performance Analysis of Intercalated Graphene Nanoribbon Based Global Level Interconnects. *Journal of Computational Electronics*, 19, 1002–1013, 2020.
100. Wadhwa, Girish, Raj, B. Design and Performance Analysis of Junctionless TFET Biosensor for High Sensitivity *IEEE Nanotechnology*, 18, 567–574, 2019.
101. Singh, Jeetendra, Raj, B. Enhanced Nonlinear Memristor Model Encapsulating Stochastic Dopant Drift. *JNO*, 14, 958–963 (2019)
102. Kumar, Mohit, Chaudhary, Tarun, Raj, Balwinder, Design and Comparative Analysis of Ferroelectric Nanowire With Dielectric HfO_2 and $Al_2O_3$for Low Power Applications *Journal of Electronic Materials*, 53, 6785–6791, 2024.
103. Thakur, Yogesh, Raj, Balwant, Raj, Balwinder, Design of Pentacene Thin-Film Transistor Based Hydrogen Gas Sensor for High Sensitivity. *ECS Journal of Solid State Science and Technology*, 13, 047005–047012, 2024.
104. Kumar, M. A., Ismail, K. B. M., Mahalingam, S., Raj, B., Kim, J., Atchudan, R. Recent Advances in Two-Dimensional Nanomaterials and Nanostructures for Healthcare Monitoring. *ACS Sensors*, 9(4), 1–8, 2024.
105. Batcha, M. I. K., Manoharan, A. K., Mahalingam, S., Balwinder, R. Carbon Fiber-Reinforced Polymers for Energy Storage Applications. *Journal of Energy Storage*, 84(Part B), 110931–110939, 2024.
106. Hussain, Shahraan, Raj, Balwinder, Sharma, Divya, Design and Parametric Optimization of Lead Free Perovskite based Tandem Solar Cell for High Efficient. *Engineering Research Express*, 6, 015315–0153122, 2024
107. Kumar, Parveen, Raj, Balwinder, Wadhwa, Girish, Singh, Balwinder, Kumar, Raj Design and Analysis of Junctionless based Gate All Around N+ Doped Layer Nanowire TFET Biosensor. *ECS Journal of Solid State Science and Technology*, 13, 017002–017010, 2024.
108. Sharma, Divya, Mehra, Rajesh, Raj, Balwinder, Enhancement in Efficiency of Methyl Ammonium Tin Iodide Based perovskite Solar Cell Using SCAPS-1 D, *Nano*, 18(14), 2350109, 2023.
109. Kumar, Parveen, Sharma, Sanjeev Kumar, Raj, Balwant, Raj, Balwinder Analysis of Device Parameter Variations in In1-xGaxAs based Gate Stacked

Double Metal Surrounding Gate Nanowire MOSFET *TEEM Springer*, 24, 570–578, 2023

110. Bhardwaj, A., Kumar, P., Raj, B., Kumar, N., Anand, S. Design and Optimization of Vertical Nanowire Tunnel FET With Electrostatic Doping. *Engineering Research Express*, 5, 045025–045033, 2023.
111. Thakur, Yogesh, Raj, B., Gill, S. S. Design and Performance Analysis of Pentacene Organic Field Effect Transistor With High-K dielectric Materials. *Journal of Optoelectronics and Advanced Materials – Rapid Communications*, 17(7–8), 2023.
112. Kumar, Parveen, Raj, Balwinder Design and Simulation of Junctionless Nanowire Tunnel Field Effect Transistor Biosensor for High Sensitivity. *Microelectronics Journal*, 137, 105826–105834, 2023.
113. Sharma, V. K., Pattanaik, M., Raj, Balwinder Leakage Current ONOFIC Approach for Deep Submicron VLSI Circuit Design. *International Journal of Electrical, Computer, Electronics and Communication Engineering*, 7(4), 239–244, 2013.
114. Kaur, Parminder, Gill, Sandeep Singh, Raj, B. Comparative Analysis of OFETs Materials and Devices for Sensor Applications. *Journal of Silicon*, 14, 4463–4471, 2022.
115. Sharma, Sanjeev Kumar, Kumar, Parveen, Raj, Balwant, Raj, B., $In_{1-x}Ga_xAs$ Double Metal Gate-Stacking Cylindrical Nanowire MOSFET for Highly Sensitive Photo Detector. *Journal of Silicon*, 14, 3535–3541, 2022.
116. Raj, B., Saxena, A. K., Dasgupta, S. Analytical Modeling of Quasi Planar Nanoscale Double Gate FinFET With Source/Drain Resistance and Field Dependent Carrier Mobility: A Quantum Mechanical Study. *Journal of Computer (JCP)*, 4(9), 1–8, 2009.
117. Bhushan, S., Khandelwal, S., Raj, B.. Analyzing Different Mode FinFET Based Memory Cell at Different Power Supply for Leakage Reduction, Seventh International Conference on Bio-Inspired Computing: Theories and Application, (BIC-TA 2012). *Adv. Intell. Syst. Comput.*, 202, 89–100, 2012.
118. Singh, Jeetendra, Raj, B. Temperature Dependent Analytical Modeling and Simulations of Nanoscale Memristor. *Journal: Engineering Science and Technology, an International Journal*, 21, 862–868, 2018.
119. Singh, Shradhya, Bala, Shashi, Raj, Balwant, Raj, B. Improved Sensitivity of Dielectric Modulated Junctionless Transistor for Nanoscale Biosensor Design. *Sensor Letters*, 18, 328–333, 2020.
120. Kumar, Vivek, Vishvakarma, S. K., Raj, B., Design and Performance Analysis of ASIC for IoT Applications. *Sensor Letters*, 18, 31–38, 2020.
121. Jaiswal, Akanksha, Sarin, R. K., Raj, B., Sukhija, Shikha A Novel Circular Slotted Microstrip-fed Patch Antenna with Three Triangle Shape Defected Ground Structure for Multiband Applications. *Advanced Electromagnetic (AEM)*, 7(3), 56–63, 2018.

Index

Pages in *italics* refer to figures and pages in **bold** refer to tables.

AI applications in circuit design, 224; *see also* machine learning
analog circuits, 14–15, 55, 57
artificial intelligence, 103, 105, 107–108, 224, 320
  integration with VLSI, 13, 55
ASICs (application-specific integrated circuits), 18, 59, 91, 114

bandgap engineering, 63
biosensors, 66–67, 118–137, 159–160, 236, *245*, 246, **247**, 309–320
bulk CMOS process, 150, 152–153

charge plasma devices, 156
CMOS circuits, 10
  fabrication process, 241
combinational logic circuits, 224
  adders, subtractors, and encoders, 225
complementary MOSFETs, 154
conductivity modulation, 126
counters, 2, 102

device modeling, 189, 240
  simulation methods, 189
digital electronics, 204, 224–225
diode characteristics, 290
drain current modeling, 237–238, 240
dynamic logic, 209, 221

electrical characteristics, 69, 71, 73–74
energy band diagram, 175, 178
error detection and correction, 101–102

ferroelectric materials, 71
flip-flops, 142, 165, 195
floating gate MOSFETs, 69
frequency response, 76

gate oxide thickness, 155
gate stack design, 176, 186
germanium heterojunctions, 292
graphene nanoribbons, 91, 115

hardware implementation, 203–204
heterojunction TFETs, 175, 181–182
high-k dielectrics, 172–174, 184, 186, *188*

IC fabrication process, 150, 242
InAs nanowires, 6, **299**, 317
input–output characteristics, 273–274
integrated circuits, 150, 154, 203, 215, 235–256

junctionless devices, 181

logic gates, 69, 203
longitudinal optical (LO) phonon modes, 7
low-power circuit design, 211; *see also* TFET

machine learning in electronics, 94, 98–99; *see also* AI
memory architecture, 101

metal–oxide–semiconductor (MOS) structures, 74, 154–155, 219
modeling tools, 103
  TCAD, Sentaurus, and Silvaco, 186
MOSFET characteristics, 175, 186, 192

nanodevice simulation, 211
nanoelectronics, 1, 10, 149, 237, 292
nanowires (NWs), 1–11, 119–121, 123–125, 131–133, 149–151, 157, 171–193, 203–226, 237–240, 249–250, 266–267, 290–300, 309, 317–320
number systems, 36, 44, 149
  conversions, 4–10

optical phonon modes, 7
optimization methods, 95–96, 98, 103, 106–108

power dissipation, 171
programmable devices, 69

quantum confinement, 93, 96, 154, 290, 297
quantum tunneling, 94, *95*, 96, 100, 102, 156

registers, 276
reliability modeling, 151, 220–221
resistive memories, 69

scaling effect, 155, 175, 180
sequential circuits, 99, 101, 103, 105
simulation tools, 183; *see also* TCAD
silicon nanowires, 1–11, 157, 237, *245*, 249, 315, 317
Silvaco TCAD, 186
subthreshold slope, 176, 179, 182, 186, 189, 191, **240**, 241
switching characteristics, 182, 207

TCAD (Technology Computer-Aided Design), 186; *see also* simulation
temperature dependence, 79, **295**, 312
ternary logic, 227
TFET (tunnel field-effect transistor), 2, 171–193
  vertical TFETs, 189
TFTs (thin-film transistors), 60–61
threshold voltage, 155–156
transverse optical (TO) phonon modes, 7

ultralow-power design, 172, 174

vertical TFETs, 189; *see also* TFET
VLSI circuit design, 18, 59, 90, 114, 145
voltage transfer characteristics, 174–175

wafer fabrication, 315–316
wide-bandgap materials, **122**

zinc oxide nanostructures, 241, 290, 292
zoning of devices (layout design), 104, 215, 224

For Product Safety Concerns and Information please contact our EU representative GPSR@taylorandfrancis.com Taylor & Francis Verlag GmbH, Kaufingerstraße 24, 80331 München, Germany